13th Hungarian Conference on Materials Science (OATK 2021)

IOP Conference Series: Materials Science and Engineering
Volume 1246

Online
9 - 12 October 2021

Editor:

Z. Weltsch

ISBN: 978-1-7138-6013-6
ISSN: 1757-8981

Printed from e-media with permission by:

Curran Associates, Inc.
57 Morehouse Lane
Red Hook, NY 12571

Some format issues inherent in the e-media version may also appear in this print version.

Institute of Physics
Dirac House, Temple Back
Bristol BS1 6BE UK

Phone: 44 1 17 929 7481
Fax: 44 1 17 920 0979

techtracking@iop.org

Additional copies of this publication are available from:

Curran Associates, Inc.
57 Morehouse Lane
Red Hook, NY 12571 USA
Phone: 845-758-0400
Fax: 845-758-2633
Email: curran@proceedings.com
Web: www.proceedings.com

TABLE OF CONTENTS

IOP Conf. Series: Materials Science and Engineering 1246 (2022) 011001 doi:10.1088/1757-899X/1246/1/011001

13th Hungarian Conference on Materials Science

Edited by

Z. Weltsch
John von Neumann University, Hungary

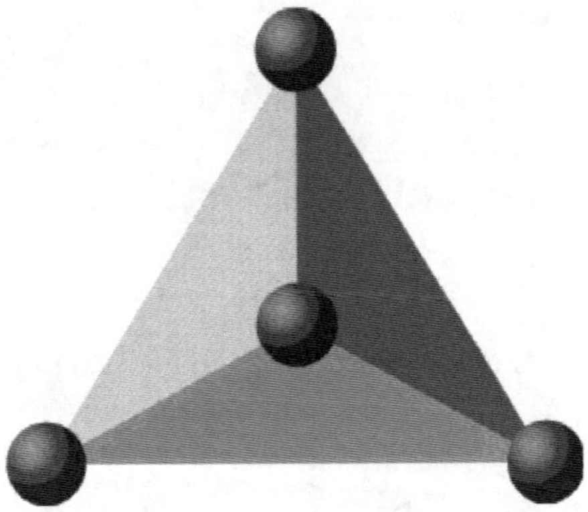

IOP Conf. Series: Materials Science and Engineering 1246 (2022) 011001 doi:10.1088/1757-899X/1246/1/011001

Scientific Committee

József Gyulai † – co-chairman

András Roósz – co-chairman

Gyula Bagyinszki	Tamás Korim
István Bársony	Zoltán Kónya
Dezső Beke	István László
Károly Belina	János Lukács
Tibor Czigány	Béla Pécz
Jánosné Fehér	János Szépvölgyi
Zoltán Gácsi	Miklós Tisza †
Jenő Gubicza	Péter Trampus
György Kaptay	Ibolya Zsoldos

Organising Committee

Péter J. Szabó – chairman

Dénes Zsámbók honorary chairman

Tamás Bárány	György A. Gémes
Mihály Réger	Balázs Verő
Valéria Mertinger	Zoltán Weltsch

Conference Secretariat

Ildikó Kónya – secretary

e-mail: oatk@oatk.hu

Sponsors of the Conference

ZEISS

Flexman
ROBOTICS

QTICS
GROUP

Hungarian Academy of Science

IOP Conf. Series: Materials Science and Engineering 1246 (2022) 011001 doi:10.1088/1757-899X/1246/1/011001

Preface

The series "Hungarian Conference and Exhibition on Materials Science, Testing and Informatics" (later simply "Hungarian Conference on Materials Science") was intended to provide a forum where Hungarian and foreign scientists and research groups interested in metals and alloys, silicates, polymers and composites get an opportunity to exchange and publish ideas and establish new and integrated partnerships. The 13th Hungarian Conference on Materials Science was held at the Balaton lakeshore in Balatonkenese, October 10-12, 2021.

The winner of the photo competition of the 13th Hungarian Conference on Materials Science was Szabó József, coloured by Adam Vida, University of Miskolc

The image with fantasy colors shows tungsten trioxide crystals that crystallized in the vicinity of oxygen bubbles during the wet chemical exploration of tungsten. Scanning electron micrographs were taken with backscattered electrons at 800x magnification.

The organisers present this volume as a peer-reviewed selection of papers submitted to publication by participants

The organisers express their gratitude to all sponsors. Thanks are due to organising societies, as the Hungarian Materials Science Society and to other Hungarian scientific and engineering societies for their support.

Budapest, May, 2022.

Peter J. SZABÓ
Chairman of Organising Committee

Peer Review Statement

All papers published in this volume have been reviewed through processes administered by the Editors. Reviews were conducted by expert referees to the professional and scientific standards expected of a proceedings journal published by IOP Publishing.

- **Type of peer review:** Single Anonymous
- **Conference submission management system:** Morressier
- **Number of submissions received:** 38
- **Number of submissions sent for review:** 35
- **Number of submissions accepted:** 35
- **Acceptance Rate (Submissions Accepted / Submissions Received × 100):** 92.1
- **Average number of reviews per paper:** 1.96
- **Total number of reviewers involved:** 15
- **Contact person for queries:**
  Name: Zoltan Weltsch
  Email: weltsch.zoltan@gamf.uni-neumann.hu
  Affiliation: John von Neumann University

IOP Conf. Series: Materials Science and Engineering 1246 (2022) 012001 doi:10.1088/1757-899X/1246/1/012001

In situ–produced polyurethane-based thermoplastic elastomers: the effect of the acrylonitrile content of the rubber phase

A Kohári[1] and T Bárány[2*]

[1]koharia@pt.bme.hu
[2]barany@pt.bme.hu
[1,2]Department of Polymer Engineering, Faculty of Mechanical Engineering, Budapest University of Technology and Economics, Műegyetem rkp. 3., H-1111 Budapest, Hungary
*Corresponding author: barany@pt.bme.hu; Tel.: (+36-1)-463-3740

Abstract. We prepared thermoplastic dynamic vulcanizates (TDVs) with an in situ–produced thermoplastic polyurethane (TPU) matrix. We used unfilled acrylonitrile butadiene rubber (NBR) with different acrylonitrile contents and investigated the effect of the polarity of the rubber phase on the properties of the TDV samples. The results show that the acrylonitrile (AN) content of the rubbers affects the tensile strength and modulus slightly but the strain at break of the TDVs only negligibly. The results of TDVs are in between those of TPU and unfilled NBRs, demonstrating the feasibility of the TPU/NBR combination.

1. Introduction

The first thermoplastic elastomers (TPEs) appeared on the market more than 60 years ago. Since then, they have become more widely used, with demand growing rapidly for various applications, such as automotive, sports and leisure goods, medical devices, technical and utility articles, and coatings with a pleasant touch. The main reason for this is that TPEs have properties similar to those of vulcanized rubbers but can be processed with processes similar to those used for thermoplastic polymers. TPEs, therefore, form a bridge between thermoplastic polymers and vulcanized rubber [1-3].

TPEs are essentially two-phase systems, with one phase being rigid at the temperature of use; the hard phase provides the strength of the material. The other is the soft phase providing flexibility and elasticity. Increasing the mass fraction of the soft phase makes the TPE softer and more elastic. Their cross-linked structure, which is responsible for their elastic properties, is not formed by irreversible chemical reactions. Instead, they are formed physically and reversibly, due to a significant reduction in the mobility of the molecular chains or molecular chain segments that form the hard phase while the melt cools (e.g. glass transition, crystallization), while their elasticity is essentially due to the mobility of the molecular chains or chain segments forming the soft phase [1, 4].

Thermoplastic polyurethanes (TPUs) belong to thermoplastic elastomers, including block copolymers. The hard segment of TPU is made up of diisocyanate and chain extender, a low molecular weight diol. The soft segment is a long-chain diol, the so-called polyol. The different segments are linked by covalent bonds in the chain, creating the structure typical of block copolymers. They can be produced in one- or two-shot processes. In the first case, all three components (isocyanate, polyol, and chain extender) are mixed, and polyaddition occurs under appropriate conditions. In the second case, the polyol is first reacted with an excess of diisocyanate in an inert atmosphere, and in the second step, the resulting isocyanate-terminated prepolymer units are linked with a chain extender. The latter is also called the prepolymer method [5, 6].

Another important group of thermoplastic elastomers is thermoplastic dynamic vulcanizates (TDVs). These materials consist of a continuous thermoplastic matrix and dynamic vulcanized rubber particles dispersed in it. The term dynamic vulcanization refers to the way the rubber is produced, as the vulcanization of the rubber and the compounding of the components take place simultaneously. The best known TDV is Santoprene, which is made of polypropylene and ethylene-propylene-diene rubber [7, 8].

For multi-phase blends, such as TDVs, the connection between the phases is vital, as it largely determines their properties. Commercially available polymers are rarely miscible or compatible, leading to poor mechanical properties. It can be improved by various compatibilisation methods. The aim of improving compatibility is to produce stable and reproducible dispersion with the desired structure and properties [9, 10].

We prepared thermoplastic dynamic vulcanizates with an in situ–synthesized polyurethane matrix. Due to the polar nature of TPU, we chose acrylonitrile butadiene rubber (NBR) as a rubber phase of the TDVs, which is also polar. During the experiment, we used unfilled NBR rubber mixes with different acrylonitrile content.

2. Experimental

2.1. Materials

The thermoplastic polyurethane matrix was prepared by the prepolymer method. For this purpose, we used the prepolymer brand Suprasec 2008, manufactured by Huntsman Co. (Texas, USA), which contained 10.2% free isocyanate groups. The chain extender was 1,6-hexanediol produced by Lanxess GmbH (Cologne, Germany). Acrylonitrile-butadiene rubbers with different acrylonitrile content were used as the rubber phase of the TDVs (Table 1).

Table 1. Types of rubber used in the experiment.

Brand name	Producer	Acrylonitrile content (%)	Mooney viscosity (ML, 1+4, 100 °C)
Perbuban 1846F		18	45
Perbunan 3445F	Arlanxeo Deutschland GmbH (Dormagen, Germany)	34	45
Perbunan 4066F		40	65

The component of the sulphur-based vulcanization systems and their suppliers were: zinc oxide (Werco Metal, Zlatna, Romania), stearic acid (Oleon, Ertvelde, Belgium), N-Cyclohexyl-2-benzothiazole sulphenamide (CBS) (Rhein Chemie, Mannheim, Germany), and sulphur (Ningbo Actmix Polymer, Ningbo, Zhejiang, China).

2.2. Preparation of thermoplastic dynamic vulcanizates

The rubber phase of the TDVs was prepared in a Brabender Plasti-Corder internal mixer (W 350 EHT, 370 cm^3) (Brabender GmbH, Duisburg, Germany). The chamber temperature was 70 °C, and the recipe is summarized in Table 2.

The TDV samples were also prepared in the above-mentioned internal mixer. The initial temperature of the chamber was set to 150 °C, the rotor speed to 50 rpm, and the mixing time to 27 min. The ratio of free isocyanate groups (NCO) in the prepolymer to hydroxyl groups (OH) from the chain extender was chosen to be 1.05. The required amounts of TPU components were measured with a syringe on a scale to the nearest hundredth of a gram. First, the matrix components (prepolymer and chain extender) were added to the internal mixer chamber, and after 8 min, the rubber mixtures were added. In all cases, the ratio of TPU to rubber mixture was 60:40 wt%.

Table 2. Formulation of the rubber mixtures.

Ingredient	Quantity (phr)
Rubber	100
Zinc oxide	5
Stearic acid	2
CBS	1.5
Sulphur	2

From the produced TDVs, we prepared 2 mm thick plates between Teflon sheets at 200 °C for 4 min after preheating for 4 min at a pressure of 2 MPa using a Teach-Line Platen Press 200E (Dr. Collin GmbH, Ebersberg, Germany). 2 mm thick plates of the rubber mixtures were also prepared with the above press. The sheets were vulcanized at 160 °C, under a pressure of 2 MPa, for the vulcanization time ((t_{90}) +1 min) of the related rubber mixes.

2.3. Test methods

The vulcanization properties of rubber mixtures were determined on a MonTech D-RPA 3000 Rubber Process Analyzer (MonTech GmbH, Buchen, Germany) under isothermal (160 °C, 1.67 Hz frequency, and 1° amplitude) conditions.

The mechanical properties of the TDV specimens were determined with a Zwick Z005 (Zwick GmbH, Ulm, Germany) tensile testing machine equipped with a 5 kN load cell. The TDV specimens were Type 1 dumb-bell shape test pieces according to MSZ ISO 37:2003. They were tensile tested at room temperature at a crosshead speed of 100 mm/min, with a grip length of 60 mm. The tear test was also carried out at room temperature on an angle test piece die (ISO 34 1:2015) at a crosshead speed of 100 mm/min and with a grip length of 56 mm. We made a 1 mm notch on the inner edge of the test pieces with a blade. In each case, the mechanical properties are the average of 5 measurements.

The Shore A hardness of the sheets was determined with a Zwich H04.3150.000 (Zwick GmbH, Ulm, Germany) hardness tester according to ISO 7619-1:2010, at 10 points per sample.

The dynamic mechanical properties of the prepared samples were determined with a TA Q800 (TA Instruments Ltd., New Castle, DE, USA) instrument. The dimensions of the samples cut out from the sheets were 2.0 x 2.5 x 10 mm (thickness x width x clamped length). The test was carried out between −90 °C and +150 °C at a heating rate of 3 °C/min, a frequency of 10 Hz, a superimposed 0.1% sinusoidal strain, and a preload of 0.01 N under tensile loading.

3. Results

3.1. Production of TDVs

In this study, we investigated the effect of the acrylonitrile content of the rubber phase. Before preparing the TDVs, we determined the isothermal vulcanization curves of the rubber mixtures (Figure 1 and Table 3). The results show that NBR 1846F and NBR 3445F rubbers with the same Mooney viscosity (45 MU) exhibited similar behaviour. In contrast, the NBR 4066F rubber, which has a higher Mooney viscosity, has a shorter scorch time (t_{10}) and a longer vulcanization time (t_{90}).

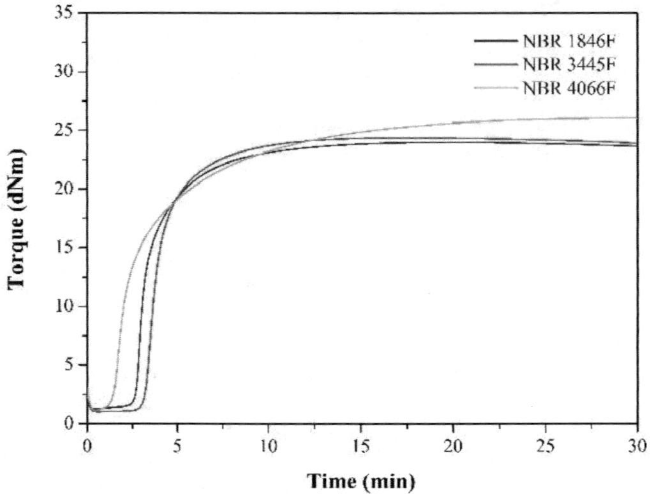

Figure 1. Isothermal vulcanization curves of the rubber mixtures.

Table 3. Isothermal vulcanization properties of the rubber mixtures.

Rubber mixtures	Scorch time, t_{10} (min)	Vulcanization time, t_{90} (min)	Max. torque (dNm)
NBR 1846F	2.75	7.00	24.01
NBR 3445F	3.28	6.81	24.40
NBR 4066F	1.56	10.7	26.04

During mixing, the internal mixer recorded the variation of the temperature of the material and the torque applied as a function of time (Figure 2). These curves provide useful information about the processes taking place in the chamber. An intense increase in the torque curves is observed after the addition of the rubber mixture, followed by a decrease. This decrease is due to the softening of the rubber mixture by kneading and high temperature. Subsequently, an intense increase in the curves is observed again due to the polymerization of the TPU and vulcanization of the rubber compounds. Finally, after reaching a maximum, the torque curves show a decrease again due to the fragmentation of the vulcanized rubber islands and the softening of the material.

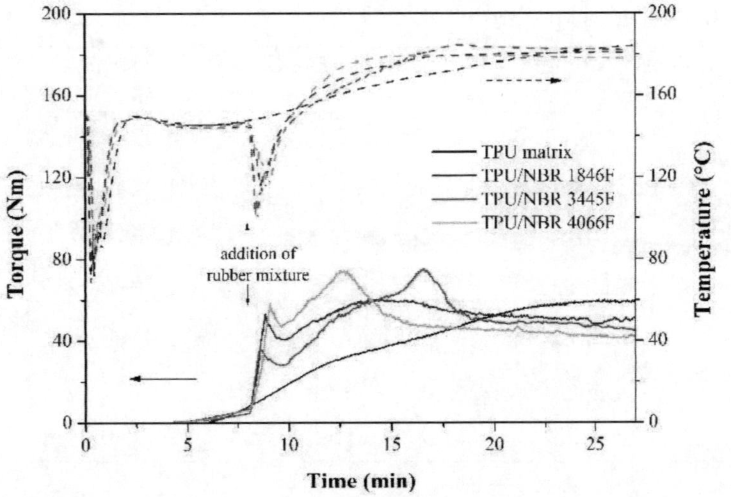

Figure 2. Torque and temperature curves recorded during the production of TDVs.

3.2. Mechanical properties

Based on the results of the tensile test (Figure 3 and Table 4), the mechanical properties of the vulcanized rubbers are far below those of the TPU matrix. Generally, the mechanical properties of TDVs are between those of the TPU matrix and (unfilled) rubber compounds. Of the rubbers, the two with higher AN content (NBR 3445F and NBR 4066F) have almost identical mechanical properties, while NBR 1846F has lower tensile strength, elongation at break, and stress at 100% and 200% elongation (M100 and M200). A comparison of the results of the TDVs does not show this, the effect of the AN content being smoothed into the scattering field. Please note that the rubber phase does not contain any reinforcement or filler. In practice, rubbers are rarely used without a filler—tests show that they have poor mechanical properties on their own.

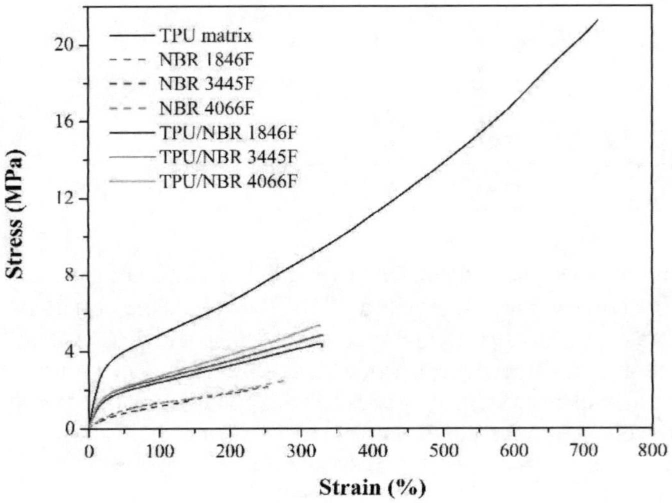

Figure 3. Typical tensile curves of the TPU matrix, the rubbers, and the TDVs.

Table 4. Mechanical properties of the TPU matrix, the rubbers, and the TDVs.

Sample	Tensile strength (MPa)	Elongation at break (%)	M100 (MPa)	M200 (MPa)
TPU matrix	20.53±1.95	722.0±9.6	4.92±0.08	6.58±0.17
NBR 1846F	1.55±0.09	150.1±9.2	1.22±0.05	—
NBR 3445F	2.40±0.19	281.5±21.6	1.34±0.01	1.81±0.01
NBR 4066F	2.52±0.19	280.2±28.8	1.30±0.02	1.89±0.03
TPU/NBR 1846F	4.45±0.15	325.7±15.9	2.44±0.04	3.33±0.08
TPU/NBR 3445F	4.92±0.37	337.3±33.2	2.55±0.03	3.47±0.07
TPU/NBR 4066F	5.45±0.33	326.0±16.5	2.74±0.04	3.87±0.09

The tear strength of the rubbers (Table 5) increased with acrylonitrile content, but this trend was again not observed for TDVs. Their tear strength is between that of the two phases, closer to that of rubber. NBR 3445F and NBR 4066F have almost identical Shore A hardness, while NBR 1846F with a lower AN content is slightly softer. This trend is also observed for the hardness of TDVs.

Table 5. Tear strength and Shore A hardness of TPU matrix, rubbers, and TDVs.

Sample	Tear strength (kN/m)	Shore A hardness
TPU matrix	45.33±1.89	84±0.3
NBR 1846F	3.97±0.14	50±0.3
NBR 3445F	4.79±0.38	54±0.2
NBR 4066F	5.79±0.41	54±0.3
TPU/NBR 1846F	14.92±0.10	68±0.5
TPU/NBR 3445F	15.98±0.41	73±0.5
TPU/NBR 4066F	15.10±0.76	72±0.4

3.3. Dynamical mechanical analysis

On the tanδ curves (Figure 4a) of the TPU matrix, there are two different glass transitions. The two glass transitions are due to the phase-separated structure of the thermoplastic polyurethane, as the soft and the hard phases have separated glass transition temperatures ($T_{g\ soft}$ and $T_{g\ hard}$). The rubbers have only one glass transition (Figure 4b), and their T_g ($T_{g\ rubber}$) increased with acrylonitrile content, resulting in a deterioration of their cold resistance. This is because as the polar fraction of the polymer chain increases, the energy required for segment movements also increases, as the chains become stiffer due to the large polar nitrile groups.

Two glass transitions are observed in the thermomechanical curves of the TDVs (Figure 4a). In the case of the TPU/NBR 1846F sample, instead of $T_{g\ soft}$ and $T_{g\ rubber}$, one glass transition temperature and the T_g of the hard segment of the TPU are visible. For the other two TDVs, the T_g of the soft segment of the TPU and of the rubber phase can be seen.

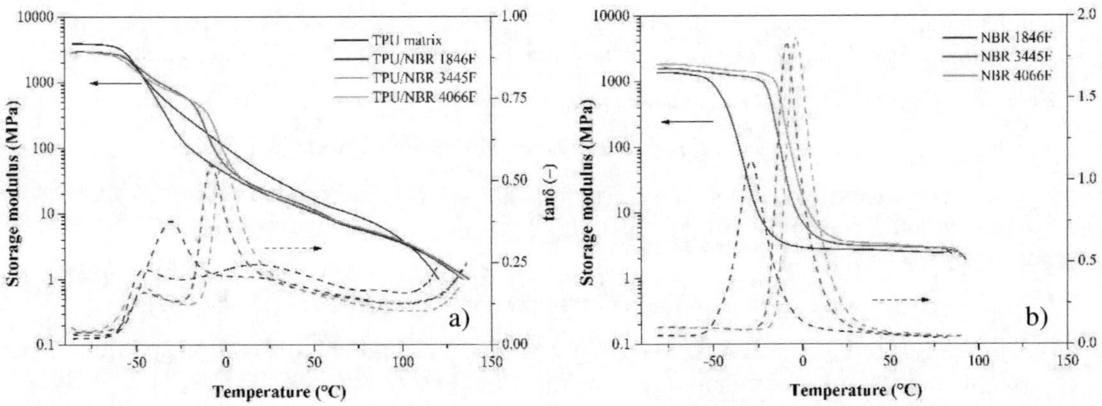

Figure 4. Thermomechanical curves of a) the TPU matrix and the TDVs, and b) the rubbers.

The room temperature storage modulus ($E'_{20°C}$) of the rubbers also increased with acrylonitrile content, for the reasons described earlier (Table 6). The modulus of the produced TDVs is lower than the modulus of the matrix TPU due to the softer rubber islands. These findings are consistent with the results of the mechanical tests.

Table 6. The glass transition temperatures (T_g) and room temperature storage modulus of the TDVs ($E'_{20°C}$).

Sample	$T_{g\ soft}$ (°C)	$T_{g\ rubber}$ (°C)	$T_{g\ hard}$ (°C)	$E'_{20°C}$ (MPa)
TPU matrix	−45.12	—	17.62	44.90
NBR 1846F	—	−28.89	—	2.74
NBR 3445F	—	−9.05	—	3.33
NBR 4066F	—	−3.42	—	3.89
TPU/NBR 1846F	-30.59		25.59	21.60
TPU/NBR 3445F	−44.68	−7.93	—	24.37
TPU/NBR 4066F	−47.62	−0.63	—	22.42

4. Conclusions

We prepared thermoplastic dynamic vulcanizates with a TPU matrix and acrylonitrile butadiene rubber. In the experiment, we investigated the effect of the polarity of the rubber phase and the acrylonitrile content of the NBR on the properties of the TDVs produced. The properties of the rubbers were influenced by their acrylonitrile content, but this was not observed for TDVs. The properties of TDVs produced with unfilled rubber compounds are between those of the TPU matrix and those of the rubber phase. We hope to improve these results further by filling the rubber phase with carbon black or with silica. In the future, we will investigate the effect of the quantity and quality of the filler used in the samples produced.

5. References

[1] Drobny J G 2014 *Handbook of Thermoplastic Elastomers* (Oxford: Elsevier)

[2] Holden G, Kricheldorf H R and Quirk R P 2004 Thermoplastic Elastomers (München: Carl Hanser Verlag)

[3] Bandyopadhyay A, Dasgupta P and Basak S 2020 *Engineering of Thermoplastic Elastomer with Graphene and Other Anisotropic Nanofillers* (Singapore: Springer Nature Singapore Pte Ltd)

[4] Kear K E 2003 *Developments in Thermoplastic Elastomers* (Shawbury: Rapra Technology Ltd)

[5] Szycher M 2013 *Szycher's Handbook of Polyurethanes* (New York: CRC Press)

[6] Piotr K 2008 *Linear Polyurethanes - Synthesis Methods, Chemical Structures, Properties and Applications* (Leiden: Koninklijke Brill NV)

[7] Karger-Kocsis J (1999) *Thermoplastic rubbers via dynamic vulcanization in Polymer Blends and Alloys*, ed Shonaike G O and Simon G P (New York: Marcel Dekker) pp 125-153

[8] Banerjee S S and Bhowmick A K 2017 High-temperature thermoplastic elastomers from rubber–plastic blends: A state-of-the-art review *Rubber Chem. Technol.* **90** 1-36

[9] Ajitha A R and Thomas S 2020 *Compatibilization of Polymer Blends* (Amsterdam: Elsevier)

[10] Thomas S, Grohens Y and Jyotishkumar P 2015 *Characterization of Polymer Blends: Miscibility, Morphology and Interfaces* (Weinheim: Wiley-VCH Verlag GmbH & Co. KGaA)

Acknowledgments

This work was supported by the National Research, Development and Innovation Office, Hungary (K128268). The research reported in this paper is part of project no. BME-NVA-02, implemented with the support provided by the Ministry of Innovation and Technology of Hungary from the National Research. Development and Innovation Fund, financed under the TKP2021 funding scheme.

Finite element analysis of closed-cell aluminum foam approximated with Weaire-Phelan unit cell structure

M File[1,3], R Cseke[1], D Huri[1,2], G Balogh[1] and T Mankovits[1]

[1] Department of Mechanical Engineering, Faculty of Engineering, University of Debrecen, H-4028 Debrecen, Ótemető u. 2-4., Hungary
[2] Doctoral School of Informatics, Faculty of Informatics, University of Debrecen, H-4028 Debrecen, Kassai u. 26., Hungary
[3] Author to whom any correspondence should be addressed.

Abstract. The finite element modelling of metal foam structures have proven to be a difficult task. Many different modelling approaches exist, both for creating the CAD model and for performing the finite element simulations. This article details the CAD and finite element modelling of an existing aluminum foam. To model the cell structure, the Weaire-Phelan unit cell was used. The different parameters of the unit cell were adjusted to fit the parameters of the existing foam. The physical properties of the finished model were then compared to the original specimen to validate its accuracy. After the comparison, it was determined, that the Weaire-Phelan unit cell can be used to model the physical properties of the structure. The validated and simplified model was then used to perform a finite element analysis under compressive loading. The results were compared to the compression test results of the existing aluminum foam. By comparing the results and calculating the relative errors, it was determined that both the physical and the mechanical properties of metal foam structures can be modelled using this exact method with significant accuracy.

1. Introduction

Closed-cell metal foams are used in a wide range of structural elements, usually the most common load case is compression. The reason of the usage is the lightness, the high compression strength and the high energy consumption under compression [1]. During the design phase, modelling the behaviour of these materials is a very important task. The behaviour of these materials under compressive loading is detailed in [2–4].

Many different geometrical modelling approaches exist, one of which is the µCT based two-or three-dimensional model generation. Two dimensional models were constructed using this method in [5,6]. Closed cell foam materials were modelled using three-dimensional models generated based on µCT images and analysed using finite element software in [7,8]. This method enables the creation of the most accurate model. However, it requires special equipment and is a time-consuming process. Another method is the usage of uniform cells to simplify the internal geometry. A combination of spherical and cruciform-shaped cells was used to model closed-cell aluminum foam in [9]. Different, already existing unit cells, such as Kelvin and Weaire-Phelan unit cells can also be used to model closed-cell metal foam materials, as presented in [10,11]. A comparison of these unit cells can be seen in [12].

This paper deals with the be investigation of the accuracy of the Weaire-Phelan unit cell, when modelling a closed-cell aluminum foam specimen for compressive loading. A specimen was modelled using these unit cells. The accuracy of the created CAD model was validated by comparing its physical properties to the original specimen. The validated model was used for the finite element simulations. By comparing the finite element simulation results to the laboratory measurement, the accuracy, and the usability of the simplified model were be determined. During the research only the elastic region was tested. The model could be expanded to the modelling of the plastic deformation zone as well in the future.

2. Materials and methods

2.1. Parameters of the modelled metal foam

The closed-cell aluminum foam was procured from Aluinvent Zrt. For the compression test five cylindrical specimens were cut according to the ISO 13314 standard [13]. The diameter and the length were 30 mm, respectively. Table 1 shows the average physical properties of the material. The porosity was calculated based on the weight of the specimens. An image about the specimen is shown in figure 1. The average cell size and wall thickness was determined using CT analysis.

Table 1. The properties of the aluminum foam

Property	Value
Weight (average)	6.6 g
Porosity (average)	89.27 %
Volume (average)	2389 mm^3
Young's modulus	2374 MPa
Poisson's ratio	0.29

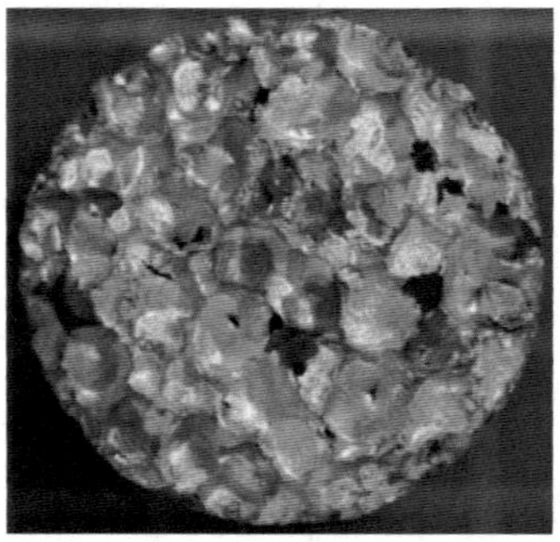

Figure 1. One of the specimens used for the modelling

2.2. Modelling the specimen using Weaire-Phelan unit cells

The 3D model was created using the SIEMENS Solid Edge software. For the cell structure approximation, the Weaire-Phelan unit cell was chosen [12]. After several steps the unit cell was modelled, see figure 2.

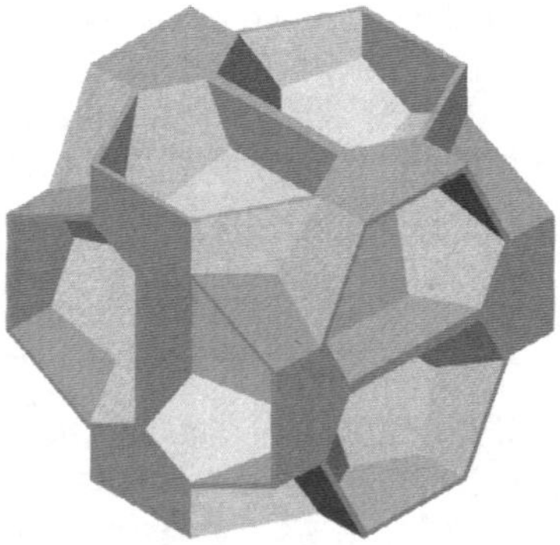

Figure 2. Weaire-Phelan unit cell modelled using CAD software

IOP Conf. Series: Materials Science and Engineering 1246 (2022) 012002 doi:10.1088/1757-899X/1246/1/012002

The unit cells were multiplied to create a cube with 30 mm edge length which was then cut to the cylinder shape of the specimen with a 30 mm diameter. The final CAD model of the aluminum foam specimen can be seen in figure 3 (a). Finally, this model was cut into a quarter model with one layer of unit cells, to prepare it for the finite element simulations, which can be seen in figure 3 (b).

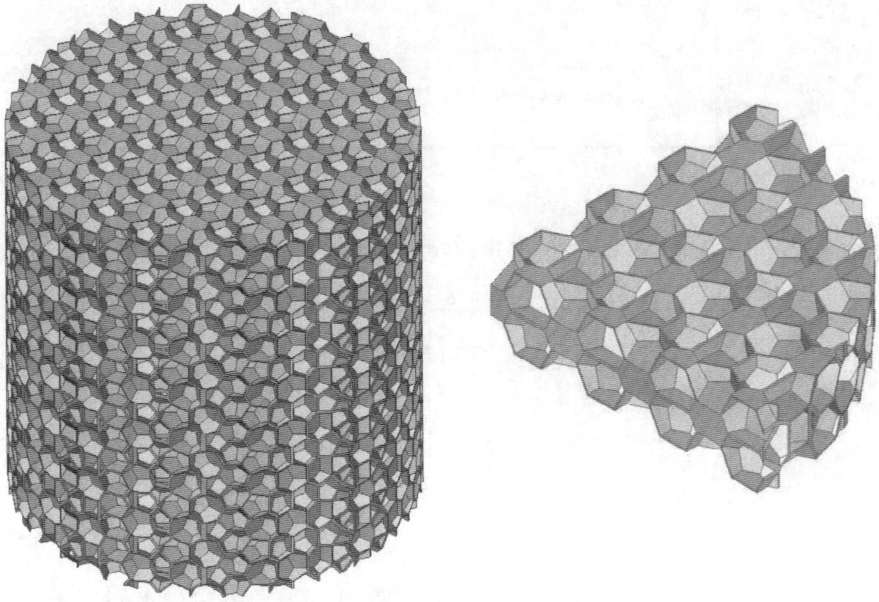

(a) (b)

Figure 3. (a) CAD model of the aluminum foam specimen; (b) one-layer high quarter model prepared for the finite element simulations

2.3. Finite element analysis

The finite element simulations were done using the Simcenter Femap with its integrated NX Nastran solver. After opening the prepared 3D model with the software, the material model was created. Since in this work only the elastic region of the material was analysed, the default linear elastic material model was used. The data required to create this model was the Poisson's ratio and the Young's modulus of the material, seen in table 1. A prescribed displacement of 0.0074 mm (0.2% strain) was applied to the top surface of the specimen as loading. Since a quarter model was used, the symmetry planes were constrained in their normal direction. The last remaining axis, the Y axis was constrained on the bottom surface and on the surface where the load was applied as well, as the software requires it for prescribed displacement.

A mesh convergence analysis was then performed to eliminate the mesh dependence of the simulated results. The force values were used as the mesh convergence criteria, relative error values were calculated between the maximum force values and convergence was detected if three deviation values were smaller than 0.5%. The maximum force values of the mesh convergence analysis are plotted in figure 4 and the calculated deviation values can be seen in table 2.

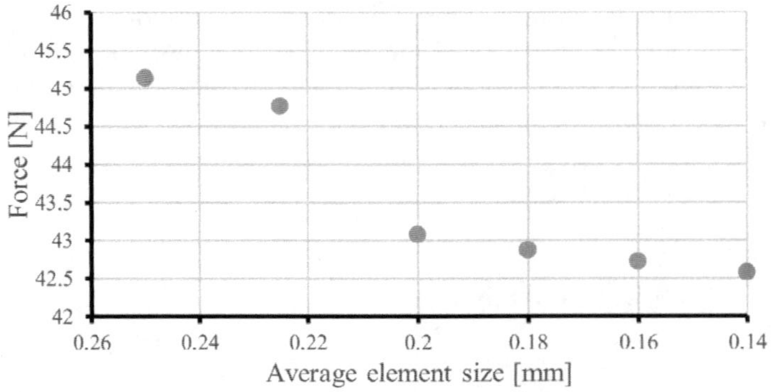

Figure 4. Results of the mesh convergence analysis

Table 2. Relative deviation values of the mesh convergence analysis

Element size [mm]	Force [N]	Relative deviation
0.25	45.132	-
0.225	44.758	0.82868 %
0.2	43.07159	4.5653 %
0.18	42.8624	0.48568 %
0.16	42.71221	0.3504 %
0.14	**42.567**	**0.33997 %**

The analysis has conducted that the 0.14 mm average element size was optimal for the tetrahedral solid elements. The parameters for the final discretisation can be seen in table 3 and the meshed finite element model can be seen in figure 5.

Table 3. The parameters of the selected mesh

Property	Value
Element type	Solid
Element shape	Tetrahedral
Nodes per element	4
Element size	0.14 mm
Number of elements	334252

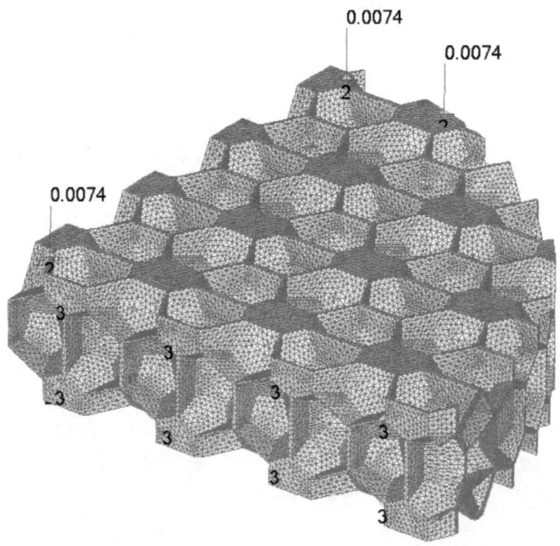

Figure 5. The meshed finite element model

2.4. Laboratory measurement

The laboratory measurement was done using the INSTRON 68TM-10 universal material testing system and the strains were measured using the INSTRON AVE2 advanced video extensometer. Similarly to the finite element simulations, the specimens were also compressed until 0.2% strain which in the case of the whole specimen is 0.06 mm. The load was applied with a 1 mm/min velocity. The results of the 5 compression tests were then summarised and the median value was selected for the comparison with the results of the finite element analysis.

3. Results and discussion

3.1. Comparison of the CAD model and the specimen

To determine the accuracy of the CAD model, its different properties were compared to the properties of the original foam specimen. Table 4 shows the comparison of the physical properties.

Table 4. Comparison of the aluminium foam specimen and the CAD model based on the physical properties

Property	Aluminum foam specimen (average values)	CAD model	Difference
Weight	6.6 g	6.901 g	4.560 %
Porosity	89.27 %	86.43 %	3.181 %
Volume	2389 mm^3	2421 mm^3	1.339 %

It is visible from the less than 5 % difference values, that the model is an accurately approximating the physical properties of the original specimen.

3.2. Results of the finite element analysis

During the evaluation of the finite element results, the force values were in focus. The simulation has conducted that compressing the specimen by 0.2% (0.0074 mm) takes 42.567 N of force. Since a one-layer quarter model was used, this force value must be multiplied by 4, meaning the force for the full specimen was 170.268 N, and the displacement must be scaled up to reflect the 0.06 mm compression of the whole specimen.

3.3. Results of the compression tests

The compression tests were also evaluated until the 0.2% strain value, which in the case of the whole specimen was 0.06 mm. From the 5 tested specimen, the median value was selected. The maximum load on this specimen was 162.608 N, the results of the 4 other specimens showed a 3.56% maximum deviation from the median value.

3.4. Comparison of the finite element analysis and the laboratory measurement

The results of the finite element simulation and the laboratory measurements were compared. The relative errors were determined, thus the accuracy of the created model was evaluated. The comparable factor was the maximum force value at the 0.06 mm (0.2%) compression. For the relative error calculation, the following equation was used:

$$e_r = \frac{v_{simulated} - v_{measured}}{v_{measured}} \cdot 100 \quad (1)$$

where e_r is the relative error, $v_{simulated}$ is the maximum force evaluated from the simulation, and $v_{measured}$ is the maximum force evaluated from the measurement. The evaluated results with the calculated relative errors can be seen in table 5.

Table 5. The calculated error value between the results of the finite element simulation and the laboratory measurement

Value	Finite element simulation	Laboratory measurement	Relative error
Maximum force	170.268 N	162.608 N	4.7 %

From the measurement and the simulation, a force-displacement curve was also evaluated, which can be seen in figure 6.

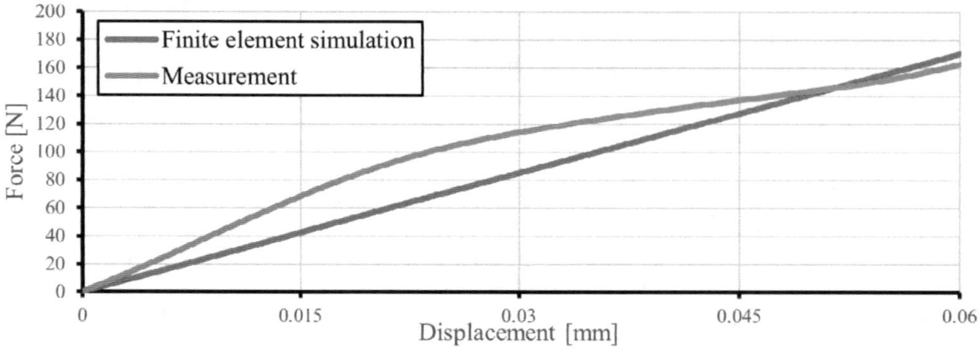

Figure 6. The force-displacement curves of the measurement and the simulation

4. Conclusion

This article introduced the modelling and finite element simulation of a closed cell aluminum foam specimen under compressive loading. The structure of the material was approximated by Weaire Phelan unit cells, which were then multiplied and shaped to match the starting foam specimen. The created CAD model was then simplified into a one-layer quarter model, to prepare it for the finite element simulations. Before the simulations, the CAD model was validated by comparing its physical properties to the original specimen. The comparison has shown that the Weaire-Phelan unit cell structure is able to model the physical properties of the material with significant accuracy, as the difference values have been less than 5 % in all cases. After the validation, the finite element simulations were performed on the simplified model. The boundary conditions and the load were given in such a way, to recreate the laboratory conditions as closely as possible. The simulated result was compared to the measured data and relative error values were calculated, to find out the accuracy of the created model. It is visible from the less than 5% relative error values, that the Weaire-Phelan unit cells and the applied finite element analysis can simulate the elastic behaviour of closed cell aluminum foam materials accurately. The use of the Weaire-Phelan unit cell can shorten the design process of structures using metal foam materials, as the creation of the structure only requires the multiplication of one unit cell, while being able to model the physical properties and the mechanical behaviour of these materials with sufficient accuracy.

Acknowledgments
The research was supported by the Thematic Excellence Programme (TKP2020-NKA-04) of the Ministry for Innovation and Technology in Hungary, within the framework of the (Automotive Industry) thematic program of the University of Debrecen.
We thank Aluinvent Zrt. for providing the closed-cell aluminum foam for the research.

References

[1] Ashby, M, Evans, A, Fleck, N, Gibson, L, Hutchinson, J, Wadley, H, et al. Metal Foams: A Design Guide. Appl Mech Rev. 2001 Nov 1;54(6):B105–6.
[2] Hu Y, Fang Q-Z, Qian J. Effect of cell structure on the uniaxial compression properties of closed-cell foam materials. Mater Today Commun. 2021 Mar;26(February):102104.
[3] Szlancsik A, Norbert Orbulov I. Compressive properties of metal matrix syntactic foams in uni- and triaxial compression. Mater Sci Eng A. 2021 Oct;827:142081.
[4] Szlancsik A, Katona B, Dombóvári Z, Orbulov IN. On the effective Young's modulus of metal matrix syntactic foams. Mater Sci Technol. 2017 Dec 12;33(18):2283–9.
[5] Caty O, Maire E, Youssef S, Bouchet R. Modeling the properties of closed-cell cellular materials from tomography images using finite shell elements. Acta Mater. 2008

IOP Conf. Series: Materials Science and Engineering 1246 (2022) 012002 doi:10.1088/1757-899X/1246/1/012002

Nov;56(19):5524–34.
[6] Kádár C, Maire E, Borbély A, Peix G, Lendvai J, Rajkovits Z. X-ray tomography and finite element simulation of the indentation behavior of metal foams. Mater Sci Eng A. 2004 Dec;387–389(1-2 SPEC. ISS.):321–5.
[7] Saadatfar M, Mukherjee M, Madadi M, Schröder-Turk GE, Garcia-Moreno F, Schaller FM, et al. Structure and deformation correlation of closed-cell aluminium foam subject to uniaxial compression. Acta Mater. 2012 May;60(8):3604–15.
[8] Hangai Y, Yamaguchi R, Takahashi S, Utsunomiya T, Kuwazuru O, Yoshikawa N. Deformation behavior estimation of aluminum foam by X-ray CT image-based finite element analysis. Metall Mater Trans A Phys Metall Mater Sci. 2013;44(4):1880–6.
[9] Hasan MA. An Improved Model for FE Modeling and Simulation of Closed Cell Al-Alloy Foams. Adv Mater Sci Eng. 2010;2010:1–12.
[10] Chen Y, Das R, Battley M, Xu Z. Compressive and shear strengths of the ductile closed-cell Kelvin and Weaire-Phelan foams along the lattice direction [100]. Thin-Walled Struct. 2018;132(November 2017):237–49.
[11] Vengatachalam B, Poh LH, Liu ZS, Qin QH, Swaddiwudhipong S. Three dimensional modelling of closed-cell aluminium foams with predictive macroscopic behaviour. Mech Mater. 2019 Sep;136(August 2018):103067.
[12] Latré SK, De Pooter S, Buffel B, Brabazon D, Seveno D, Desplentere F. Comparative study of a cubic, Kelvin and Weaire-Phelan unit cell for the prediction of the thermal conductivity of low density silica aerogels. Microporous Mesoporous Mater. 2020 Jul;301(March):110206.
[13] ISO. Mechanical testing of metals - Ductility testing - Compression test for porous and cellular metals. Int Stand. 2011 [cited 2022 Jan 24];ISO 13314(1).

IOP Conf. Series: Materials Science and Engineering 1246 (2022) 012003 doi:10.1088/1757-899X/1246/1/012003

Investigation of aluminium clinched joints by FEA

Sz. Jónás[1], M. Tisza[2,†], P. Z. Kovács[3]

[1] Simulation Engineer, Knorr-Bremse Rail Systems Budapest, Hungary
[2,†] Professor Emeritus, University of Miskolc, Institute of Materials Science and Technology, University of Miskolc, Hungary
[3] Associate Professor, University of Miskolc, Institute of Materials Science and Technology, University of Miskolc, Hungary

E-mail: szabolcs.jonas@knorr-bremse.com

Abstract. The aim of the present paper is to compare the results of clinch joining simulations of two finite element software. The comparison is taken in between the resultant force-displacement curves, and in between the most important geometrical parameters of the joints, as the undercut (C) and neck thickness (t_N). The joining process was performed between two AA6082 sheets in two different heat treatment conditions by a TOX clinching tool. For the simulations the flow curves of material have determined by the so-called Watts-Ford compression test for both conditions. The applied FE software packages were the MSC.Marc&Mentat 2019 and the DEFORM2D 10.2.

1. Introduction

In the recent decades the usage of aluminium is highly increased due to the industrial demands of mass reduction, but also considering the strength issues. Figure 1. shows an excellent example for the distribution of the applied materials in a car body according to [1].

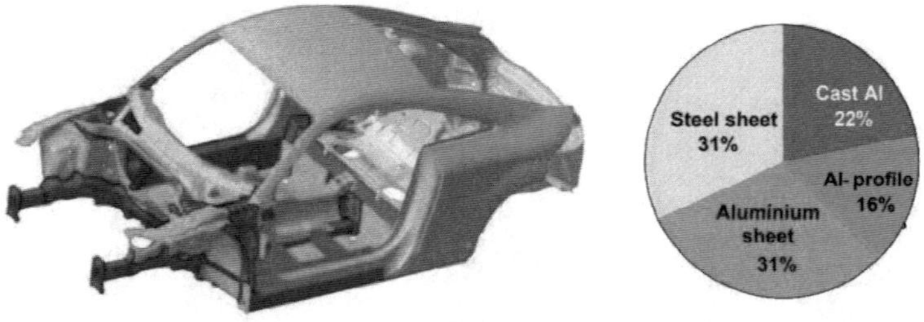

Figure 1. Aluminium content of a car body [1]

According to the data reported in [2] the aluminium content of car bodies nowadays is more than four times compared to the year 1975, and the increasing usage will not stop, especially due to the continuous improvement of cars, such as electrification. It can be stated the lightweight manufacturing is a key area of the car manufacturers. It can be seen in figure 2. that in these years and the near future based on the actual needs, the aluminium content is about 200 kg in a car. The percentage of sheets from the total mass of aluminium in a car body according to [2] is around 23% (~46 kg), and it will double in 2030.

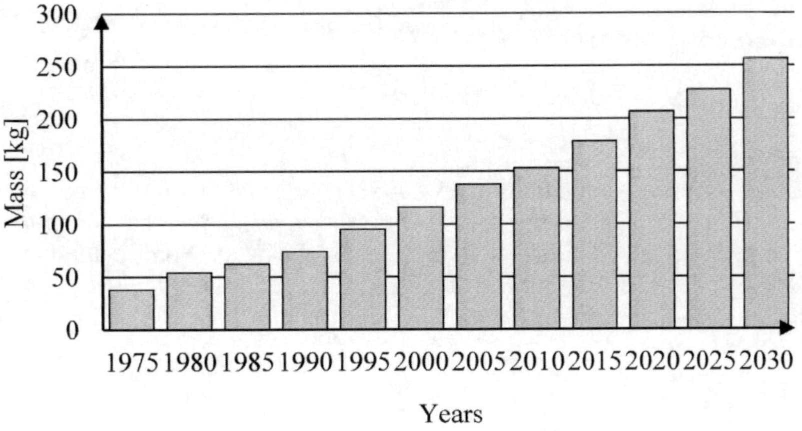

Figure 2. Increasing aluminium content in cars [2]

The further need of mass reduction, CO2 emission regulations and the amount of aluminium (and the concept of the so-called multi-materials) predicts the importance of joining techniques which are environmentally friendly, fast, effective, cheap, have relatively high strength and can join dissimilar materials, also. Clinch joining or clinching is one of these innovative joining techniques.

Clinch joining is a type of mechanical joining processes where the sheet metal parts are joined through an intense local plastic deformation by a special tool without use of any external elements such as bolts. It is suitable for the connection of steel and aluminium. The quality of the clinched joint is influenced by the material of the joined sheets, the thickness of the sheets, and last, but not least the geometrical parameters of the clinching tools [3].

Several researchers analysed aluminium clinched joints by several aspects such as presented in [4], [5]. From the point of view of structural strength of multi-material joints, the investigation of the dissimilar joints is a particularly important topic. Numerous articles are dealing with the dissimilar material joints between high strength steel and high strength aluminium sheets e.g., [6][7].

In the present work the finite element modelling of AA6082 type of aluminium alloy in two heat treated conditions is analysed by two different FE systems, and the results are compared.

The most important resultant geometrical parameters of round clinch joints are depicted on figure 3. These parameters are namely the size of undercut (C), the size of neck thickness (t_N) and the size of the residual bottom thickness (t_B).

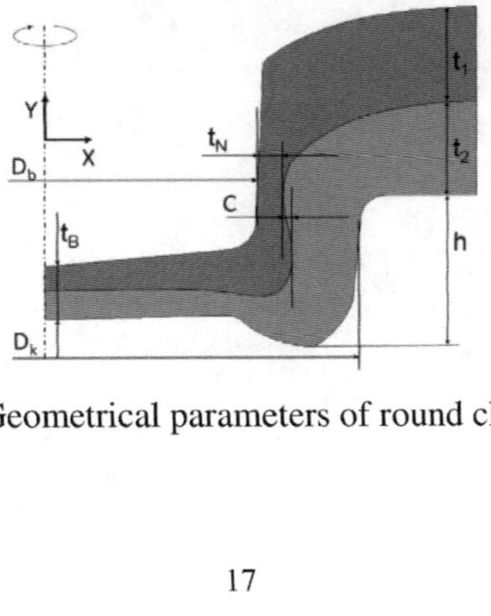

Figure 3. Geometrical parameters of round clinch joints

The residual bottom thickness is a measurable value, so this value is important from the point of view of quality assurance. In this study, every simulation result was considered that the value of the residual bottom thickness is t_B=0.5 mm. The other two parameters were analysed and compared. Both parameters are depending on the bottom thickness, and both parameters are highly affecting the strength of the joints.

2. Material and method

2.1. Applied material

The applied material for the simulations is AA6082 type of aluminium alloy in two heat-treated conditions (T0 and T6). The applied thickness of the sheets is 1 mm. The flow curve of the alloy is measured for both heat-treated conditions (T0 and T6) by the Watts-Ford method at the laboratory of the University of Miskolc. The solution heat-treatment procedure was done at 525°C for 30 minutes. The results of the measurement data of the aluminium sheets for both conditions can be seen in figure 4.

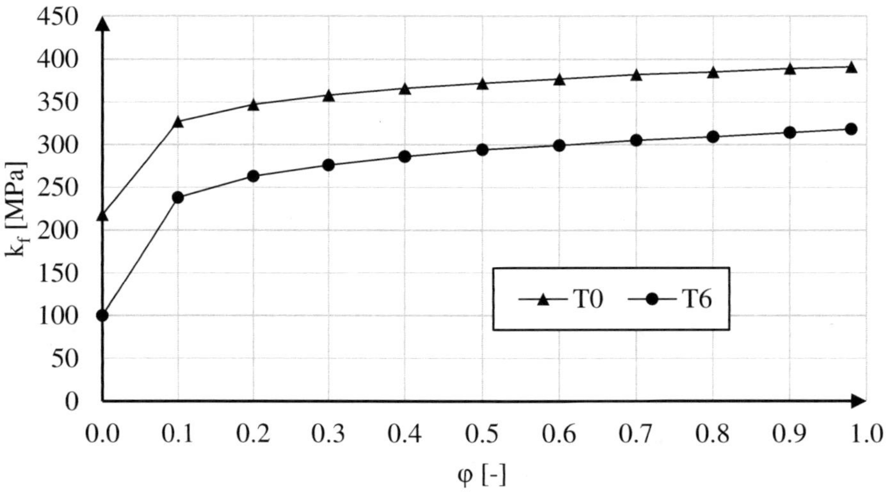

Figure 4. Measured flow curves of AA6082 in different heat-treated conditions

As it can be seen the maximal measured value is at φ=0.98, but according to [8] during clinching the strains can reach up to φ=2÷3, which value cannot be measured, however different extrapolation techniques are available to describe the higher strain regimes, such as Voce, Hockett-Sherby, Swift etc.

According to the previous experiments, the material is easier to join in T6 condition, in T0 condition cracks often occur during the manufacturing process.

2.2. FE model

For the FE modelling two software were used. On the one hand, the special, dedicated to forming processes simulation FE software, the DEFORM 2D was used, and on the other hand, the MSC.Marc was used which is a general-purpose one.

There are several parameters which could be the same in both systems, but several ones depend on the software and cannot be modified by manually (e.g., meshing).

The similarities in the two FE models are the following:

- both solvers are implicit,
- element type (4 node quadrilateral),
- applied remeshing during the process,
- geometry of the clinching tool and sheet thicknesses,
- elastic properties (E=70 GPa and ν=0.33),
- non-extrapolated flow curves (according to the measurement, figure 4.),
- friction parameter (shear friction, m=0.12),
- dimension of the problem (2D axisymmetric),
- elastic-plastic sheets and rigid tools,
- displacement driven punching tool.

The differences in the two FE models are the following:

- definition of the remeshing criteria,
- slightly different boundary conditions,
- software specific flow curve extrapolation during calculation.

The two software are extrapolating the true stress-true plastic strain curves differently when the strain reaches up the last data point of the available ones. The MSC.Marc is using linear extrapolation method, in contrast the DEFORM 2D is using exponential method ($k_f = C\varepsilon^n\varepsilon^m + k_{f,0}$). In this study the built-in extrapolation methods are used to analyse the process of clinching, i.e., the applied flow curves are not extrapolated according to any material laws. The applied material data can be seen in figure 4.

In the MSC.Marc model the nodes on symmetry axis are prescribed in horizontal move and two nodes at the end of the modelled sheets were prescribed in vertical movement, the DEFORM 2D model does not consider these conditions. Although this difference has no or easily eliminable effect on the results. However, both software uses remeshing during the forming process, but the details are different, as per the case of MSC.Marc every single detail of the remeshing criteria have to be manually described, the DEFORM 2D use a fully automatic remeshing algorithm. These differences are mainly caused by the goal of the software. Since the DEFORM 2D is a dedicated one for forming analyses, the MSC.Marc is a general purpose one, therefore the users often different, i.e., the DEFORM is mainly applied by the industrial, and the MSC.Marc is used by academical users.

The meshed FE models can be seen in figure 5 in the initial configuration of the simulation.

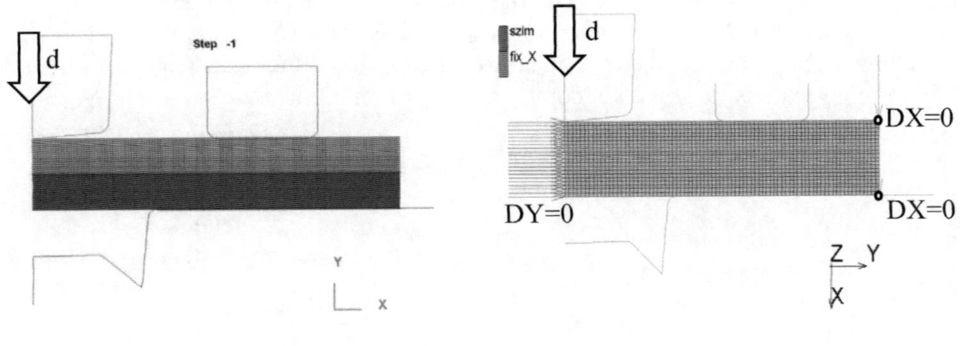

Figure 5. Meshed FE models (left: DEFORM 2D, right: MSC.Marc)

In both cases linear, four-noded quadrilateral elements were used to set up the models, and the frictional behaviour is also the same, based on shear friction with the coefficient of m=0.12.

3. Results

The most important results are the effective strain distribution of the models, the force need of the process, and the geometrical parameters (particularly t_N and C) of the joints.

The comparison of the effective strain distribution of the two models can be seen in figure 6. The distribution of the strain is highly similar, but as it can be seen the values of the maximum strain is slightly different, in case of MSC.Marc calculation result is higher. This phenomenon is caused by the different internal extrapolation methods. The MSC.Marc uses linear extrapolation based on the last two data points of the available data set, in contrast DEFORM uses exponential extrapolation method. According to the results the value of k_f is higher in case of linear extrapolation than the exponential extrapolation, which leads to a different stress-strain state at higher strains. The signed zones (see figure 6) are also differing slightly, in case of DEFORM 2D the upper sheet is thinning a little more, close to the axis of symmetry. In case of MSC.Marc the shape of the interface is straight at the bottom zone of the joint, in contrast the result of the DEFORM 2D, the interface is curved. The zone of interest from the interface point of view is framed, see in figure 6.

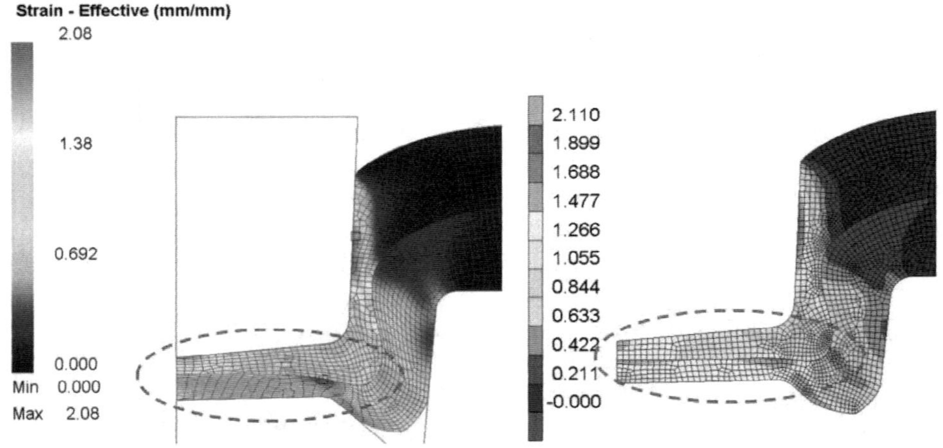

Figure 6. Effective strain distribution – T6 condition (left: DEFORM 2D, right: MSC.Marc)

The differences can also be examined on the forming force-punch displacement (F-d) curves. Figure 7. show the simulated F-d curves for both conditions. The results of MSC.Marc simulations shows higher forming forces (~2 kN) in both cases at the peak values. The difference between the simulated curves starts to increase at d≈1,5 mm when the lower sheet reaches the bottom of the die, and the strains are getting higher than the maximal measured data. Due to the heat-treatment in condition T6 the maximal forming force is less by 2 kN compared to T0 condition.

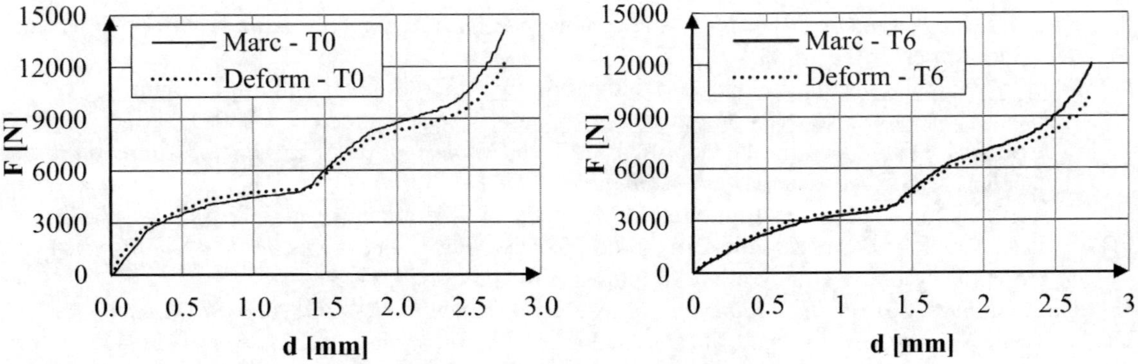

Figure 7. Forming force – punch displacement (left: T0, right: T6)

The values of t_N and C parameters are also investigated (figure 8.) and found that the value of neck thickness is higher in T6 case, and the value of undercut is higher in T0 condition. These differences cannot be neglected from the point of view of the joint strength, for instance the neck thickness is highly affected the tensile-shear strength of the joints as its reported in [8]. The trends seem to consequent comparing the results of the two software, the MSC.Marc results are higher.

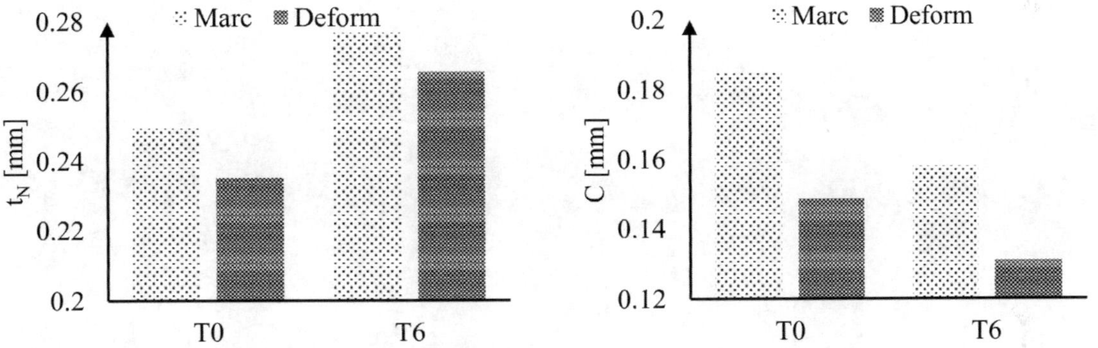

Figure 8. Main geometrical parameters (left: t_N, right: C)

4. Conclusion

Both software are applicable to simulate the clinch joining process. The internal extrapolation methods are differing from each other, which leads to non-negligible differences in the results, such as the predicted forming force or predicted geometry. The underestimated forming force can cause tooling failure during experiments, or the overestimated geometrical properties can cause inaccurate joint strength predictions.

The most important conclusion is that the flow curves must describe the behaviour of the materials as good as possible, and according to the results the extrapolation is a necessary step before the simulation of clinch joining, however the proper extrapolating technique is depending on the type of material. Due to the unknown values in the high strain regime, it is highly recommended to use multiple methods (e.g., Voce, Hockett-Sherby, Swift etc.) and choose the proper one based on experimental results.

Acknowledgement

This research was supported by the Hungarian Higher Education Excellence Programme (FIKP).

References

[1] M. Tisza, Zs. Lukács 2018, *High strength aluminum alloys in car manufacturing*, IOP Conf. Ser.: Mater. Sci. Eng. 418 012033

[2] https://www.lightmetalage.com/news/industry-news/automotive/aluminum-continues-unprecedented-growth-in-automotive-applications/ (last visited: 17.10.2021)

[3] Y. Song, L. Yang, G. Zhu, L. Hua, R. Liu, 2019, *Numerical and experimental study on failure behavior of steel-aluminium mechanical clinched joints under multiple test conditions*, International Journal of Lightweight Materials and Manufacture, Vol. 2, Iss. 1,pp. 72-79

[4] M. K. Sabra Atia, M. K. Jain, 2017, *Die-less clinching process and joint strength of AA7075 aluminum joints*, Thin-Walled Structures, Vol. 120, pp. 421-431

[5] F. Lambiase, 2015, *Clinch joining of heat-treatable aluminum AA6082-T6 alloy under warm conditions*, Journal of Materials Processing Technology, Vol.225, pp. 421-432

[6] Y. Ge, Y. Xia, 2020, *Mechanical characterization of a steel-aluminum clinched joint under impact loading*, Thin-Walled Structures, Vol. 151, 106759

[7] C.-J. Lee, J.-M. Lee, H.-Y. Ryu, K.-H. Lee, B.-M. Kim, D.-C. Ko, 2014, *Design of hole-clinching process for joining of dissimilar materials – Al6061-T4 alloy with DP780 steel, hot-pressed 22MnB5 steel, and carbon fiber reinforced plastic*, Journal of Materials Processing Technology, Vol. 214, Iss. 10, pp. 2169-2178.

[8] S. Coppieters, 2012, *Experimental and numerical study of clinched connections*, PhD dissertation, KU Leuven

Optimized metal foam-filled Crushbox structure to improve the impact energy absorption

J.Kertesz[1,2], T.A. Kovacs[3]

[1] University of Debrecen Faculty of Engineering, Air- and Road Vehicles Dep., Hungary
[2] Óbuda University, Doctoral School on Safety and Security Sciences, Hungary
[3] Óbuda University, Bánki Donát Faculty of Mechanical and Safety Engineering, Hungary

E-mail: [1,2]kertesz.jozsef@eng.unideb.hu, [3]kovacs.tunde@bgk.uni-obuda.hu

Abstract. According to the transport statistics, there are many personal injury affected road fatalities on the EU's roads. The increasing urbanization results in huge daily crowded paths in the cities. Drivers try to recoup for lost time in traffic jams with increased speed and reduced tracking distance, and the inattentive behavior to using mobile phones is a common cause of rear-end collision.The most common personal injury of the rear-end collision would be decreased by an optimized crumple zone to reduce the impact load force. With mass optimization not only the operational efficiency would be improved but the impact energy can be reeducated also. To achieve this aim integrated and advanced materials must be applied. This study is focusing on the optimization of the bumper-crushbox, and apply an improved structure in which more volume of the metal foam can be utilized, than the recent solutions.

1. Introduction

Analysing the reports of the ACEA (European Automobile Manufacturers' Association) we can realize in the last decade the number of cars is in use have been increased by more the 42%. According to their published report in 2010 more than 278 million vehicles were in use in Europe, while this number in 2020 has been almost exceeding 400 million [1]. Due to the increasing urbanization, the cities and their roads have become crowded with passenger cars day by day [2]. The road infrastructure (intersections, traffic lights program) can not follow this pace of the increase of traffic. The crowded roads result in impatient and uptight behaviour in the member of the traffic. Due to this impatient and careless driver behaviour generates upgraded tendency of the rear-end collisions. Analysing the statistics about the road fatalities, the rear-end collision gives a considerable proportion of the whole crash occasions [3]. Usually, the impact energy level of a rear-end collision is lower than a head-of-collision, notwithstanding the fatality can involve serious personal injury and bodywork damage as the effect of it. Therefore this study is focusing on the improvement of the outer passive safety system of road vehicles.

2. Damaged parts of the bodywork of the rear-end collision

Figure 1. shows the most frequently damaged vehicle body parts by rear-end collision (REC). The investigation is focusing on the city-road type rear-end collision, therefore the distinction is made between three main separations. The extent of the physical injury of the vehicle and the personal injury of the driver or passenger is influenced by the difference in the kinetic energy of the participants. The kinetic energy depends on the mass of the vehicle and the velocity of the vehicle. However analyzing the effect of the collision, we have to compare the kinetic features of each vehicle simultaneously. In the first group, where the low speed or low mass caused impact energy is low, only some front and rear optical parts will be damaged or destroyed. Broken bumper shells, broken mouldings, grilles, PDC sensors are the most common damaged vehicle body parts during low impact energy. If we increase the speed or/and mass of each participant vehicle, this results in higher impact energy during the collision. Not only the optical bumper shell will be destroyed, but –being the first member of the crumple zone- the bumper beam will be also damaged. We frequently experienced the sub- or full breaking of the glass-fibre-reinforced polymer manufactured front wall-beam. Owing to the bumper beam damaging, behind the beam, the coolers (air, water, oil, air-condition) are often idamaged by this collision. In favour of the mass optimization, the support of the head and rear light, support of the coolers is made by polymer, which results in easily damaging of these components. Over about 30 km/h or with heavy massed vehicles the impact energy will be multiplicated and result in serious technical damaging in the participants' vehicles. Over the bumper beam, the front and rear rails, the rear wall is damaged by the hood and tailgate destroying. The front hood conducts the impact energy through the support to the A-pillar, where the impact force will have appeared in windshield breaking. Naturally, this type of collision involves the activation of the airbags and the seat-belt tensioners, therefore the dashboard, inner upholstery expand on the list of parts to be replaced.

Figure 1. Most common bodywork damages of the REC. [own source]

3. Description of the safety level of the vehicles

Basically, the safety system of the vehicles can be divided into passive and active safety technologies. Active technologies try to prevent collision to improve the safety of the traffic. Nevertheless, the passive safety system tries to reduce the personal injury and technical injury effect of the collision [4]. Furthermore, the distinction is made between inner and outer passive solutions. However nowadays, in the case of modern vehicles, there is no strict boundary between the passive and active safety system, since their common operation of influence the level of the vehicle's safety. Owing to this we can use the integrated safety notion to describe the safety of the vehicles. The integrated safety is in fact a description based on the Haddon matrix model expressing the complete safety of vehicles. With a systematic approach to road transport, we can divide the human-vehicle environment into three. These elements in the pre-, during-, and post-accident states continuously determine and actively influence the circumstances of the accident and its consequences. Consequently, technological developments based on traffic accident research are based on this system-based principle [5].

4. Kinematics of the collision

The aim of this study is to reduce the effect of the collision focusing on the outer passive safety system development. However, the effect is influenced by the impact energy more exactly the impact force, which is relevant with the kinetic energy. According to the energy conversation statement, the kinetic energy must be equal to the impact energy during the collision [6].

$$E_{kinetic} = E_{impact} \tag{1}$$

The dimension of the crumple zone compression is designated as S. According to Newton's 3rd axiom in the case of two bodies interaction, they apply forces on each other that are equal in magnitude and opposite in direction. This force creates work on the S-displacement, resulting in the compression of the crumple zone.

$$E_{impact} = F \cdot S \tag{2}$$

$$F = m \cdot a \tag{3}$$

$$E_{kinetic} = \frac{1}{2} m v^2 \cdot \tag{4}$$

Take equal the (2) (3) and (4) equations:

$$\frac{1}{2} m v^2 = m \cdot a \cdot S \tag{5}$$

˙where the *m* is the impact mass, the *v* is the kinetic velocity and the *a* is the declaration of the vehicle.

Analysing the (5) formula, there are more options to reduce the amount of the F force, which loads the drivers and the passengers during the collision. According to the correlation, with an extended crumple zone, we can expand the compression distance results in an F force reduction. In another way, the amount of the F force that influences the collision effect can be reduced if the temporal changes of the velocity are prolonged. Furthermore, with mass optimization, not only the efficiency but safety can be improved simultaneously. Figure 2. shows the Crushbox of the bumper structure, which provides the above mentioned one of displacement of the crumple zone.

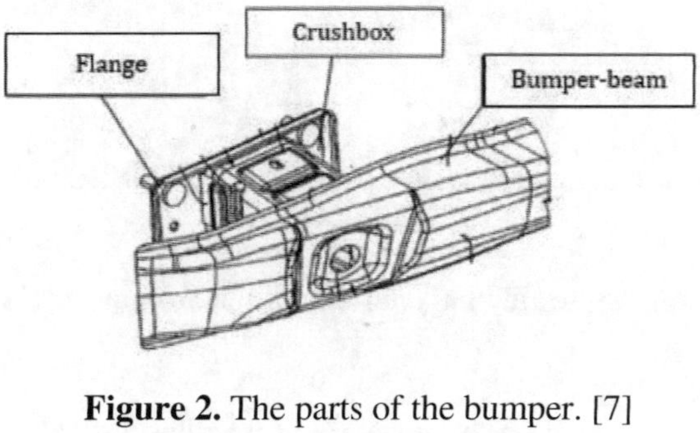

Figure 2. The parts of the bumper. [7]

5. Foam-filled crushbox to enhance the energy absorption

There are many recent investigations that confirm the statement, according to which the impact energy absorption capacity of each carrying body element can be improved with the application of the metal foam [8][9][10]. However, during the literature review, it can be experienced that the applied metal foam frequently is not utilized entirely. Figure 3. shows a foam-filled tube after impact investigation. As it can be seen the applied metal foam volume can be divided into three types of groups. Alongside the crsuhbox wall, the metal foam cells reach the maximum deformation option. Next to the extremely densified zone, in the normally densified region, the metal foam could not reach the maximum deformation state, resulting in the applied foam could suit for more energy absorption. Furthermore under there is a region, in which the foam after the collision stayed in undeformed form since it could not take part in the energy absorption. Figure 3. introduced FEM analysing also which confirms that a significant volume of the applied foam is not utilized in the absorption during the collision.

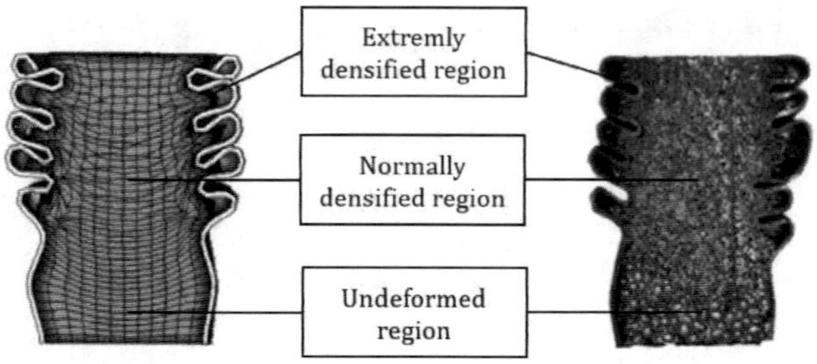

Figure 3. Impact loaded foam-filled tube. [5]

6. Utilization of the advantageous characteristics of metal foam

Before introducing the optimized Crushbox, the stress-strain characteristics of the metal foam must be discussed, since it gives the fundamental of the metal foam utilization. As it can be seen in figure 4. after the linear elasticity (cell wall bending) of the curves there is the ideal energy absorber zone. In this section the largescale strain involves almost permanent stress value, therefore one of the most proper materials for the impact absorbing. Numbers of open literature have reported the dynamic behaviours and features of the metal foams [11][12][13]. After the ideal zone, when the cells become fully deformed, there is the safety back-up zone. The aim with the application of the metal foam is to utilize the maximum horizontal characteristics (Crush Plateau) of the curves or to further extend this horizontal line.

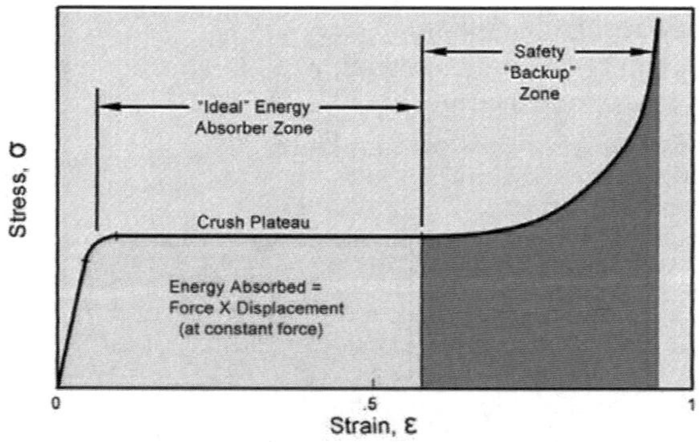

Figure 4. Stress-strain characteristic of the aluminium foam. [14]

7. Optimized Crushbox

For us, the optimized crusbox is based on the structure of the telescopic cylinder. As in figure 5. can be seen contrast with the standard one-element Crushbox, we separated it into more sections. Each section is separated by a metal foam plate. When the collision acts force on the top of the Crushbox, the first upper member slips down into the below part meanwhile it compressed the metal foam and the cells begin to densify thereby absorbing the impact energy. Since the actuating load distributes on the entire surface of the metal foam, the efficiency could be more favourable. During the operation, we try to avoid or postpone the negative vertical line of the stress-strain characteristics, therefore before the cell

IOP Conf. Series: Materials Science and Engineering 1246 (2022) 012004 doi:10.1088/1757-899X/1246/1/012004

achieves the overloaded state the flange of the first upper element lays on the upper flange of the second element which begins to shink into the lower element. In the second section, the metal foam is also loaded on the entire surface by the compressing force. Moreover owing to the separated sections we can apply different characterized metal foam, or different material foams can be applied simultaneously that can result in new stress-strain characteristics due to their serial arrangement. The sliding components are mounted with a certain amount of clearance to provide the displacement, with a proper clearance the energy absorption can be improved since more energy will be dissipated to overcome the friction between the components. With a different variety of clearance and with different features metal foams applying we can create modulated Crushbox.

Figure 5. Structure of the developed telescopic based crsuhbox. [own source]

8. Conclusions

The significant growth of the road vehicles in use results in frequently crowded paths on the city roads. In order to retrieve the wasted time in traffic jams, drivers often neglect the following distance and the speed limits during the transport which results in the increment of the number of rear-end collisions occasions. To prevent the serious personal injury effect of this type of facility the impact energy must be reduced during the collision involving the improvement of the outer passive safety system of the vehicles. Recent investigations verify that the foam-filled Crushbox of the bumper-beam results in enhanced energy absorption efficiency. However, the amount of absorbed energy would be increased by our developed optimized structure, which involves lower impact forces on the drivers and passengers. In contrast with the standard Crushbox, the telescopic structure provides the possibility to apply different featured foam simultaneously thereby creating a modulable structure. Owing to the enhanced absorption capacity, conventional metals (steel, aluminium) can be replaced by lighter but advanced materials, for example, CFRP(Carbon Fiber Reinforced Polymer).

References

[1] ACEA (European Automobile Manufacturers Association) report – Vehicles in use Europe, January 2021

[2] Hamidreza Salaripoor, Mohammadbagher B Azimi and Masoud Asgari - Optimized foam filling configuration in bi-tubular crush boxes; a comprehensive experimental and numerical analysis Engineering Research Express, Volume 2, Number 1; Published 15 January 2020

[3] A I Radu - Analysis of driver's seat rigidity in the case of rear-end collisions using a virtual multibody model; 2021 IOP Conf. Ser.: Mater. Sci. Eng. 1169 012021

[4] Devendra Vashist, Naveen Kumar – A review of active and passive automotive safety systems; International Journal of Engineering Sciences & Management Research, ISSN 2349-6193

[5] Kőfalvi Gy. Ignácz F. – Utasmozgás vizsgálata gépjárművek ütközésénél (2011) Biomechanica Hungarica IV. évfolyam, 1. szám pp. 19-29, DOI: 10.17489/biohun/2011/1/03

[6] Davoud Khorasani-Zavareh , Maryam Bigdeli , Soheil Saadat c, Reza Mohammadi - Kinetic energy management in road traffic injury prevention: a call for action; J Inj Violence Res. 2015 Jan; 7(1):36-37. ; DOI: 10.5249/jivr.v7il.458

[7] United States Patens - Michael R., Ulrich Lütke-Bexten, Wilhelm A., Elmar M. – Bumper for a motor vehicle (Patent No.: US7357432B2) Benteler Automobiltechnik GmbH, Paderborn (DE) (Apr. 15 2008)

[8] Hong-WeiSong, Zi-JieFan, GangYu, Qing-ChunWang, A.Tobotac - Partition energy absorption of axially crushed aluminium foam-filled hat sections, International Journal of Solids and Structures, Volume 42, Issues 9–10, May 2005, Pages 2575-2600

[9] H.R. Zarei, M. Kröger - Optimization of the foam-filled aluminium tubes for crush box application; Thin-Walled Structures Volume 46, Issue 2, February 2008, Pages 214-221

[10] P. D. Jeyakumar, G. Devaradjane - Energy Absorption by Square, Circular, Pentagon and Hexagonal Steel Tubes Under Impact Loading for Automotive Crush Box Applications; International Review of Mechanical Engineering (I.RE.M.E.), Vol. 7, N. 7; ISSN 1970 - 8734 November 2013

[11] V. Deshpande, N. Fleck - High strain rate compressive behaviour of aluminium alloy foams; Int J Impact Eng, 24 (3) (2000), pp. 277-298

[12] K.A. Dannemann, J. Lankfords - High strain rate compression of closed-cell aluminium foams; Mater Sci Eng A, 293 (1) (2000), pp. 157-164

[13] T. Mukai, H. Kanahashi, T. Miyoshi, M. Mabuchi, T. Nieh, K. Higashi - Experimental study of energy absorption in a close-celled aluminium foam under dynamic loading; Scr Mater, 40 (8) (1999), pp. 921-927

[14] ERG materials & aerospace® - http://ergaerospace.com/technical-data/duocel-foam-energy-absorption/ dowload : 01.11.2021

IOP Conf. Series: Materials Science and Engineering 1246 (2022) 012005 doi:10.1088/1757-899X/1246/1/012005

Investigation of the torque transmission time constant of a multi-disc magnetorheological clutch in the case of ramp function excitation

S Mester[1], B Horváth[2], A Lukács[1] and I Szalai[1,2]

[1] Institute of Mechatronics Engineering and Research, University of Pannonia, H-8900 Zalaegerszeg, M. Gasparich St. 18/A, Hungary
[2] Research Centre for Engineering Sciences, Functional Soft Materials Research Group, University of Pannonia, H-8201 Veszprém, PO Box 158, Hungary

Abstract. In this work the torque transmission time constant of a multi-disc magnetorheological (MR) fluid clutch is presented. The self-built clutch contains six layers of MR fluid and it is excited by an external electromagnetic coil. The transmitted torque is measured by a torque sensor using a self-developed LabVIEW software. Torque transmission is measured using ramp excitation with different slopes. The time constant of the clutch is derived from the time delay of the torque response. Our measurements showed the time constants as a hyperbolic function of the slope of the ramp excitation.

1. Introduction

Magnetorheological (MR) fluids are intelligent materials that consist of high permeability particles dispersed in a fluid with low magnetic permeability [1]. They contain mostly mineral or silicone oil as the carrier liquid and ferromagnetic particles (e.g. iron) as the dispersed phase (1 μm to 10 μm in diameter) [2,3]. Their special behaviour is that their rheological properties change under the influence of an external magnetic field: their apparent viscosity increases. Under the influence of the external magnetic field, magnetic dipole moment of the dispersed particles is induced and they are organized into pairs and then into chains and columns, which resist the shear effects on the liquid [4]. The effect can be reversibly changed by turning off the field. Due to the apparent increase in viscosity, magnetorheological fluids can be used in clutches [5], vibration dampers [6,7], and superfinishing [8,9], among others.

Two basic types of magnetorheological clutches can be distinguished: the disc type, where the fluid is located between the discs, and the cylindrical type, where the fluid is between the circumferential surfaces of concentric cylinders [10]. In addition to the basic types, mixed types and couplings with several fluid layers are also being tested, thus increasing the amount of the transmitted torque [11].

2. Experimental

In order to maximize the transmitted torque, we designed and built a multi-disc MR clutch, the internal structure of which is shown in figure 1. Only the magnetorheological fluid, which is located between the inner discs (9) and the outer discs (6), creates a connection between the input shaft (1) and the output shaft (10).

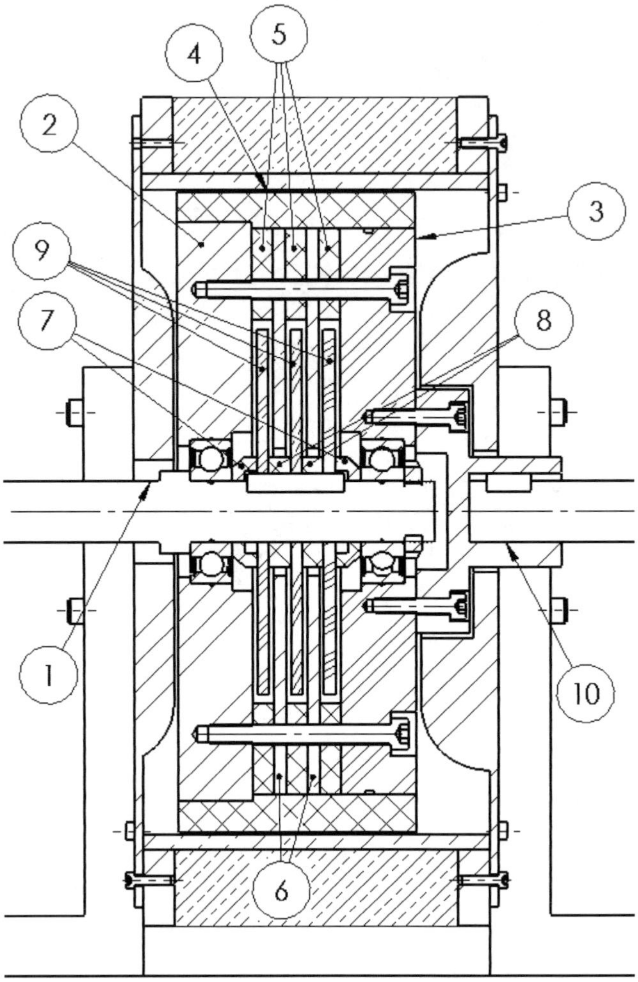

Figure 1. Internal structure of the multi-disc magnetorheological clutch. 1 – Input shaft, 2 – End disk one, 3 – End disk two, 4 – disk clamp, 5 – Outer spacers, 6 – Outer disks, 7 – End spacers, 8 – Inner spacers, 9 – Inner disks, 10 – Output shaft

The inner discs are held by the inner spacers (8) and end spacers (7), while the outer ones are held in the correct position by the outer spacers (5) relative to the end discs (2 and 3). The magnetic field generating coil surrounds the discs and flux guides direct the magnetic field to the discs from both sides.

To determine the torque transmission time constant of the clutch, a series of measurements were performed, where we measured the transmitted torque during a ramp function excitation. Measurements were made at five different speeds (30 min^{-1} to 70 min^{-1} in 10 min^{-1} increments) and at six different slopes (20 $kAm^{-1}s^{-1}$, 27 $kAm^{-1}s^{-1}$, 34 $kAm^{-1}s^{-1}$, 38 $kAm^{-1}s^{-1}$, 43 $kAm^{-1}s^{-1}$ and 47 $kAm^{-1}s^{-1}$). During the experiments the clutch was filled with the MRF-122EG fluid from Lord Corporation.

3. Results

Examples of the measured time domain torque responses during ramp excitation with different slopes and at 70 min^{-1} are shown in figure 2.

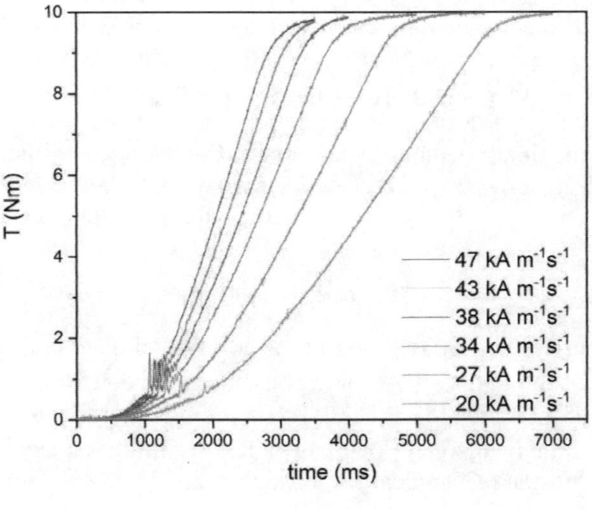

Figure 2. Torque responses at n=70min^{-1} speed and different slopes.

The time domain torque response of the MR clutch to the ramp excitation $H(t) = At$ can be described by the response of a first-order linear system as

$$T(t) = KA\left[t - \tau\left(1 - e^{-\frac{t}{\tau}}\right)\right] \quad (1)$$

where t is time, τ is the time constant, $A = \frac{\Delta H}{\Delta t}$ is the slope of the ramp, and K is the steady state gain of the system. Due to the exponential term's fast decay, the response can be approximated as $T(t) \approx KA[t - \tau]$ when $t \gg \tau$. The response time is defined as the time delay in the steady state region, if both the excitation and the response is normalized to have a slope of unity, $\frac{H(t)}{A}$ and $\frac{T(t)}{KA}$ respectively [12]. In the region of the applied magnetic field strength the system's magnetic behaviour is linear. After the normalization the linear regions of the response and the excitation are fitted, and the time constants of the whole system and the electronics are gained as the time delays, respectively. The asymptotic linear fit to the normalized curves is shown in figure 3.

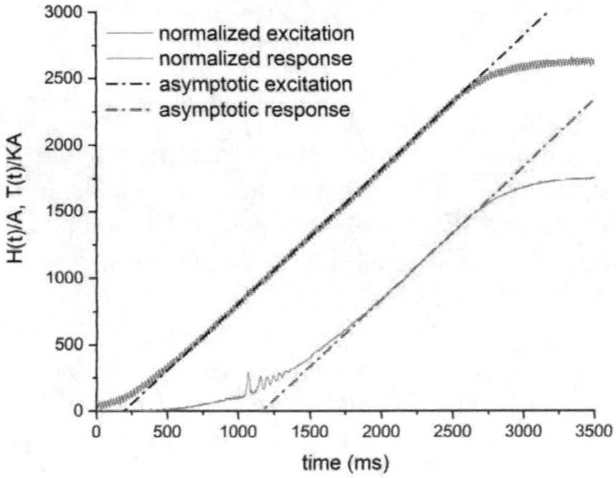

Figure 3. Fitted function to the excitation and response at n=70min^{-1} speed and slope of 47 kAm^{-1}s^{-1}.

A summary of the time constants for the entire system is shown in figure 4 and figure 5. As a function of speed, the time constant is independent of speed with a good approximation. At low speeds, a slight increase can be observed at all slopes, however, there is no significant deviation after the 40 min^{-1} speed, which still exists, are already within the limits of the measurement inaccuracy. As a function of field strength, the time constants show a decreasing trend as the slope increases. If plotted as a function of the reciprocal of the slope, the time constants at each speed fit on a straight line. It follows that the time constants shown in figure 5 are hyperbolically dependent on the slope.

The gained time constants are representative to the whole system, and are the sum of three components:

$$\tau = \tau_f + \tau_m + \tau_e \tag{2}$$

where τ_f is the time constant of the fluid, τ_m is the time constant of the clutches mechanical parts (e.g. bearings, timing belt, inertia etc.) and τ_e is the time constant of the electromagnetic coil and the power supply. The time constant of the fluid (MRF-122EG) has been measured beforehand, $\tau_f = 29{,}8\ ms$ at 47 kAm^{-1}s^{-1}. The response time of the electronics is gained by fitting an asymptotic to the excitation (figure 3) at each setting. Thus the mechanical response time at 47 kAm^{-1}s^{-1} and 70 min^{-1} is

$$\tau_m = \tau - \tau_f - \tau_e = 1164{,}5\ ms - 29{,}8\ ms - 197\ \ ms = 937{,}7\ ms \tag{3}$$

At other slopes and speeds the difference between τ and τ_m is similar, showing the dominance of the mechanical component of the response time due to the timing belt and the high inertia of the system.

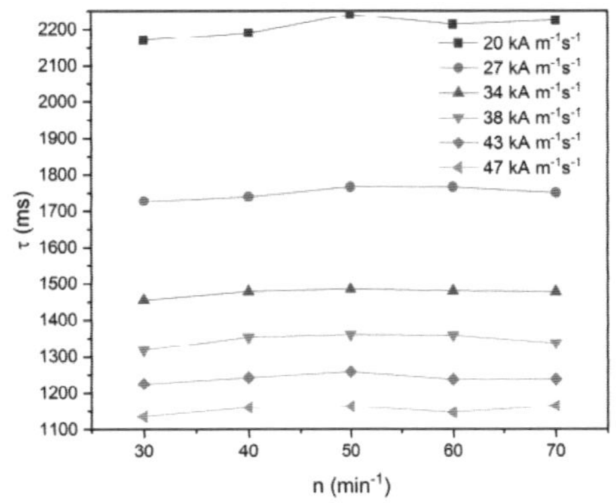

Figure 4. Time constants as a function of speed at different slopes.

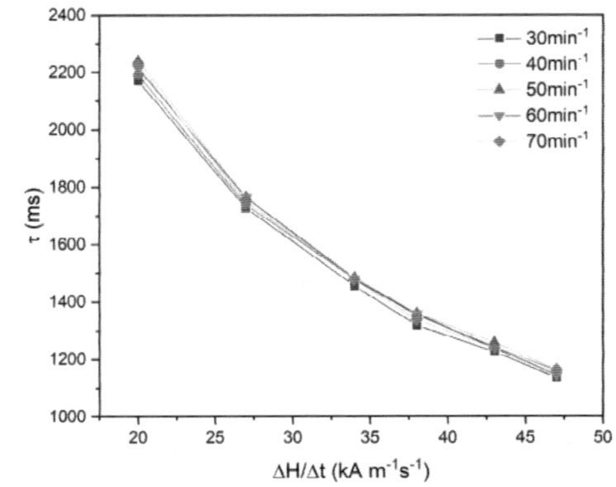

Figure 5. Time constants as a function of slopes at different speeds.

4. Conclusion

After the design and construction of the multi-disc magnetorheological clutch, I performed a series of measurements to determine the torque transmission time constants using ramp function excitation.

The time constants are independent of speed, while they are hyperbolically dependent on the slope of the excitation. In the future, I want to examine the time constants in a different measurement procedure.

References

[1] Rabinow J 1948 The Magnetic Fluid Clutch *Trans. Am. Inst. Electr. Eng.* **67** 1308–15
[2] Rankin P J, Ginder J M and Klingenberg D J 1998 Electro- and magneto-rheology *Curr. Opin. Colloid Interface Sci.* **3** 373–81
[3] Asiaban R, Khajehsaeid H, Ghobadi E and Jabbari M 2020 New magneto-rheological fluids with high stability: Experimental study and constitutive modelling *Polym. Test.* **87** 106512

[4] Patel R 2011 Mechanism of chain formation in nanofluid based MR fluids *J. Magn. Magn. Mater.* **323** 1360–3
[5] Tsai N C, Yeh C T and Chiu H L 2018 Linear magnetic clutch to automatically control torque output *ISA Trans.* **76** 224–34
[6] Yao G Z, Yap F F, Chen G, Li W H and Yeo S H 2002 MR damper and its application for semi-active control of vehicle suspension system *Mechatronics* **12** 963–73
[7] Yoon D S, Park Y J and Choi S B 2019 An eddy current effect on the response time of a magnetorheological damper: Analysis and experimental validation *Mech. Syst. Signal Process.* **127** 136–58
[8] Peng W, Li S, Guan C, Li Y and Hu X 2018 Ultra-precision optical surface fabricated by hydrodynamic effect polishing combined with magnetorheological finishing *Optik (Stuttg).* **156** 374–83
[9] Hashmi A W, Mali H S, Meena A, Khilji I A, Chilakamarry C R and Saffe S N binti M 2022 Experimental investigation on magnetorheological finishing process parameters *Mater. Today Proc.* **48** 1892–8
[10] Böse H, Gerlach T and Ehrlich J 2013 Magnetorheological torque transmission devices with permanent magnets *J. Phys. Conf. Ser.* **412**
[11] Hema Latha K, Usha Sri P and Seetharamaiah N 2017 Design and Manufacturing Aspects of Magneto-rheological Fluid (MRF) Clutch *Mater. Today Proc.* **4** 1525–34
[12] Horváth B, Decsi P and Szalai I 2021 Measurement of the response time of magnetorheological fluids and ferrofluids based on the magnetic susceptibility response *J. Intell. Mater. Syst. Struct.*

Interaction of zero energy levels with crystal defects in graphene. A network theory study

I László, S Kugler

Department of Theoretical Physics, Budapest University of Technology and Economics, H-1521 Budapest, Budafoki út 8. Hungary

Abstract. Applying network theory, we present relations concerning the degeneracy and localization of the zero energy modes in graphene. We demonstrate that in most of cases network properties of the lattice are responsible for existence and nonexistence of these topological determined energy levels.

1. Introduction

Graphene is a two-dimensional crystalline form of carbon in a honeycomb lattice. It can be imagined as a single layer of graphite. The band structure of a single layer graphene has a two-dimensional character and a linear dispersion relation of electronic wave functions with a perfect electron-hole symmetry, in which Fermi surface consists of two cones touching at one singular point, the so-called Dirac point. At the Dirac point density of states is zero. There are other zero-energy modes which are produced by vacancies and/or by other crystal defects [1-7]. Properties of the vacancy zero-energy modes are usually explained in one orbital one site approximation by the help of a theorem based on the bipartite property of the graphene sheet [8-10]. In the case of pure bipartite structure, where the number of atoms on the sublattice A is marked by $\mathrm{N_A}$ and those on the sublattice B is marked by $\mathrm{N_B}$ with the relation of $\mathrm{N_A} > N_{\mathrm{B}}$, the multiplicity of the zero-energy value is equal to $(\mathrm{N_A}-\mathrm{N_B})$. Here first neighbor interactions and zero diagonal matrix elements in the Hamiltonian matrix are used. If a Stone-Wales transformation [11] is applied in the graphene sheet, two pentagons and two heptagons will be formed, and the bipartite property of the lattice will be destroyed. It causes that the above-mentioned method cannot be used to determine the zero-energy modes.

In our previous work [12] we developed a method to determine the zero-energy modes in non-bipartite lattices, like the graphitic amorphous carbon structures where a one orbital one site approximation was applied for each atomic site. Later, we developed this method for cases where more than one atomic orbital is used for atomic sites [13-16]. We called these energy levels "topologically determined energy levels".

Based on our previous results we are presenting a method to determine the number of zero-energy modes in some atomic arrangements including presences of Stone-Wales transformations and/or vacancies.

2. Method

2.1. Energy levels of the perfect graphene lattice

Two-dimensional hexagonal graphene structures are described by unit vectors $\mathbf{a}_1 = \mathrm{a}(\sqrt{3},-1)/2$ and $\mathbf{a}_2 = \mathrm{a}(\sqrt{3},1)/2$. These vectors are given in a two-dimensional orthonormal basis. Each unit cell contains $r = 2$ atoms at relative positions of $(\mathbf{a_1} + \mathbf{a_2})/3$ and $2(\mathbf{a_1} + \mathbf{a_2})/3$. Using the lattice translation vector $\mathbf{t} = t_1\mathbf{a_1} + t_2\mathbf{a_2}$ we can describe the hexagonal lattice with two integers t_1 and t_2. The reciprocal unit cell vectors $\mathbf{b_1}$,$\mathbf{b_2}$ come from the vectors $\mathbf{a_1}$,$\mathbf{a_2}$ by the relations

$$\mathbf{b_1} = 2\pi(\mathbf{a_2} \times \mathbf{z})/(\mathbf{a_1} \cdot \mathbf{a_2} \times \mathbf{z}) \quad (1)$$

$$\mathbf{b_2} = 2\pi(\mathbf{z} \times \mathbf{a_1})/(\mathbf{a_1} \cdot \mathbf{a_2} \times \mathbf{z}) \quad (2)$$

The electronic wave function of the wave vector $\mathbf{k} = k_1\mathbf{b_1} + k_2\mathbf{b_2}$ has the form

$$|\mu\mathbf{k}\rangle = \sum_{\nu=1}^{r} c_{\mu\mathbf{k}}^{\nu} |\mathbf{k}\rangle_\nu \quad (3)$$

were

$$|\mathbf{k}\rangle_\nu = \frac{1}{\sqrt{M}} \sum_{\mathbf{t}} e^{i\mathbf{k}\cdot\mathbf{t}} |\mathbf{t}\rangle_\nu \quad . \quad (4)$$

The number of unit cells is M, and $|\mathbf{t}\rangle_\nu$ is the atomic basis function centred at the site ν of the unit cell $\mathbf{t}$. The $c_{\mu\mathbf{k}}^{\nu}$ coefficients are the μ -th eigenvectors of the $\mathbf{H}(\mathbf{k})$ Fourier transform of the Hamiltonian and $E_\mu(\mathbf{k})$ is the corresponding eigenvalue. This Hamiltonian is given by the relation

$$\mathbf{H}(\mathbf{k}) = \sum_{\mathbf{t}} e^{i\mathbf{k}\cdot\mathbf{t}}\, \mathbf{h}(\mathbf{t}) \quad (5)$$

were

$$\mathbf{h}(\mathbf{t})_{\mu\nu} = \langle \mathbf{0}^\mu | \mathbf{h}_{\mathbf{eff}} | \mathbf{t}^\nu \rangle \quad . \quad (6)$$

Here $\langle \mathbf{0}^\mu | \mathbf{h}_{\mathbf{eff}} | \mathbf{t}^\nu \rangle$ is different from zero, only if the μ -th and ν -th atoms of unit cell translation vectors $\mathbf{0}$ and $\mathbf{t}$ are first neighbors. The non-zero off-diagonal matrix elements or the hoping integrals are the followings: $\langle \mathbf{0}^\mu | \mathbf{h}_{\mathbf{eff}} | \mathbf{t}^\nu \rangle = -\gamma$.
The $E_\mu(\mathbf{k})$ eigenvalues of the Hamiltonian $\mathbf{H}(\mathbf{k})$ can be written as:

$$E_\mu(\mathbf{k}) = \pm\{3\gamma^2 + 2\gamma^2 \cos 2\pi k_1 + 2\gamma^2 \cos 2\pi k_2 + 2\gamma^2 \cos 2\pi(k_2 - k_1)\}^{1/2} \quad (7)$$

The Fermi energy is at the value of $E_\mu(\mathbf{k}) = 0$ and the positive and negative bands of Eq. (7) touch at the $\mathbf{K}$ points with the coordinates $(\mathbf{b_1} + 2\mathbf{b_2})/3$ and $(2\mathbf{b_1} + \boldsymbol{b_2})/3$. After developing the energy $E_\mu(\mathbf{k})$ in Taylor series at these two $\mathbf{K}$ points we obtain the Dirac cones. Here the energy depends linearly on the wave number $\mathbf{k}$ at the $\mathbf{K}$ points. Dispersion relation $E_\mu(\mathbf{k})$ of Equation (7) is obtained by the help of a very simplified tight-binding Hamiltony. It was supposed that near the Fermi level only the π - electrons are important and each atomic site contained only this one atomic orbital. We supposed further first neighbour interaction in the Hamiltonian.

2.2. Network theoretical method to determine the degeneracy and localization of the topologically determined electronic energy levels.

Let us suppose that we want to find the eigenvalues and eigenvectors of a Hamiltonian matrix $\mathbf{H} = [H_{ij}]$. This is a symmetrical square matrix of size $N \times N$. Here N is the number of atomic orbitals. Each of these symmetrical square matrices can be seen as networks represented by a graph $G = (V, E)$, where $V = \{v_1, v_2, \ldots, v_N\}$ is the set of N vertices and (v_i, v_j) is an edge of graph $G = (V, E)$ if and only if $H_{ij} = H_{ji} \neq 0$. Here E is the set of edges, which is $(v_i, v_j) \subset E$ and for the number of vertices in a set of V we shall use the notation $|V|$. That is in our case $N = |V|$.

The characteristic polynomial $D(\varepsilon)$ of $\mathbf{H}$ can be written as

$$D(\varepsilon) = \det|\varepsilon\mathbf{I} - \mathbf{H}| = \sum_{n=0}^{N} a_n\, \varepsilon^{N-n} \quad (8)$$

here $\mathbf{I}$ is the unit matrix and the coefficients of $D(\varepsilon)$ are given by

$$a_n = (-1)^n \sum H(i_1 i_2 \ldots i_n) \quad (9)$$

where the summation runs over all $\binom{N}{n}$ principal minors of H of order n. These minors are given by

$$H(i_1 i_2 \ldots i_n) = \sum_{(j_1 j_2 \ldots j_n)} (-1)^{t(p)} H_{i_1 j_1} H_{i_2 j_2} \ldots H_{i_n j_n} \qquad (10)$$

where $t(p)$ is the parity of the permutation $j_1 j_2 \ldots j_n$ with respect to the permutation $i_1 i_2 \ldots i_n$.

If any matrix element $H_{i_k j_k}$ equals to zero, the product $H_{i_1 j_1} H_{i_2 j_2} \ldots H_{i_n j_n}$ equals to zero as well. In the other case if each term of the product $H_{i_1 j_1} H_{i_2 j_2} \ldots H_{i_n j_n}$ different of zero, the graph $G = (V, E)$ of matrix $\mathbf{H} = [H_{ij}]$ contains a subgraph of n vertices having only the components of loop, complete graph of two vertices or cyclic graphs (Figure 1).

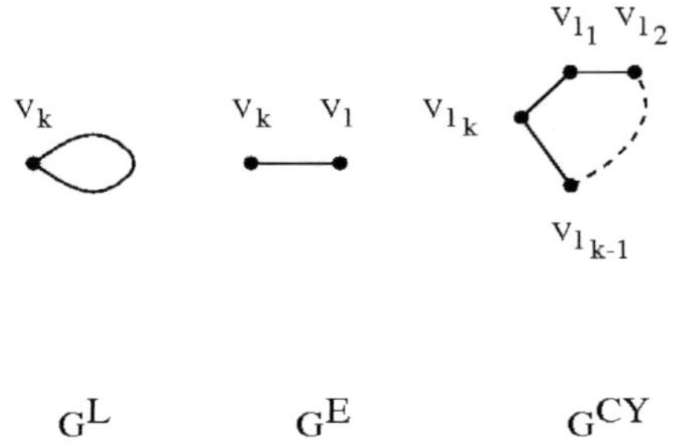

Figure 1. Definition of graph components: loop (G^L), complete graph of two vertices (G^E) and cyclic graph (G^{CY}).

The name of a graph is Sachs graph in chemical graph theory, or 2-matching and q-factor in mathematics if its components are those of Figure 1. Thus Sachs-graph contains only cycle, edge and/or loop components. If there are no Sachs-graphs with $N, N-1, \ldots, N-\nu+1$ vertices in G than from Eq. (8) follows that

$$a_N = a_{N-1} = \ldots = a_{N-\nu+1} = 0, \qquad (11)$$

and the $\varepsilon = 0$ is a ν -fold degenerated eigenvalue of H. Here we remark that the study of $\varepsilon = \alpha = H_{kk}$ eigenvalue can be carried out with the study of the $\varepsilon = 0$ eigenvalue of the matrix $\mathbf{H}' = \mathbf{H} - \alpha\mathbf{I}$. The eigenvalues are topologically determined as their values do not depend on the actual values of the off-diagonal matrix elements. They belong to the appropriate distribution of the off-diagonal matrix elements of zero values.

Let G' be a graph of the matrix $\mathbf{H}'$. It was shown [10,12] that $\varepsilon = \alpha = H_{kk}$ is a ν -fold not accidentally degenerated eigenvalue of $\mathbf{H}$ if and only if the vertices of graph G' can be partitioned into three disjoint sets V_1, V_2 and V_3 with the following properties. Set V_1 contains $|V_1|$ independent (not neighboring) loop-free vertices, $V_2 = \Gamma(V_1)$, that is in V_2 are all the neighbors of V_1 and

$$\nu = |V_1| - |V_2| > 0. \qquad (12)$$

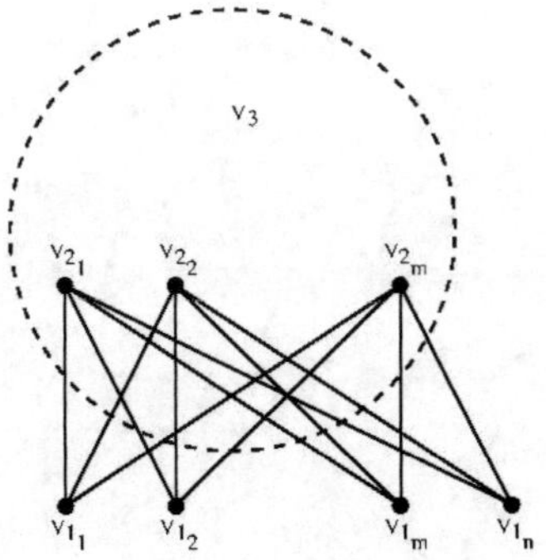

Figure 2. The partition of a graph G, into the sets V_1, V_2 and V_3. Set V_1 contains $|V_1|$ independent (not neighboring) loop-free vertices, $V_2 = \Gamma(V_1)$, that is in V_2 are all the neighbors of V_1. In set V_3 are the other vertices.

Other vertices of G' are in V_3. (see Figure 2). It was proved further [10] that the non-zero coefficients of the corresponding eigenvectors are only on the set V_1. That is, they are localized on the atoms of the set V_1 and eliminating any of the vertices of set V_1 decreases the number ν of the degeneracy. Thus, it is a global property of the network. The before mentioned V_1, V_2 and V_3 partitioning of the vertices is a necessary and sufficient condition for the non-existence of Sachs-graphs [16,17]. We remark here that in general cases the graph of Hamiltonian matrix describes not only the atomic network but also the network of atomic orbitals [10]. In the case of one orbital one site approximation these two networks are the same.

3. Application

The atomic arrangement of our first example for topologically determined energy values can be seen in Figure 3.

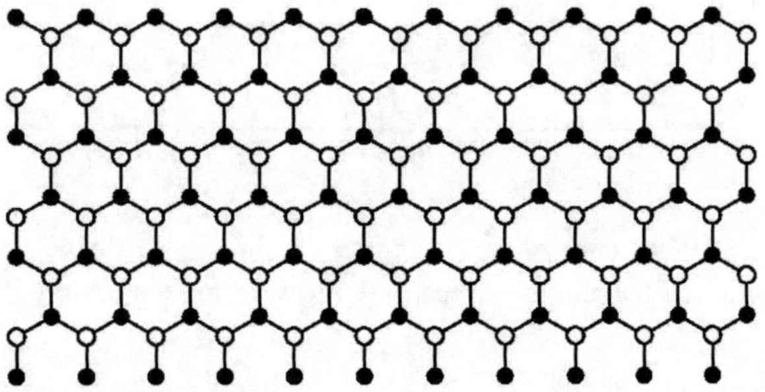

Figure 3. A perfect graphene cluster. Filled circles belong to the set V_1, atoms in V_2 are displayed by open circles. $|V_1| = 77$, $|V_2| = 66$ $|V_3| = 0$. Multiplicity of the $\varepsilon = 0$ eigenvalue is equal to $|V_1| - |V_2| = 11$.

Atoms in graphene marked by filled circles belong to the set V_1 while atoms in V_2 are displayed by open circles. Numbers in these two sets equal $|V_1| = 77$ and $|V_2| = 66$. Thus, the multiplicity of the $\varepsilon = 0$ eigenvalue becomes $|V_1| - |V_2| = 11$ and the corresponding eigenvectors are localized on the black atoms. Important to remark that the Dirac-cones are touching only if the number of carbon atoms

IOP Conf. Series: Materials Science and Engineering 1246 (2022) 012006 doi:10.1088/1757-899X/1246/1/012006

in the grapheme is infinite. Values of these topologically determined eigenvalues in graphene are always zero.

Figure 4. A graphene sheet with full of defects. Eliminating the vertices marked with arrows, the graph will have Sachs graph and the multiplicity of the $\varepsilon = 0$ eigenvalue is equal to zero.

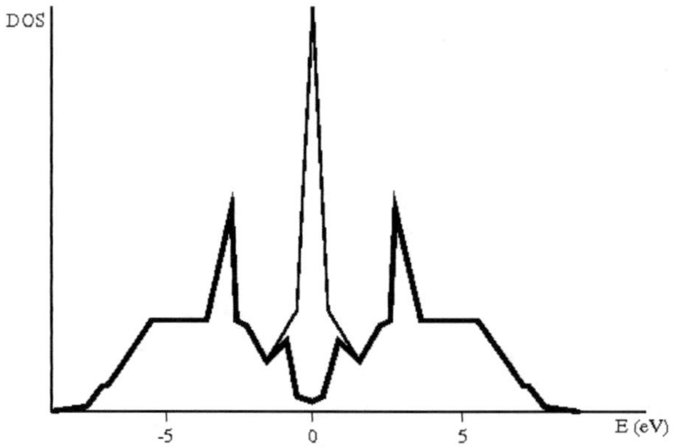

Figure 5. Calculated density of states of the structure in Figure 4. The peak at zero (thin line) disappears after eliminating of the atoms marked with arrows and the multiplicity of the $\varepsilon = 0$ eigenvalue changes to be zero.

A graphene sheet with several defects is shown in Figure 4 as our second example. It is often called haeckelitelike [18] structure. This atomic arrangement has four topologically determined $\varepsilon = 0$ energy levels. In ref. 8 we calculated the density of states for amorphous carbon structure which was displayed in Figure 4. The calculated density of states is presented in Figure 5. It has a large peek at $\varepsilon = 0$ (thin line). This peak disappeared after eliminating $\nu = |V_1| - |V_2|$ π -orbitals in each haeckelitelike graphene by bounding – for example – hydrogen atoms. Such kind of peeks are often explained by the number of pentagons and heptagons in the structures.

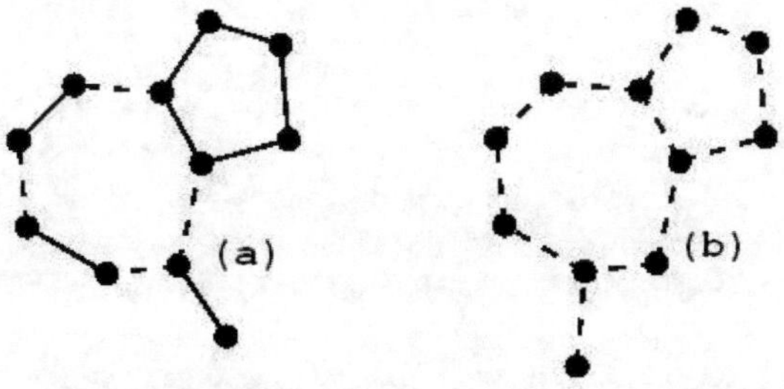

Figure 6. Two simple structures having the same number of atoms, pentagons and heptagons and an additional atom. The structure (a) does not have topologically determined energy level, but structure (b) does one. Solid lines mean edges in isolated edges or in cycles for the Sachs graph. Dashed lines mean edges do not belonging to Sachs graph.

Example 3. In Figure 6 we present two structures having the same number of atoms, a pentagon, and a heptagon plus an additional atom. These two structures are near the same atomic arrangements, but only one of them has topologically determined energy level [15]. The corresponding peeks can be found also in ab-initio calculations as well [15]. It means that this phenomenon is not an artifact of the tight-binding method.

Figure 7. A perfect graphene lattice with $N_{AB} = N_A = N_B$ number of atoms in the sublattices A and B. Periodic boundary condition is applied.

Figure 8. Graphene lattice after introducing defects in the structure of Figure 7. In this case $N_f = 1$ and $N_o = 4$ are the number vacancies and $N_{SW} = 1$ is the number of Stone-Wales transformations.

The next example is the cases of vacancies and Stone-Wales transformations. We introduce defects in an originally perfect graphene lattice with periodic boundary condition, which contains $N_{AB} = N_A = N_B$ atoms in the sublattices A and B (see Figure 7). In our last example we take away N_f atoms from the lattice A and N_o atoms from the lattice B and introducing N_{SW} number of Stone-Wales transformations. Figure 8 shows an example of $N_f = 1$, $N_o = 4$ and $N_{SW} = 1$. If $N_o - N_f > 0$ than we define the atoms in set V_3 as in Figure 8 the four vertices with Y inside a circle symbol around each Stone-Wales transformation. Thus $|V_3| = 4N_{SW}$, $|V_2| = N_B = N_{AB} - N_{SW} - N_o$ and $|V_1| = N_A = N_{AB} - 3N_{SW} - N_f$. According to Equation (12) the ν degeneracy of the eigenvalue $\varepsilon = 0$ is

$$\nu = |V_1| - |V_2| = N_o - N_f - 2N_{SW} > 0 \quad (13)$$

and they are localized on the atoms of set V_1.

$$\nu = 0 \quad (14)$$

in other cases.

If $N_o - N_f < 0$ we change the coloring of the lattices and use the same idea for determining the ν degeneracy of the eigenvalue $\varepsilon = 0$ as above.

4. Conclusion

We have presented a simple method how to derive the multiplicity of the $\varepsilon = 0$ eigenvalue if we know network of atomic orbitals. The vertices of atomic orbital network can be divided into three sets V_1, V_2 and V_3 with the following properties: (1) set V_1 contains $|V_1|$ independent (not neighboring) loop-free vertices, (2) $V_2 = \Gamma(V_1)$, that is in V_2 are all the neighbors of set V_1. (3) $\nu = |V_1| - |V_2| > 0$, then the multiplicity of the $\varepsilon = 0$ energy value is ν and the corresponding eigenvectors are localized on the set V_1. In the case of one orbital one site approximation the atomic orbital arrangement and the atomic arrangement is the same. From this theorem follows the statements, that if a graphene sheet with boundary conditions contains N_f vacancies on lattice A and N_o atoms on lattice B and it

IOP Conf. Series: Materials Science and Engineering 1246 (2022) 012006 doi:10.1088/1757-899X/1246/1/012006

contains further N_{SW} Stone-Wales transformations than the multiplicity of the $\varepsilon = 0$ energy value is ν , if $\nu = N_o - N_f - 2N_{SW} > 0$ and $\nu = 0$ for other cases.

This method is valid for any two dimensional and quasi two-dimensional carbon systems like fullerenes or nanotubes. Here we should remark, that in these conditions we are not speaking about the accidental degeneracies.

5. References

[1] Cot P, Parnell J, Habibian S, Straßer C, Ostrovsky P M, Ast C R, Phys Rev B 2020, 101:235116

[2] Zhang Y, Gao F, Gao S, He L, Science Bulletin 2020, 65:194

[3] Linhart L, Burgdörfer J, Libisch F, Phys. Rev. B 2018, 97:035430

[4] Tian W, Li W, Yu W, Liu X, Micromachines. 2017, 8:163

[5] Cresti A, Ortmann F, Louvet Th, Tuan D V, Roche S, Phys. Rev. Lett. 2013, 110:196601

[6] Pereira V M, Guinea F, Lopes dos Santos J M B, Peres N M R, and Neto A H C, Phys. Rev. Lett.2006, 96:036801

[7] Pereira V M, Lopes dos Santos J M B, and Neto A H C, *Phys. Rev. B* **2008**, 77:115109

[8] Brouwer P W, Racine E, Furusaki A, Hatsugai Y, Morita Y, and Mudry C, Phys. Rev. B 2002, 66:014204

[9] Lieb E H, Phys. Rev. Lett. 1989, 62:1201

[10] Sutherland B, Phys. Rev. B 1986, 34:5208

[11] Stone A J, Wales D J, Chemical Physics Letters. 1986, 128:50

[12] Kugler S, László I, Phys. Rev. B 1989, 39:3882

[13] László I, Menyes Cs, Phys. Rev. B 1991 44:7730

[14] László I, Int. J. Quantum Chem. 1993, 48:135

[15] Kádas K, László I, Kugler S, Solid State Communications 1996, 97:631

[16] László I, J. of Mol. Struct. (THEOCHEM) 2000, 501-502:501

[17] Lovász L and Plummer M D, 1986 *Matching Theory,* (Akadémia Kiadó – North-Holland Budapest-Amsterdam, Hungary-Netherlands)

[18] Terrones H, Terrones M, Hernández E, Grobert N, Charlier J-C, and Ajayan P M, Phys. Rev. Lett. 2000, 84:1716

IOP Conf. Series: Materials Science and Engineering 1246 (2022) 012007 doi:10.1088/1757-899X/1246/1/012007

Examining the reliability of the data analysis system for continuous steel casting

K Wizner[1] and A Kővári[1]

[1,2] University of Dunaújváros, Dunaújváros, Hungary

[1] wiznerk@uniduna.hu, [2] akovari@uniduna.hu

Abstract. The application of industry 4.0 guidelines is also an integral part of the life of ISD Dunaferr Zrt. Working closely with technology professionals, we have created a data analysis system that processes data from continuous steel casting. The life cycle of slabs can be traced in a complex way from raw steel production to the end of hot rolling. In an industrial environment, the program provides an opportunity to study the effects on slabs (e.g. change in casting speed, change in mould level, change in the position of tundish stopper rod (TSR), etc.). The results provided by the system are suitable for both qualifications made by experts and automatic one during the visual inspection using the web display system. In this paper, we examined the correlation between the slabs burdened with known significant changes in casting parameters in the cast sequence and the slabs with significant changes classified by the data processing system to verify the reliability of the data analysis system. The results confirmed the reliability of the data analysis system and pointed to the greater reliability of the study of cast length-based changes versus time-based changes.

1. Introduction

The application of industry 4.0 guidelines is also an integral part of the life of ISD Dunaferr Zrt. [1]. To meet these challenges, several highly automated systems need to be put in place [2] to optimize production [3]. To achieve this a data analysis system which processes the data of the continuous steel casting for the Steelworks of ISD Dunaferr Zrt. was developed in close cooperation with the technological experts [4]. It makes the life cycle of slabs traceable from raw steel production to the end of hot rolling [5]. In an industrial environment, the program provides an opportunity to study the effects on slabs during the manufacturing process [6] (e.g. change in casting rate [7. 8], change in mould level [9, 10], change in position of TSR indicating inclusion attachment/detachment [11-13], etc.). The results provided by the system are suitable for both qualifications made by experts and automatic one during the visual inspection using the web display system [14].

Due to the above conditions, the correlation between significant changes in time and cast length (casting speed, mould steel level, TSR position) and the fact of sequence start (0: no; 1: yes) was examined for the sequence starting slabs (Table 2). At the start of casting, there is no risk of the inclusion cluster attachment/detachment, so in the TSR position at the start of casting stage (even if it exceeds the critical level) the slab is not rated based on changes above the limit value. For these reasons, there is no significant correlation between the sequence-starting slabs and those burdened with significant TSR position change either for time- or cast length-based changes.

2. Materials and methods

During the examination of the measured data, it was observed that the reliability and accuracy of data have improved year by year with the continuous development of the measurement systems and the technical conditions. As a compromise solution, the studies were based on the period from 01.01.2018 to 30.06.2020, which contains sufficient data to perform the studies safely with little data uncertainty (25,429 sequences → 122,712 heats → 977,827 slabs).

The data of steelmaking, secondary metallurgy and continuous casting as well as the conditions for the compositional classification of slabs are derived from the Oracle database of the steelworks. Downgrade data for rolled products (sheet, coil) are derived from the SAP / R3 database, which is part of the corporate governance system.

When processing the data, it should be borne in mind that there may be several empty (null) and zero (0) values between the parameters during data collection. An empty value is clearly manageable as it clearly indicates data loss. However, in the case of zero it can be decided only knowing the surrounding data whether the value is real or it has been recorded only due to a data error. During the processing of the data of the continuous steel caster, we tried to replace and correct the missing and erroneous data mentioned above, however, due to the peculiarities of the process, in many cases the clearly erroneous data sections in the given data environment proved to be correct in other data environments. Time and length data are vital for later analysis, so we excluded the strand data where critical data uncertainty was observed in a multi-step analysis. Due to critical data uncertainty, ~ 1.1% of the strand data were excluded from all sequences examined. Table 1 shows the incidence rate of critical data errors. Critical data problems with data with a 6-second storage cycle prevent further testing due to the uncertainty of the length data. The uncertainty of the data with a 20-second storage cycle "only" makes the determination of the actual heat start uncertain, and therefore prevents "only" the associated calculations (e.g. localization of the actual heat start and mixed strand section).

Table 1. Proportion of critical data problems during processing.

Data error phenomenon	Proportion*
Data with a storage cycle of 6 seconds	
Data storage rate differs from 6 seconds	0.610%
Length change is greater than maximum length change	0.726%
There is a change in length while the mould is empty (probably strand breakout)	0.624%
Strand length data is null	0.006%
Data is missing at the beginning of the strand	0.008%
The difference between the cast length and the cut length is too large	0.351%
Data with a storage cycle of 20 seconds	
The weight of the ladle has fallen below the minimum value	1.181%
Data storage rate differs from 20 seconds	0.204%
There are more heat start than heats	0.310%
The start of each heat cannot be determined	1.420%

*One strand can be affected by multiple data problems

The reliability of the results provided by the created data processing system can only be judged if they can be compared with authentic data and results that are independent of it. Due to the nature of continuous casting technology, numerous data can be relied upon for this study.

For example, the significant casting speed and mould steel level change at the beginning of casting can be clearly attributed to the sequential starting slabs, while the event of deceleration (casting speed below the minimum value) always affects the sequential starting and ending slabs. Both the start and end of the sequence and the deceleration event can be easily defined, and the involvement of the slabs can be clearly identified, and is therefore comparable to the involvement of a significant non-steady state identified by the data processing system.

The extent of the expected correlation proving the goodness of the data processing system always depends on the compared events, but it can be stated that no perfect match ($R^2 = 1$) can be expected for any of the events, even with the perfect operation of the data processing program. The degree of correlation (R^2) was considered weak from 0.2, moderate from 0.4, and strong from 0.7.

3. Results and discussion

The location of non-stabilized events (e.g. significant mould level fluctuations) in many cases occupies a prominent place on the cast strand (e.g. casting start, casting end). Therefore, the strands in this section are affected by the particular casting event due to the technology [15]. This correlation was used to determine the correlation of the involvement of slabs with the casting event according to the technology as well as the data processing system.

3.1. Examining the reliability of changes based on time and cast length

The time-based monitoring and evaluation of the processes mainly provides information on the production process, but only provides approximate information on the effects on a given stage of the product (cast strand / slab) that affect its quality. With the help of the data processing system, it was also possible to examine the changes during casting based on the cast length [16]. For the first slab of the cast strand, the casting conditions are far from constant. During a short time and cast length, it is affected by a significant change in casting speed (increasing the drawing speed from a standing position to the recommended casting speed) and a change in mould level (due to filling and the nature of the control). In addition to these conditions, the secondary cooling zone of the casting machine does not even have a steady state of cooling and support.

Due to the above conditions, the correlation between significant changes in time and cast length (casting speed, mould steel level, TSR position) and the fact of sequence start (0: no; 1: yes) was examined for the sequence starting slabs (table 2). At the start of casting, there is no risk of the inclusion cluster attachment/detachment, so in the TSR position at the start of casting stage (even if it exceeds the critical level) the slab is not rated based on changes above the limit value. For these reasons, there is no significant correlation between the sequence-starting slabs and those burdened with significant TSR position change either for time- or cast length-based changes.

Table 2. Significant time- and cast length-based changes and correlation of sequence starting slabs (R^2).

		Casting speed change (based on)		**Mould steel level change (based on)**		**TSR position change (based on)**		**Sequential start**
		Time	*Length*	*Time*	*Length*	*Time*	*Length*	
ΔSpeed	*Time*	1	0.678	0.002	0.425	0.100	0.100	0.410
	Length	0.678	1	0.003	0.705	0.151	0.151	0.704
ΔLevel	*Time*	0.002	0.003	1	0.009	0.001	0.001	0.001
	Length	0.425	0.705	0.009	1	0.132	0.132	0.860
ΔTSR	*Time*	0.100	0.151	0.001	0.132	1	0.977	0.134
	Length	0.100	0.151	0.001	0.132	0.977	1	0.134
Sequential start		0.410	0.704	0.001	0.860	0.134	0.134	1

Based on the comparison of the correlations shown in the table, the following conclusions can be made:

- In case of casting speed and mould steel level, cast length-based changes show a stronger correlation with the sequence starting slabs than time-based changes. Based on this, more reliable studies are possible by examining cast length-based changes than by time-based changes.
- At the casting speed, it can be observed that the changes based on time and cast length are moderately correlated with each other, so it is worth examining both aspects further in later studies.
- Significant level fluctuations typically occur with significant changes in casting speed. Again, a stronger correlation is observed for cast length-based changes.

- The time-based change in mould steel level does not show a significant correlation with either the other changes or the sequence starting slabs. Based on this, it can be stated that it is not reliable for slab tests.
- In the case of a significant change in the TSR position, it can be observed that there is a strong correlation between the changes based on time and the cast length. This suggests that a significant change in plug position indicative of inclusion detachment typically occurs at a casting stage in which the casting speed does not change (significantly), i.e. the quotient of the cast length and the elapsed time is (nearly) constant.

Based on the results, it can be stated that the significant changes in casting speed based on the length as well as the significant changes in the mould level show a strong ($R^2 \geq 0.7$) correlation compared to the reference changes (sequence start) and can therefore be considered reliable. The relevance of the time-based significant change in casting speed is questionable due to its medium correlation ($R^2 \geq 0.4$), so further studies are needed to verify its usability. The time-based significant change in mould level cannot be considered reliable based on the results.

3.2. Examining the reliability of casting speed changes

The cast slabs at the start and end of casting were affected by a significant change in casting speed due to the nature of the technology, so we examined the correlation between the casting speed change parameters and the sequential start / end slabs.

Table 3. Correlation of casting deceleration conditions (R^2).

		Casting speed change (based on)		**Decelerated**	**Start or end**
		Time	*Length*		
Casting speed change base	*Time*	1	0.678	0.201	0.188
	Length	0.678	1	0.291	0.327
Decelerated		0.201	0.291	1	0.452
Start or end		0.188	0.327	0.452	1

Based on table 3, a weak ($R^2 \geq 0.2$) correlation is observed for deceleration-affected slabs with a significant casting speed change and the sequential start / end slabs. No significantly higher correlation is expected because deceleration often occurs even in cases where the other two circumstances do not. Table 4 shows that the proportion of the slabs decelerated below the minimum casting speed is 3-4 times that of the slabs affected by a significant change in casting speed. The weak correlation in the light of the circumstances confirms the finding that the data processing system reliably identifies slabs subject to significant change in casting speed.

Table 4. Production rates of slabs affected by changes in casting speed from different perspectives.

Slab property		**Proportion in production**
Burdened with significant casting speed change	*Time-based*	4.69%
	Cast length-based	6.96%
Decelerated below minimum casting speed		19.47%
Sequential start or end		10.31%

4. Conclusion

The data processing and display system presented in former publications, operating at the continuous casting machine of ISD Dunaferr Zrt. provides an opportunity to study the effects of non-steady-state casting events during casting, examining the slab in a complex way from raw steel production to the end of hot rolling. During the verification of the reliability of the data processing system, it was observed that the product-related changes in cast length are more reliably correlated with the changes in the cast sections of the cast strand used in the reference (e.g. sequence start and end) changes over time than the process-related changes in time. The correlations observed during the subsequent tests (the relationship of downgrades with the results of the data processing system) confirm the reliability of the qualification of the slabs and thus the results of the data processing system.

5. References

[1] Neef C, Hirzel S and Arens M 2018 Industry 4.0 in the Europiean Iron and Steel Industry *Fraunhofer Institiute for System and Innovation Research ISI Germany* p 32

[2] Harald P 2017 How could Industry 4.0 transform the Steel Industry? *Future Steel Forum, Warsaw*

[3] Portász A and Nagy-Rácz I 2019 Together for the future! Application of AI in the production of competitive steel products *Microsoft Envision Forum, Budapest*

[4] Wizner K and Kővári A 2019 Analysis of the steel production / casting process *21th Mining, metallurgy and geology conference, Hungarian Technical Scientific society of Transylvania* pp 153-157

[5] Wizner K and Kővári A 2021 Qualification of the differences between the cast and cut strand length in the continuous casting machine of ISD Dunaferr Zrt. *Hungarian Mining and Metallurgical Society (Metallurgy)* **154 (1)** pp 1-4

[6] Wizner K and Kővári A 2020 Root cause analysis of metallurgical defects in continuous cast steel slabs at ISD DUNAFERR Zrt. *IOP Conference Series: Materials Science and Engineering* **903** 012043 p 8

[7] Zhang Q, Wang L and Wang X 2006 Influence of casting speed variation during unsteady continuous casting on non-metallic inclusions in IF steel slabs 2006 *ISIJ International* **46 (10)** pp 1421-1426

[8] Wang Y and Zhang L 2010 Transient fluid flow phenomena during continuous casting: Part II – Cast speed change, Temperature fluctuation, and steel grade mixing *ISIJ International* **50 (12)** pp 1783-1791

[9] Sengupta J, and Thomas B G 2006 Effect of sudden level fluctuation on hook formation during continuous casting of ultra-low carbon steel slabs *Modeling of casting, welding and advanced solidification process* **XI** pp 727-736

[10] Shin H-J, Thomas B G, Lee G-G, Park J-M, Lee C-H and Kim S-H 2004 Analysis of hook formation mechanism in ultra low carbon steel using CON1D heat flow – solidification model *Materials Science & Technology* New Orleans, LA. p 16

[11] Tehovnik F, Burja J, Arh B and Knap M 2015 Submerged entry nozzle clogging during continuous casting of Al-killed steel *Metalurgija* **54 (2)** pp 371-374

[12] Braun T B, Elliott J F and Flemings M C 1979 The clustering of alumina inclusions *Metallurgical Transactions* **B 10B** pp 171–84.

[13] Girase N U, Basu S and Choudhary S K 2007 Development of indices for quantification of nozzle clogging during continuous slab casting *Iron and Steelmaking* **34 (6)** pp 506-512

[14] Wizner K and Kővári A 2020 Novel concepts for establishing expert support systems to investigate the defect occurring in metallurgical phases in the technology of ISD DUNAFERR Zrt. *Conference Series: Materials Science and Engineering* **903** 012001 p 6

[15] Wizner K and Kővári A 2020 Effect of the production conditions of continuously cast steels on the degree of hot rolled product downgrading *Acta Materialia Transylvanica* **3 (1)** pp 55-59

[16] Kővári A, Katona J, Wizner K, Ujbányi T, Nagy B, Berki B and Sudár A 2021 Application possibilities of human-computer, display and analysis interfaces *Dunakavics, Online Journal of University of Dunaújváros* **9 (5)** pp 45-60

IOP Conf. Series: Materials Science and Engineering 1246 (2022) 012008 doi:10.1088/1757-899X/1246/1/012008

A possible medieval recycling technique – smelting iron using hammerscale

Á Thiele[1] **and B Török**[2]
[1] Department of Materials Science and Engineering, Faculty of Mechanical Engineering, Budapest University of Technology and Economics, Műegyetem rkp. 3., H-1111 Budapest, Hungary
[2] Institute of Metallurgy, University of Miskolc, H-3515 Miskolc-Egyetemváros, Hungary

Abstract. Hammerscale is a common by-product of the forging process of iron and it may appear in a large amount in the ancient smithy workshops as well. Its recycling (i.e. re-smelting) in early bloomery furnaces might have been an economical and reasonable process, although it is not known whether it works or not, and also there is no archaeological evidence for this technique. In our recent investigation, flaky hammerscale was smelted using the copy of the 8-10th century embedded furnaces of the Avars and the conquering Hungarians. The resulting bloom was processed to a billet and its chemical composition was further examined. Hammerscale proved to be an excellent raw material for the early bloomery iron smelting technology: good iron yield could be achieved due to the high reaction surface and iron content. Moreover, the extracted bloomery iron had low carbon content and had a very low amount of undesirable impurities, e.g. P, As. Cu, S. etc.

1. Introduction

Hammerscale is a common by-product of the forging process of iron and its modern recycling by the steelmaking industry is well known: hammerscale or mill scale is reduced to sponge iron powder that can be used as raw material in the electric furnaces for steelmaking, eg. [1-3]. However, hammerscale may appear in a large amount also in ancient smithy workshops and its recycling (i.e. re-smelting) in early bloomery furnaces might have been an economical and reasonable process theoretically. It is not known whether it works or not, and there is no archaeological evidence for this technology. This paper shows the results of smelting experiments of hammerscale to describe its use in the archaeometallurgy of iron.

1.1. General characterization

Most studies basically distinguish between two forms of appearance, the flaky and the spheroid hammerscale. Flaky hammerscale forms in the forge fire during the heating of the workpiece in oxidizing atmosphere and at high temperature. The oxidation of the surface becomes faster after the hot workpiece is removed from the forge fire for further hot working, e.g. hammering, so flaky hammerscale forms mostly in this stage of the process. Most sources agree that spheroidal hammerscale is produced mainly during forge-welding [4,5]. Here it should be noted that the formation method of flaky and spheroid hammerscale depends on the temperature. Under its melting point (ca. 1300°C) a thin layer of solid hammerscale covers the surface of the workpiece due to the frontal oxidation (the oxidation of the workpiece can be frontal and inter granular). This brittle, solid hammerscale cracks and falls off the workpiece during its plastic deformation. Over the melting point of the hammerscale the steel workpiece starts to burn and in the beginning small sparks (which are burning steel particles or grains) are rejected. Later if the temperature is increased and/or the atmosphere turns more oxidizing, the sparking of the workpiece intensifies. When the hammerscale is molten, it can drip down into the forge fire during the heating process, or slip off from the surface of the workpiece during hammering, e.g. in case of forge welding. Molten, slipped off hammerscale takes the form of a sphere due to the surface tension (see figure 1. later). In a smithy workshop, flaky hammerscale appears in much larger quantities than spheroidal hammerscale.

Several studies deal with the typology of hammerscales mainly on the basis of their morphology and shape. McDonnel examined samples of hammerscale with SEM-EDS and identified two general shapes: flakes and spheroids. The microstructures of both types were similar, however the spheroids were often hollow and contained more phosphorus. It is suggested that the spheroid hammerscale formed during the forge welding process where flux (mainly sand) has an important role in their formation [6]. Hammerscales from smithy workshops of the Roman period in Britain were examined and classified by Allen (Awre, Gloucestershire) [7] and by Sim (Silchester, Hampshire) [8]. Allen identified twelve types of hammerscales on the basis of their shape and size. Although numerous examples of spheroid or droplet-shape hammerscales have been found as an evidence of the purification process of the bloom or semi-finished iron product, the most common shape was the flake. Dungworth and Wilkes discuss in detail in their summary study, illustrated by a large number of SEM images and results of EDS examinations, the characteristics of the hammerscales in terms of both the archaeological finds and the hammerscales produced by reconstruction experiments. In addition to the two general types of hammerscales mentioned above, a third type is also defined, the so-called miscellaneous hammerscale. It accounts to less than 10% of the total amount of hammerscale collected by them, has very variable shapes and sizes but many consist of half spheroids or sheets with a rather bubbly texture on one surface. It was found that hammerscale samples usually have relatively large equiaxed iron oxide grains with a small amount of silica along grain boundaries. The layers of flaky hammerscales indicated several periods of heating before the flake became detached. The spheroidal hammerscale has a more variable microstructure than the flakes but most commonly this consists of a hollow spheroid [4].

1.2. Chemical and mineral composition

Hammerscale has a mineral composition of mainly iron oxides in the mineral form of magnetite (Fe_3O_4) which causes a notable magnetic character [1, 4]. The formation of hammerscale starts at temperatures between 500-600°C and becomes more rapid at around 800°C. Its structure is generally characterized by a layer of wüstite (FeO), sticking to the hot iron workpiece, a thicker central zone of Fe_3O_4 (magnetite) and sometimes an outer, very thin, easily flaked layer of Fe_2O_3 (hematite).

The chemical composition of hammerscale depends on the stage of the iron purification process and no sharp line can be drawn between slag and hammerscale. During the medieval bloomery iron smelting process high amount of slag forms from the unreduced iron oxides and gauge of the ores (mainly SiO_2, little CaO, Al_2O_3, etc), this slag may contain very little amount of magnetite, it consists mainly of fayalite ($2FeO{\bullet}SiO_2$) and glassy phase. After the smelting process the spongy iron bloom is compressed and forged. During this compressing process slag (it is still rather slag than hammerscale) melts off from the bloom into the forge fire and also splashes off during compressing and later at the forging phase as well. The relative iron content of the slag increases during this process, i.e. its wüstite (FeO) and magnetite (Fe_3O_4) content increases, while the SiO_2-content and the amount of other alkaline compounds decreases. Finally, in case of the forging of a compact iron bar almost pure iron oxide, hammerscale forms. Hammerscale may contain residues of flux that may be applied for forge-welding (e.g. pure quartz, SiO_2) and elements that burned off from the surface of modern alloyed steel materials (e.g. Mn). In Sim's extensive survey (1998), 17 different samples of hammerscales from bloom purification, forging and welding were demonstrated. These were archaeological finds and results of reproduction experiments and modern processes. Chemical analyses of the experimentally produced hammerscales revealed that silica content decreased as iron was refined from the state of bloom towards a workable product and higher iron oxide content could be detected in the stage when well-refined billets were manufactured [8].

1.3. Archaeological background

Hammerscale is usually unearthed in iron-related archaeological excavations and may provide important information regarding the function of the excavated archaeological feature. According to Pleiner, iron losses of 3-4 % could be expected in a single heating, and even a higher percentage in longer and higher temperature heating [9]. The ferromagnetic property of hammerscale may help to identify forging operations during the archaeological excavation. In some sites of ancient and medieval forges, hundreds or even thousands of grams of hammerscale have been found. On the excavation site, the location of hammerscale can also help to locate the smithing activity precisely because it is often found in the immediate vicinity of the smithing hearth and anvil.

In the archaeological excavations, hammerscale is one of the most important evidence of iron smithing activity. However, it is often too small to be noticed during unearthing but can be detected using a magnet. The surface of the floor of a smithy workshop should be sampled throughout at 0.2–0.5m intervals in order to examine the distribution of hammerscale. 0.2 litres of sample is adequate for magnetic the susceptibility screening and quantification of hammerscale [10].

In the literature of archaeometallurgy there are numerous studies of hammerscale found mainly in Roman smithy workshops. One of the earliest (1941) identifications of hammerscale in the archaeological papers comes from the excavation report ("…the material is the magnetic oxide of iron, Fe_3O_4, containing ferrous oxide in considerable amount and some metallic iron. There can be no doubt that it is a mill-scale or smithy scale…") of the Roman fort of Benwell on Hadrian's Wall [11]. A good example is the Roman smithy workshop at Nailly in France, where the magnetite layer of the examined samples of hammerscale pieces was covered by a thin layer of hematite or fayalite [9].

An analysis of a sample of hammerscale from a Roman iron smelting workshop of Ashwicken (England) showed more than 85 % magnetite and the rest of the material was fayalite in the form of a thin cementing layer [12]. Tylecote identified hammerscale as a by-product of forging including different iron oxides (wüstite, magnetite and hematite) covered by a cementing film of silica originating from the slag inclusions of the bloom and from the sand used as flux [13].

2. Methods and results

2.1. Investigations of hammerscale

Hammerscale was collected in the smithy workshop of Ádám Thiele, where ca. 40 kg of hammerscale accumulates each year as a by-product of forging axes, knives, etc. This hammerscale was flaky for the most part and it contained spheroidal hammerscale only in traces. The chemical composition of the hammerscale was analysed by the means of wavelength-dispersive X-ray spectroscopy (type S8 TIGER) in the Dunaferr Labor Nonprofit Ltd. The result for the main oxides are summarized in table 1.

Table 1. Main oxides of the hammerscale (wt%).

Fe_2O_3	SiO_2	Al_2O_3	CaO	MnO	MgO	K_2O	Na_2O	P_2O_5
87.9	6.11	2.16	0.54	0.7	0.21	0.15	1.08	0.28

The hammerscale collected from the floor of the smithy workshop was not pure iron oxide, its iron content calculated to hematite (Fe_2O_3) was 87.9 wt% (which means 56 % of elemental iron). High SiO_2 and Al_2O_3 content may be caused by the clay minerals coming from the mud brought on the shoes into the building of the workshop. High Na_2O content could be explained by the use of borax, sodium tetraborate ($Na_2B_4O_7 \cdot 10H_2O$) as flux for forge welding. The relatively high MnO content could come from the Mn alloyed structural and tool steels that are mainly forged in the workshop (S235 and 90MnCrV8, both have a Mn content of ca. 1.5-2 wt%). The weight percent of other oxides were under 0.5 wt%.

Macro photos and SEM micrographs were made of flaky and spheroid hammerscale. The spheroid hammerscale samples (cf. figure 1) were produced for this investigation only. The burning steel workpiece was removed from the fire, and the showering sparks were collected into a simple tin can. The hollow spheroid hammerscale grains had a diameter of ca. 20-80μm (see figure 1/B).

Figure 1. Marco photo (A) and SEM micrograph of hollow spheroid hammerscale (B).

A more detialed SEM-EDS examination has also been carried out on some samples of flaky hammerscale (figure 2/A) on which Zeiss EVOMA10 electron microscopy equipped with EDAX energy dispersive spectroscopy was used. The measured thickness values of 150-350 μm of the cross sections (figure 2/B and C) are in line with the thickness of the layers of hammerscales mentioned by Dungsworth and Wilkes [4]. In figure 2/D a layered structure of the cross section can be observed clearly. In most cases of the examined samples the two sides could be distinguished by their colour: on the grey side (figure 2/E and F) only iron oxide could be measured by EDS while on the black side hollow microstructures (figure 2/G) and special shaped formations (figure 2/H) could be observed at a higher magnification in some places. In the black side EDS examination resulted in 2.0-4.5 wt% Na. 1.5-2.5 wt% Si. 0.5-0.6 wt% Ca and 3.2-3.6 wt% Mn probably originated from the flux and/or slag.

2.2. Experimental smelting of hammerscale

The hammerscale was smelted using the copy of the 8-10th century embedded furnaces of the Avars and the conquering Hungarians. The experimental furnace was built in a steep hillside of a clay-sand mixture. Its height was 75 cm above the level of the twyer (shaft) and its depth was 25 cm under the twyer (hearth). The biggest inner diameter of the furnace at the level of the twyer was 35cm. The furnace was closed with a breast wall during smelting. For thermal insulation, earth dust was heaped into the lower part of the furnace to increase the temperature of its hearth. Air was supplied (ca. 150 litre/min) through the twyer using a centrifugal fan chocked with a piece of paper. A slag tapping hole was punched through the breast wall under the twyer for slag tapping if necessary. Fig 3 shows the structure and main parts of the experimental furnace.

In the beginning of the smelting experiment the furnace was preheated for 30mins by burning smaller wood branches. Then the furnace was filled up with charcoal and preheating was going on for an additional 15 mins. The charcoal used for the experiment was made from oak and beech wood and its grain size was between 5-40 mm (charcoal was broken into smaller pieces and then the powder was shifted). In the next 2 hours and 45 mins hammerscale and charcoal was charged into the preheated furnace in 300g portions, so their ratio was 1:1 during the smelting process. 12.5 kg hammerscale was smelted in total. Before the first charge of the hammerscale, 14 kg of charcoal was used (ca. 11kg to fill up the furnace and 3kg for preheating). After the opening of the furnace, 6 kg glowing charcoal was collected and cooled down, thus saved for a next smelting. 20.5 kg charcoal was used altogether. During the smelting process the twyer was clean, and no solid slag block choked the air supply in front of the twyer.

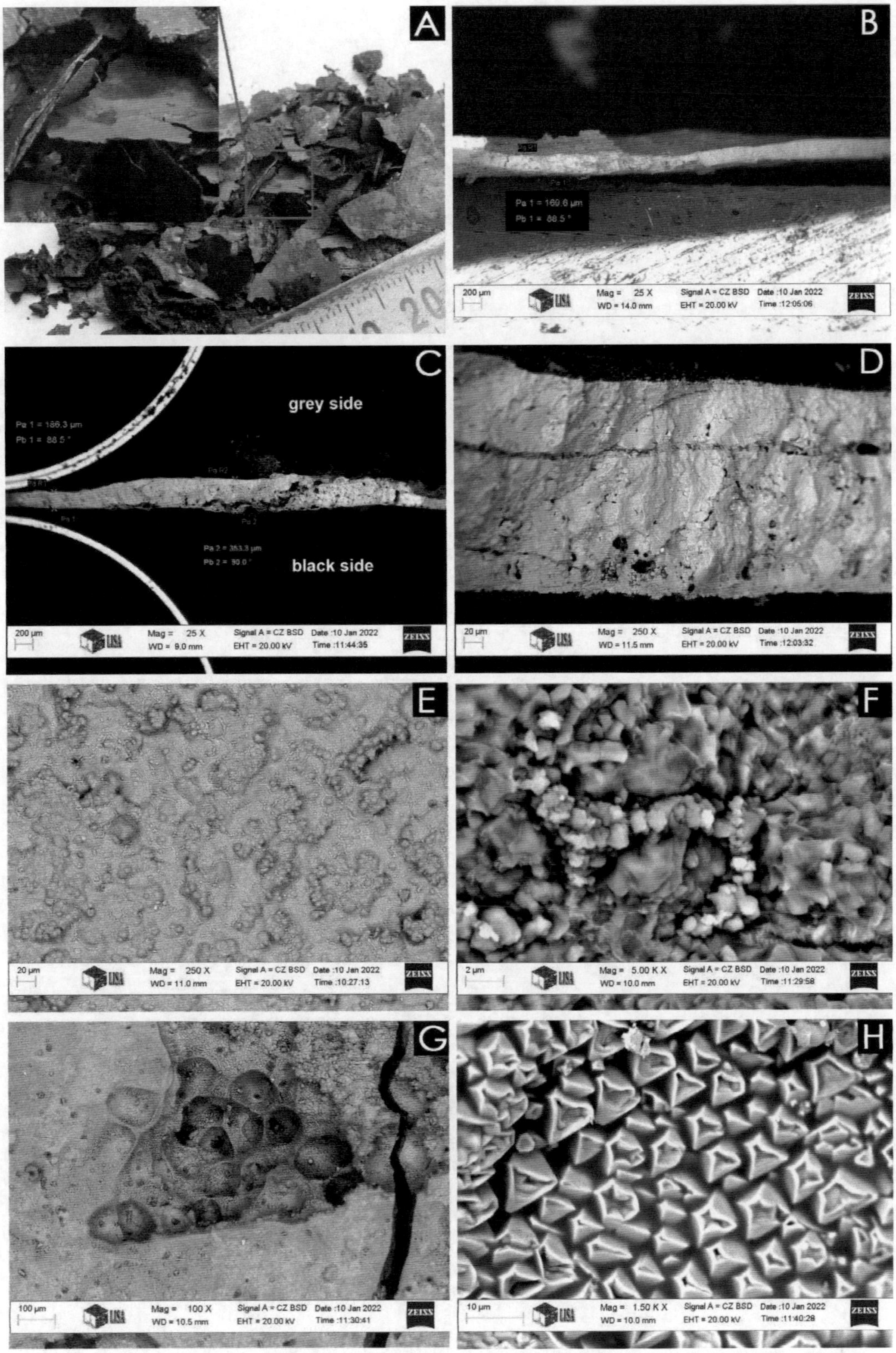

Figure 2. Marco photo (A) and SEM micrographs of flaky hammerscale (B – H).

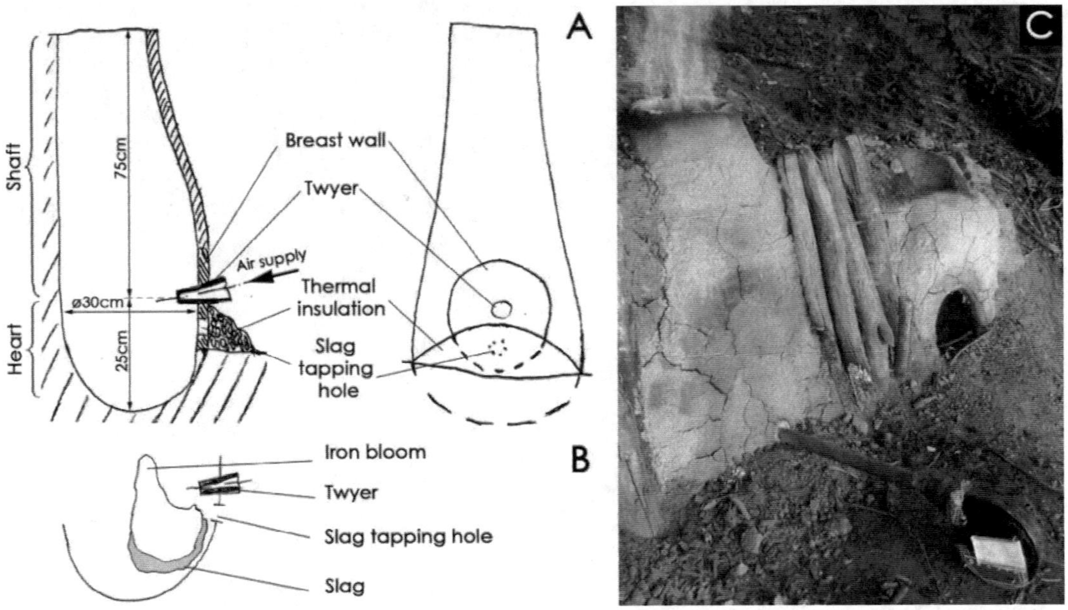

Figure 3. The structure and main parts of the experimental furnace. Schematic sketch of the furnace (A). Shape and position of the bloom at the end of the smelting process (B). Photo of the furnace (C).

The first slag tapping was needed 15 mins after the last hammerscale portion was charged and the slag was tapped three times ca. every 10 mins. The tapped slag weighed 2.3 kg in total. The tap slag had low viscosity at lower temperature (on red glow) as well, it had a narrow solidification range, so it was not glassy and sticky, its fracture surface was greyish black with a few bubbles. So, it was very easy to tap the slag of the hammerscale. For comparison, the tap slag of iron ores with low iron and high silica content is usually glassy, has high viscosity, a wide solidification range, its fracture is greenish black, contains much more bubbles and sometimes also pure quartz particles (white, marked with a red arrow, cf. Fig 4/B) visible with the naked eye. This kind of slag is hard to tap and usually marks a poor-quality bloom. Figure 4 shows the fracture surface of a tap slag sample of the hammerscale and a slag sample of a kind of bog iron ore (from Bátorliget, Hungary, Fe_2O_3=57%, SiO_2=28.4%, Al_2O_3=2.72%, CaO=2.81%, MnO=3.29% and P_2O_5=3.67%) smelted under the same conditions.

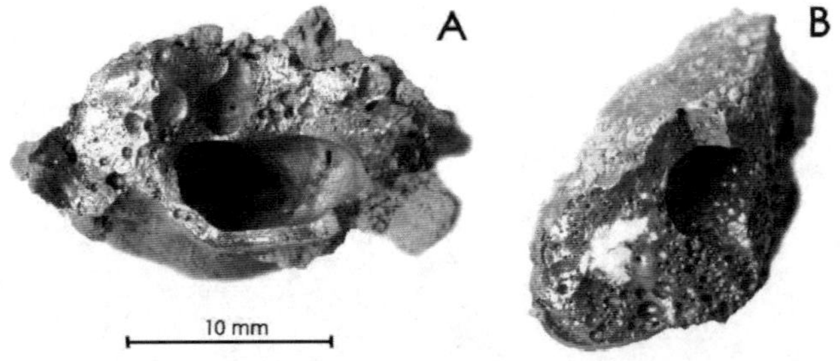

Figure 4. Fracture surface of a tap slag sample of hammerscale (A) and of bog iron ore with high silica content (B).

The smelting was ended, and the breast wall of the furnace was opened 45 mins after the last hammerscale portion was charged (so the smelting took 4 hours and 15 mins altogether). The iron bloom was unusually high, it towered ca. 10-15cm above the level of the twyer and it had an "L" shape with a spongy top and denser bottom. There was a little slag block glued to the bottom of the bloom (the slag block is usually bigger if the smelted iron ore has a lower iron content or is under-reduced in the furnace). Then the iron bloom was removed and compressed on a wood log with a wooden hammer. The bloom weighed 3.9 kg after this first compression. The slag block that broke to pieces weighed ca. 1.4kg.

A short video about this smelting experiment: https://youtu.be/UP8Dwi8JD30

2.3. *Processing the iron bloom*

After the first compression, the bloom was reheated in a charcoal fuelled smithy hearth (similar to the 8-10th century ones unearthed in excavations), and compressed further with wooden tools five more times. The compressed bloom weighed 2.8 kg. Using a power hammer and a modern coke-fuelled smithy forge fire, this bloom was forged into a 210 mm long and 60x20mm cross sectioned billet, which finally weighed 1.85 kg. There was no metallic iron loss during this forging step and the iron bloom was easy to forge, brittleness was not observed (bloomery iron smelted from bog iron ores rich in phosphorus or arsenic or copper-rich mineral ores usually shows hot-shortness and also cold-shortness).

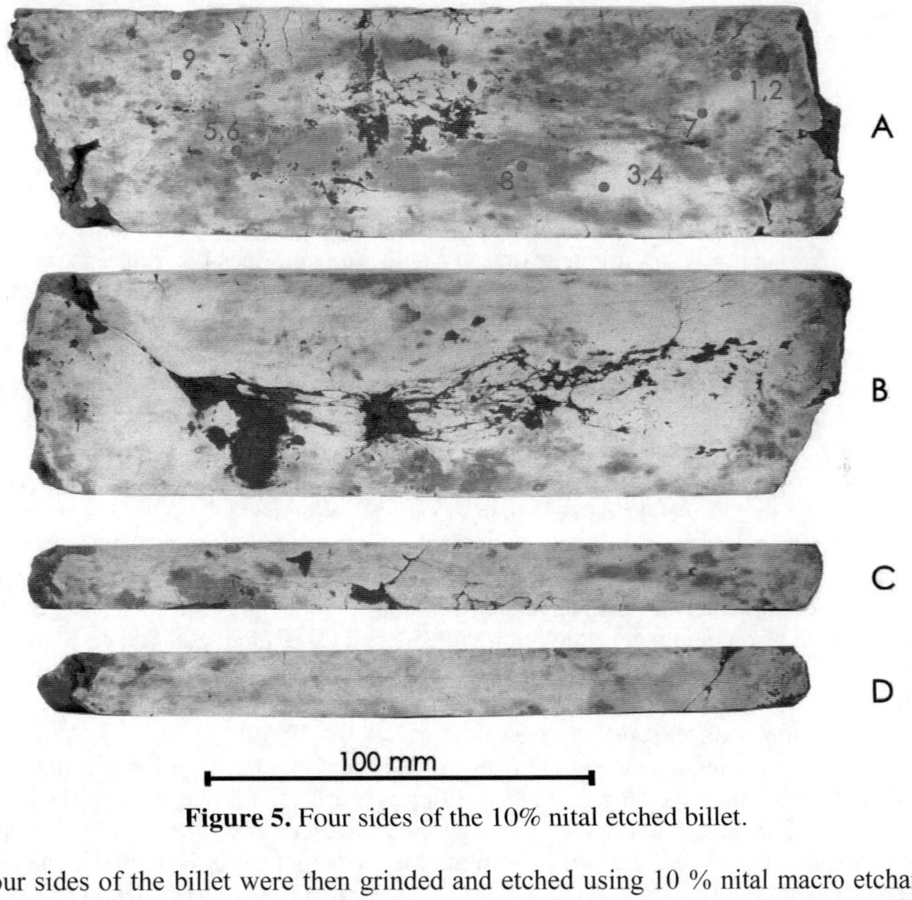

Figure 5. Four sides of the 10% nital etched billet.

The four sides of the billet were then grinded and etched using 10 % nital macro etchant to make the carbon distribution visible. The etched billet can be seen in figure 5. Based on the etched surface, it can be stated that the carbon distribution is very heterogeneous and carburized parts appear mainly on the first side (figure 5/A). So, the bloomery iron material extracted from hammerscale had low carbon content, it can be qualified rather iron than steel (several similarly etched bloomery iron billets were examined earlier by the means of Glow Discharge Optical Emission Spectrometry (GDOES) method, and the carbon content corresponds to the darkness level of the etched surface [14]).

An Oxford Instrument portable X-ray fluorescence spectrometer (50 kV, Rh anode, Silicon Drift Detector, Measurement time: 30 sec.) has been used for the ED-XRF examination (the analyzed points are marked in red in figure 5/A). Fundamental parameter (Alloy FP) methods use a complex mathematical analysis to calculate the concentrations of elements (from 0 wt% to 100 wt%) in the sample. For metals with inherently unknown composition, such as excavation finds and products of reproduction experiments, it is recommended to use this method. Low alloy method includes the light elements as well. ED-XRF analysis results (summarized in table 2) suggest that the bloomery iron material contains a very small amount of alloys, it is unalloyed iron with a definitely low carbon content.

Table 2. P-XRF analysis results (wt%).

Nr.	Class	Al	Si	P	S	Ti	Cr	Mn	Fe	Ni	Cu	Mo
1	Low alloy	0.13	0.21	0.05	0.08	0.04	0.01	0.02	98.85	0.11	0.33	0.04
2	Alloy_FP					0.00	0.05	0.03	99.36	0.13	0.29	0.03
3	Low alloy	0.18	0.18	0.05	0.09	0.04	0.01	0.02	98.85	0.16	0.49	0.06
4	Alloy_FP					0.00	0.06	0.02	99.16	0.18	0.43	0.06
5	Low alloy	0.39	0.50	0.06	0.10	0.03	0.06	0.03	98.81	0.17	0.49	0.05
6	Alloy_FP					0.00	0.10	0.04	99.04	0.20	0.47	0.05
7	Low alloy	0.18	0.32	0.09	0.10	0.00	0.01	0.02	98.93	0.23	0.40	0.04
8	Low alloy	0.19	0.27	0.04	0.11	0.01	0.02	0.02	98.75	0.35	0.59	0.05
9	Low alloy	0.41	0.60	0.06	0.10	0.02	0.02	0.02	98.96	0.12	0.41	0.05

3. Discussion

Experimental smelting results suggest that hammerscale is not only a suitable but an excellent raw material for the early bloomery iron smelting technology. Firstly, flaky hammerscale has a high reaction surface due to its low thickness. However, the several mm width of the flakes prevents its fast sinking in the furnace and provides a longer time for reduction and does not choke the air flow in the furnace (dusty iron ores are not suitable for iron smelting). Secondly, hammerscale has a high iron content so a good iron yield could be achieved. This statement is supported by the small amount of slag (small slag block, little tapped slag) and the heavy bloom. The iron yield calculated from the weight of the compressed bloom is 22.4%, while from the iron billet it is 14.8%. And thirdly, hammerscale has low P, As, Cu, S, and other undesirable impurities that would have a detrimental effect on the mechanical properties of the bloomery iron.

Ancient smiths sometimes worked with bloomery iron that had high P and As content, also Cu might appear in much higher weight percent than in case of modern steels. All these elements caused cold and/or hot shortness, so their presence reduced the workability of the bloomery iron. However, we did not examine the hammerscale of these kind of bloomery iron materials with impurities (the enrichment of these elements in hammerscale could be a good question), but we can state that bloomery iron smelted from hammerscale with low P, As and Cu content could have been a very good raw material for the ancient smiths. This kind of bloomery iron could have had more or less carbon content depending on the smelting technology and could possibly have been a suitable raw material for a wide range of tools and weapons.

The hammerscale we used in our experiment provided just enough slag for a compact bloom to form. The suitable amount of the slag is important during iron smelting because in the hearth of the furnace, it helps the diffusion welding of the iron particles reduced from the iron oxides in the shaft. Smelting experiments with pure industrial 0.3-5 mm grain sized hematite (Brazilian hematite from the ISD Dunaferr company, Fe_2O_3 = 97.9 wt%, SiO_2 = 1.3 wt%, Al_2O_3 = 0.62 wt% and CaO = 1.01 wt%) resulted in huge but very spongy iron blooms without slag. It was not possible to compress and forge these blooms because it broke into pieces due to the poorly diffusion welded iron particles. But charging sand (consisting mainly of quartz) to the pure hematite in a ratio of 1:6 provided enough slag, i.e. sand acted as flux for the diffusion welding of the iron particles. The flaky hammerscale collected from the floor of the smithy workshop had just enough flux (mainly simple mud powder) to get the suitable amount of slag and a dense bloom without decreasing the iron yield too much. But in case of smelting pure hammerscale that contains only iron oxides, charging some kind of flux (e.g. sand. dusty clay. wood ash. broken tap slag. etc.) might be necessary.

As hammerscale may have appeared as a large amount of by-product in a smithy workshop where bigger iron workpieces were processed for a longer period of time, the recycling (i.e. re-smelting) of the hammerscale in bloomery furnaces might have been an economical and reasonable process. There is no archaeological evidence of the extent to which this raw material was used in any of the historical iron smelting workshops, but the re-smelting of the hammerscale was theoretically and practically feasible. On the other hand, this recycling process suggests that the site of forging was close to, or at least in contact with, the smelting sites.

4. Conclusions

The results of reconstruction experiments and material examinations reveal that flaky hammerscale is an appropriate and self-fluxing raw material for the iron smelting process of the ancient bloomery. Good iron yield could be achieved due to the high reaction surface and iron content. The extracted bloomery iron had low carbon content and had a very low amount of undesirable impurities, e.g. P, As, Cu, S, etc.

References

[1] Martin M I, Lopez F A and Torralba J M 2012 Production of sponge iron powder by reduction of rolling mill scale *Ironmak. Steelmak.* **39(3)** 155–162
[2] Camci L, Aydin S and Arslan C 2002 Reduction of ion oxides in solid wastes generated by steelworks *Turk. J. Eng. Environ. Sci.* **26** 37–44
[3] Park J-W, Ahn J-C, Song H, Park K, Shin H and Ahn J-S 2002 Reduction characteristics of oily hot rolling mill sludge by direct reduced iron method *Resour. Conserv. Recycl.* **34(2)** 129–140
[4] Dungworth D and Wilkes R 2009 Understanding hammerscale: the use of high-speed film and electronmicroscopy *Hist. Metall.* **43** 33–46
[5] Veldhuijzen H A 2009 Of Slag and Scales; Micro-Straigraphy and Micro-Magnetic Material at Metallurgical Excavations UCL Institute of Archeology
[6] McDonnell G 1984 The study of early iron smithing residues *Crafts of the Blacksmith* ed Scott B G and Cleere H (Belfast: UISPP) pp 47–52
[7] Allen J R L 1986 Interpretation of some Romano-British smithing slag from Awre in Gloucestershire *Hist. Metall.* **20** 97–104
[8] Sim D 1998 *Beyond the Bloom. Bloom refining and iron artefact production in the Roman world* BAR International Series 725 (Oxford: Archaeopress)
[9] Pleiner R 2006 Iron in Archaeology. Early European Blacksmiths. *Archeologicky Ústav* AV ČR. Praha pp 110–1
[10] Bayley J, Dungworth D and Paynter S 2001 *Archaeometallurgy* (London: English Heritage Publications) p 6
[11] Simpson F G and Richmond I A 1942 The Roman Fort on Hadrian's Wall At Benwell *Archaeologia Aeliana* **19** 1–43
[12] Tylecote R F and Owles E 1960 A second-century iron smelting site at Ashwicken. Norfolk. *Norfolk Archaeology* **32** 142–162
[13] Tylecote R F 1962 *Metallurgy in Archaeology* (London: Edward Arnold) p 45
[14] Thiele Á, Török B Controlling possibilities of the carbon content in the iron bloom during the iron smelting. 5th International Conference on Archaeometallurgy. Hungary. Miskolc. 19-21th June. 2019 (poster)

Acknowledgement

The research reported in this paper and carried out at BME has been supported by the NRDI Fund (TKP2020 NC, Grant No. BME-NC) based on the charter of bolster issued by the NRDI Office under the auspices of the Ministry for Innovation and Technology.
We are grateful to Prof. György Kaptay and Árpád Kovács (Institute of Physical Metallurgy, Metalforming and Nanotechnology) for their support in the SEM-EDS examinations.

Design and manufacture of closed die multi-axial forging tool

Zs Juhász[1], T Bíró[1] and J B Renkó[1]

[1] Department of Materials Science and Engineering, Faculty of Mechanical Engineering Budapest University of Technology and Economics, Budapest, Hungary, H-1111 Budapest

Abstract. In the last decades many solutions were developed to achieve multi-axial forging. Outstanding among these is the two-way process that can be implemented on the Maxstrain unit of the Gleeble thermophysical simulator. Although the experiments performed on the Gleeble system were well suited for characterize mechanical models, this system had some serious issues, such as the outflow of material from the forging zone. To solve these problems, a new forming tool was designed, in which the total volume of the workpiece is deformed, and the shanks used for fastening can be omitted.

Keywords: multi-axial forging, design, material science, manufacture, testing, simulation

1 Introduction

In the last decades the production and testing of ultrafine grain materials has undergone a significant development [1, 2]. The average grain size of these materials is approx. 70-500 nm, which results in increased strength, higher fatigue resistance, increased toughness, superplasticity compared to conventional coarse-grained bulk materials [3, 4].

In most cases this particle structure is achieved by severe plastic deformation (SPD) in bulk materials. The coarse-grained, as received bulk material undergoes a structural transformation under high shear stress, which results in an ultrafine-grained state [5].

Several solutions have been developed for the practical implementation of this technology, like high pressure torsion, equal channel angular pressure, friction-assisted lateral extrusion, alternating extrusion-compression, or multi-axial forging [6-11]. Multi-axial forging stands out from the listed technologies with its easy toolability and implementation. It can be used to forge specimens up to 50-100 mm or even larger [12], that allows to step out of the usual size range of laboratory tests.

In previous studies, the MaxStrain unit of a Gleeble 3800 thermophysical simulator was used to physically implement the two-way multi-axial forging [13]. The equipment makes it possible to control the parameters of forging, such as load, tool movements and temperature, with high accuracy [14]. The only setback of this system is the fastening of the workpiece. The workpiece must have two relatively big stems, of which the volume is significantly larger compared to the volume of the deformation zone (figure 1).

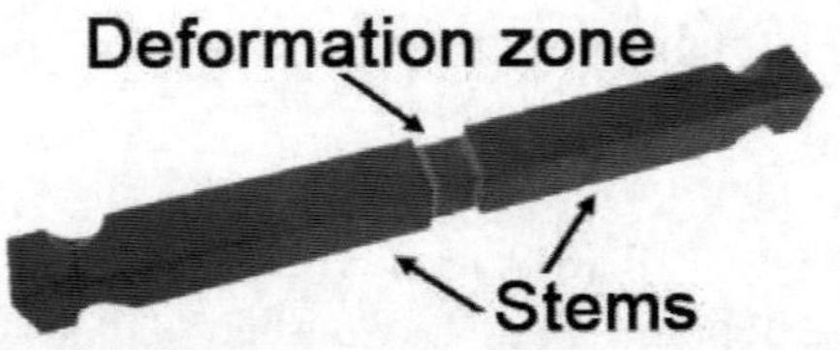

Figure 1. Gleeble specimen with a significantly smaller deformation zone in the middle compared to the size of the stems.

The size of the formed zone between the individual forming steps has showed a clear decrease, which was the result of a material flow during forging in the direction of the stems [15]. As a result, the forgeable volume of the specimen decreased with each forming steps, leading to a volume loss of approx. 25% during a 10-step forging cycle (figure 2).

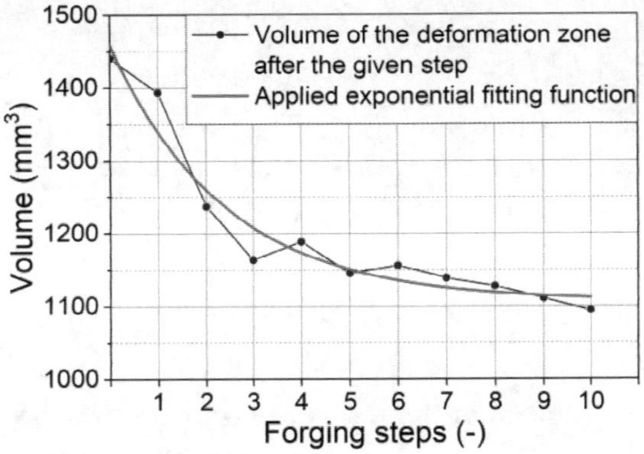

Figure 2. Volume change of the deformation zone after each forming step.

Due to such a decrease in the volume of the forged zone, a problem with reliability and general validity of the material properties determined by this technique arises. In order to eliminate this problem, we aimed to design and manufacture a new closed-die multi-axial forging tool, which can be used to study non-monotonic material behaviour.

2 Design and manufacture

2.1 Constructional considerations

At the beginning of the design process a list of requirements was made. The tool was designed for the already available MTS 810 universal material testing equipment, so the geometry of the tool as well as the maximum load capacity of the machine, which was 250 kN, had to be considered. Other aspects were the locking of the shafts, the centering of the specimen, the provision of manual or machine tool rotation, and ergonomic design in line with occupational safety considerations [16]. The mechanical properties of the CuE to be tested had to be taken into account to dimension certain parts of the tool.

To eliminate the material flow into the stems, the workpiece was modified so that its entire volume will be forged. To do this, a closed die cavity had to be created that could prevent the uncontrolled flow of volume in any directions.

2.2 Tool movements

In order not to have out the workpiece from the tool during a multi-step forging process, it is necessary to form two pairs of perpendicular tools. Thus, there is always an active, moving tool pair and a passive, fixed position tool pair for the given forming step. The tool pairs are bounded on two sides by flat closing plates to form a 20mm×20mm×10mm tool cavity. The nominal dimensions of the specimen are 20mm×10mm×10mm.

During a forging step, the specimen is compressed parallel to its 20 mm side length so that the new nominal dimensions of the originally 20mm×10mm×10mm workpiece will be 10mm×20mm×10mm (figure 3). At the end of the given step, the stamps are pushed to their rear end position by means of the ejector pins, and then the side is locked by leaving the pins inside. By removing the ejectors from the passive tool pair of the previous step, the active and passive tool roles are swapped. Before starting the next step, the tool must be rotated by 90 °.

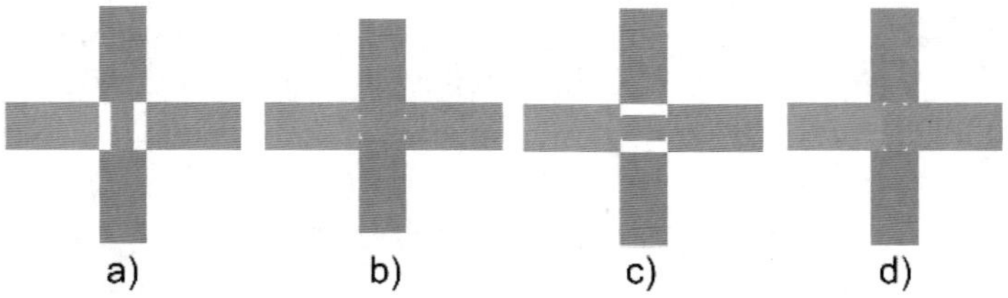

Figure 3. Initial state of a two-way forging cycle (a), final state of the first forging step (b), resetting of the tool position before the next step (c) and final state of the second forging step (d). The workpiece is grey, and the tools are green.

2.3 Tool design

The tool can be divided into three main units. These are the center block, the linear actuators, and the tool housing (figure 4).

The tool housing provides a stable frame for the tool and ensures the uniform movements of the tools. The center block is where the specimen is forged, so its primary function is to position the stamps and to form a closed die cavity. The main function of the linear actuators is to move the stamps and to compensate any asymmetric movements by allowing the center block move.

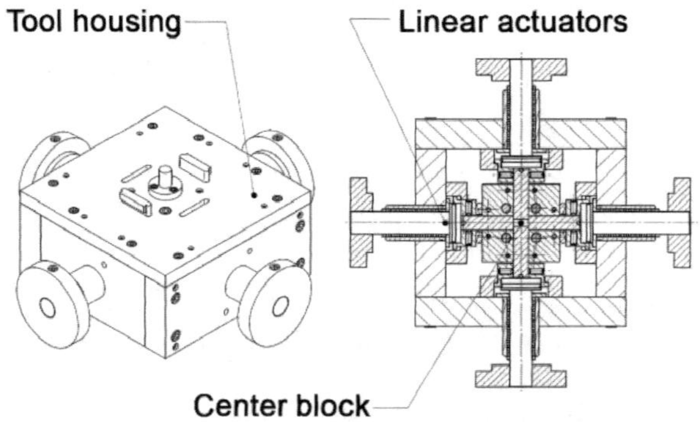

Figure 4. The main structural elements of the tool

If one of the two opposing tools moves faster relative to the other, the stamp holder will collide with the boundary ring. Since the boundary ring is attached to the tool housing, thus the entire tool moves with it and begins to pull over the stamp on the opposite direction. This way, when the end position is reached, the workpiece will always be in the centre.

2.4 Watts-Ford experiment

To size some components, the load force was estimated using QForm3D software. To improve the accuracy of the finite element simulation, we measured the planar deformation flow curve of the CuE material to be tested later with the tool. To determine the flow curve Watts-Ford method was used [17]. The result of the measurement and the flow curve fitted to it are shown in figure 5.

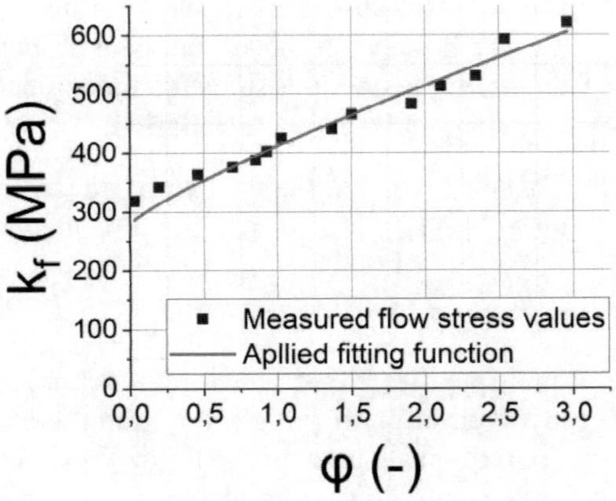

Figure 5. Flow curve obtained from Watts-Ford experiment

To determine the flow curve, a simplified Johnson-Cook equation was used. The effect of forming rate and temperature were negligible due to the nature of the Watts-Ford experiment [18]. The form of the function thus simplified is shown in Eq. 1.

$$k_f = C_0 + C_1 \cdot \varphi^n \tag{1}$$

The flow stress (k_f) for the given deformation (φ) was determined by fitting three parameters on the measured points (C_0, C_1, n), of which $C_0 = 280\ MPa$ is the yield strength of the material.

2.5 Finite element simulation

For the sizing of the components, a 3D finite element model was used to estimate subsequent tool loads. Although the forging process can be simplified to a 2D planar deformation problem, it could not provide the necessary information about the loads of the two planar plates closing the die cavity.

The flow curve determined during the Watts-Ford experiment allows to simulate the forging of the workpiece. Based on the simulation, the loading of the tools for four forging steps is shown in figure 6.

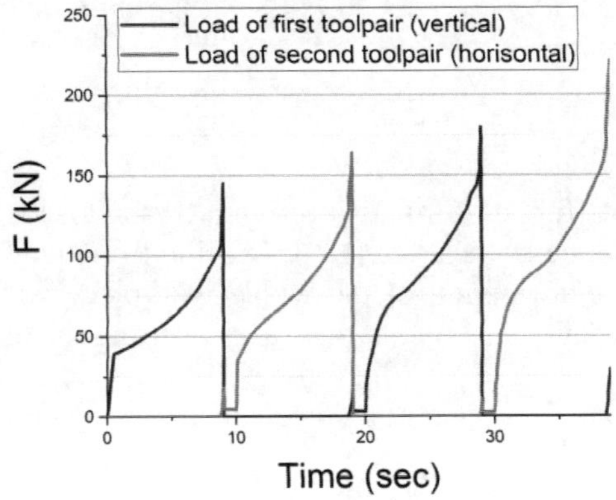

Figure 6. Tool load during finite element simulation of the first 4 forging steps

The force values shown in figure 6 increase drastically by the end of each individual forging steps. This is because tool movements were used as boundary conditions during the simulations instead of tool load. When the available space in tool cavity, which decreases with the tool movement, is almost zero, the workpiece can no longer deform substantially. This results in a nearly vertical jump in the force curves. Under real measuring conditions, however, this will not be a problem, as the forces in this way mark the end of the forming step, and then the operator stops the movement of the tools immediately.

It is also important to mention, that the applied material model assume a monotonous hardening material, although previous studies have shown that the load always decreases as a result of a change of direction during multi-axial forging [11], [13], [15]. Thus, the tool load will never reach the critical level.

2.6 Manufacturing
The tool must be able to move in multiple directions, which requires precise machining and assembly of the individual parts. The parts involved in the forging and guiding them must therefore be finalized in sub-assemblies after roughing. To avoid inaccuracies due to the friction of the tool movements, very strict form and position tolerances are required during production.

3 Results
The designed tool was manufactured and shown in figure 7. The tool fully complies with the requirements list.

Figure 7. Finished multi-axial forging tool

4 Summary
Based on our knowledge of tools for multi-axial forging, a new, closed die cavity forging tool was designed. Its structural design eliminates the uncertainties arising from the volume loss of the previously used Gleeble system due to the deformation of the whole workpiece volume. The construction also allows to examine plane strain deformation on a larger range. During the experiments with this tool the cumulative plastic deformation can exceed 10. Measurements on a wide scale like this can help to better understand the mechanical properties and operating mechanisms of ultrafine grain materials.

Acknowledgement
Supported by the ÚNKP-21-2 New National Excellence Program of the Ministry for Innovation and Technology from the source of the National Research, Development and Innovation Fund.

References

[1] Langdon T G 2007 „The principles of grain refinement in equal-channel angular pressing", *Mat. Sci. and Eng.* A **462** 3-11

[2] Huang Y and Langdon T G 2013 „Advances in ultrafine-grained materials" *Mat. Tod.* vol. **16**, Nr. 3, March, 85-93

[3] Krallics G, Fodor A and Agena A 2006 "Anisotropic mechanical properties of an ultrafine-grained aluminium alloy", *Ult. Gr. Mat.* **IV.** TMS, San Antonio, USA

[4] Valiev R Z, Islamgaliev R K and Alexandrov I V 2000 „Bulk nanostructured materials from severe plastic deformation", *Prog. in Mat. Sci.* **45** 103-189

[5] Krallics G, Fodor A 2007 "Determining the anisotropy coefficients of an ultrafinegrained Al6082 alloy subjected to ECAP" *Mat. Sci. For.* **537-538** 215-221

[6] Korbel A, Richert M and Richert J 1981 „The effects of very high cumulative deformation on structure and mechanical properties of aluminium" *Met. and Mat. Sci.*, Roskilde, Denmark, 445–450

[7] Nagy P and Gonda V. 2013 „Könyöksajtolás végeselemes modellezése" *Acta Mat. Trans.* **XVIII**. Kolozsvár 273-276

[8] Segal V M, Reznikov V I, Drobyshevkij A E and Kopylov V I 1981 *Rus. Met.* **1** 115

[9] Vu V, Beygelzimer Y, Kulagin R and Toth L 2018 "The New Plastic Flow Machining Process for Producing Thin Sheets" *Adv. in Mat. Sci. and Eng.* 10.1155/2018/8747960.

[10] Zrnik J, Dobatkin S V and Mamuzi I 2008 „Processing of metals by severe plastic deformation (SPD) – structure and mechanical properties respond", *Metalurgija* **47** 211-216

[11] Szabó P J, Bereczki P and Verő B 2011 „The Effect of Multiaxial Forging on the Grain Refinement of Low Alloyed Steel" *Per. Pol. Mech. Eng.* **55** 63-66

[12] Naser T S B and Krállics G 2012 „The effect of multiple forging and cold rolling on bending and tensile behavior of Al 7075 alloy", *Mat. Sci. For.* **729** 464-469

[13] Bereczki P, Krallics G and Renkó J 2019 "The effect of strain rate under multiple forging on the mechanical and microstructural properties" *Proc. Man.* vol. **37** 253-260, ISSN 2351-9789

[14] Szabó P J, Bereczki P and Verő B 2011 "The Effect of Multiaxial Forging on the Grain Refinement of Low Alloyed Steel", *Period. Polytech., Mech. Eng.* Vol. **55/1** pp. 63-69

[15] Bereczki P, Szombathelyi V and Krállics G 2014 „Determination of flow curve at large cyclic plastic strain by multiaxial forging on MaxStrain System", *Int. Jour. of Mech. Sci.* **84** 182-188.

[16] Aurélie R 2012 "Functional design method for improving safety and ergonomics of mechanical products" *Jour. of Biomed. Sci. and Eng.* **05** 457-468.

[17] Krállics G, et al. 2013 "Limitations in Measuring the Flow Stress." *Mat. Sci. For.* vol. **752** 85–94

[18] Shrot A and Bäker M 2012 "Determination of Johnson–Cook parameters from machining simulations" Com. Mat. Sci. vol **52** 298-304

IOP Conf. Series: Materials Science and Engineering 1246 (2022) 012010 doi:10.1088/1757-899X/1246/1/012010

Inherent strain based estimate of the residual deformations for printed MS1 and 316L parts

V Gonda, I Felde, R Horváth and M Réger

Óbuda University, Népszínház u. 8, 1081 Budapest, Hungary

gonda.viktor@bgk.uni-obuda.hu

Abstract. In this study, the effect of the magnitude of inherent strain is examined for the simulated deformations of printed specimens, where the material grade is either of MS1 or 316L steel powders. While both material qualities are steel grades, and the same sets of inherent strain values with different magnitude and directional variations are defined in the simulations, allowing isotropic and orthotropic behaviour too, the resulting deformation in the printed part may also be strongly influenced by the mechanical properties of the different steel grades.

1. Introduction

Common processes for additive manufacturing technology for metal components are laser heating/melting processes, such as selective laser melting (SLM), which can be executed on steels [1,2] as well as on advanced alloys [3]. In this process, a metal powder layer is applied to the heated surface of the base plate of same material quality, and the corresponding layer of the model geometry is irradiated to melting by a laser. It continues layer-to-layer until the part is completely built. The unused metal powder is removed from the working area, then the base plate is unmounted, if necessary heat treatments are performed, and the printed part is separated from the build plate. Finally, the supporting structure is removed from the workpiece.

During the process, different thermal loads are exposed to different parts of the workpiece: in some locations there is less constrain towards thermal expansion than in others, resulting in a distribution of inhomogeneous residual stresses over the piece, causing the printed geometry may deviate from the designed one. It is worth estimating the deformations before printing by simulation, which makes it possible to adjust the geometry for the workpiece and/or supports based on these estimated deformation results.

Simulations of additive manufacturing processes require high computational power, therefore simplifications and assumptions are established about modeling, to the extent that the accuracy of the estimate is still acceptable. Related to the geometry, meso/macro scale simulations are considered for the simulations of the printing process, and from a physical point of view it is possible to use solely a mechanical model to estimate the deformations within a relatively short computational time. The key to the physical simplification is the application of inherent strains [4], that is the source of the residual stresses, and it can be considered as a calibration parameter for a specific material quality, machine, and process parameter set [5, 6]. Determination of inherent strains require complex procedures [7], or special specimens which are used with a simulation feedback loop for iterative refinement of results [8].

In this study, the effect of the magnitude of inherent strain is examined for the deformations of a printed specimen, where the material grade is either of MS1 or 316L steel powders. Even both material qualities are steels, and the same sets of inherent strain values with different magnitude and directional variations are chosen, allowing isotropic and orthotropic behavior too, the resulting deformation in the printed part may also be strongly influenced by the elastic/plastic mechanical properties of the different steel grades.

2. Inherent strain method

This method was originated in determination of welding deformations with an efficient way [4]. It is very time consuming to simulate large welded structures – as well as multi layered additive manufactured parts by coupled thermo-mechanical analysis. Inherent strains method is an approximate alternative for the simulation of welding and additive manufacturing simulations where residual deformations are in the main focus. Many studies were performed on welding [9-17]. Recently, it has been successfully applied on modeling additive manufacturing [5-7].

In additive manufacturing, when cooling after laser irradiation is reached, the resulting strain ε may have elastic ε_e, plastic ε_p, thermal ε_t, creep strains ε_{cr} and phase transformation induced strain ε_{ph} components as:

$$\varepsilon = \varepsilon_e + \varepsilon_p + \varepsilon_t + \varepsilon_{cr} + \varepsilon_{ph}. \quad (1)$$

Inherent strain ε_i, is defined as the difference between the deformation strain and the elastic strain, therefore it can be written as:

$$\varepsilon_i = \varepsilon_p + \varepsilon_t + \varepsilon_{cr} + \varepsilon_{ph}. \quad (2)$$

Inherent strains are typically applied as mechanical boundary conditions in the calculations for the layer which is subjected to high temperature during the additive manufacturing simulation process to capture the structural response at the end of the deposition step. The structure can be treated as elastic-plastic with the inherent strains applied either over a number of increments or as a one-increment total value in the finite element analysis. The actual layer thickness in the simulations can be larger than the real layer thicknesses to reduce calculation costs, following a meso/macro scale approach to estimate deformations with reasonable accuracy.

The actual values of inherent strain components ε_{i-x}, ε_{i-y}, and ε_{i-z} in the simulations are taken as calibration parameters. For the calibration, dedicated comb test structures might be used in correspondence with the special calibration routine of the Simufact Additive simulation software [8] or other procedures described in [5, 7, 10].

3. Simulation Methods

In the present study, the finite element software of Simufact Additive 2020 [8] was used for the simulations.

3.1. Specimen geometry and orientation

The workpiece was a U-Charpy specimen with a dimensions of 55 mm × 10 mm × 10 mm with a U-notch of 2 mm wide and 5 mm deep. Orientations of the workpieces were 0°, 45°, and 90° with respect to one axis of the build plate as shown in figure 1. A measurement point is defined on each specimen at the corner shown in figure 2, where x and y displacement data were collected at the end of the simulations.

3.2. General process definitions

Pure mechanical simulations were executed of the manufacturing process stages of build, release of the build plate and removal of the supports. Base plate deformations were considered in the simulations.

Generic machine definitions with a build space of 250 mm × 250 mm × 250 mm were used, where the build plate occupied the planar dimensions with a thickness of 30 mm of same material grade as the specimens. Specimen geometric parts are imported from stl files, support structures of 2 mm height were generated and the assembly was positioned over the build plate.

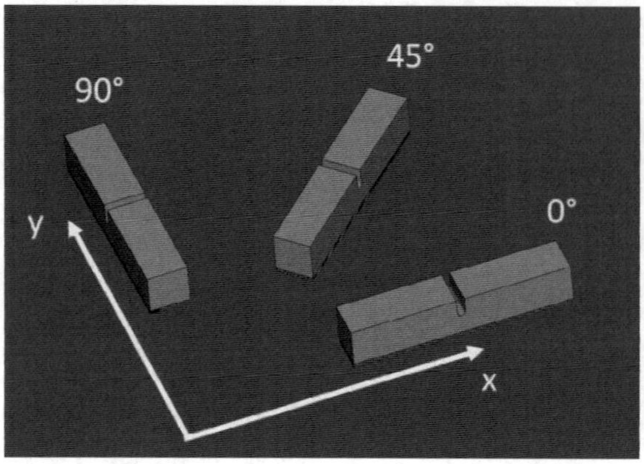

Figure 1. Orientations of the specimens on the build plate.

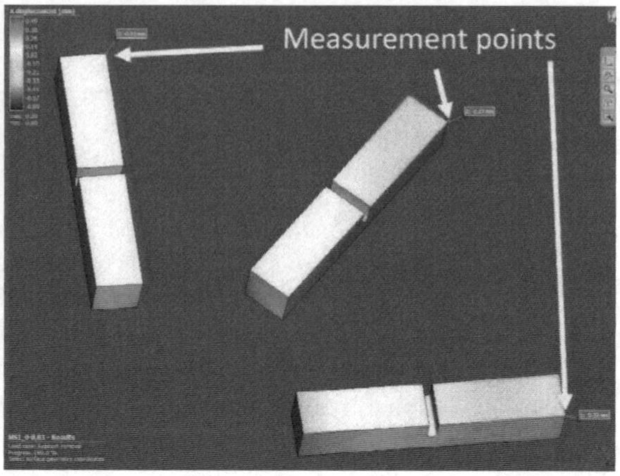

Figure 2. Measurement points for x and y displacements.

For the build parameters, a layer thickness of 0.03 mm was chosen. Uniform voxel size was set to 1 mm for the build space, and 4 mm for the base plate. All together 31292 voxels with 42277 nodes were generated in the simulation space.

3.3. Material properties

Material properties were taken from the Simufact Material interface, where MS1 and 316L steel powder material property sets are available. MS1 is a maraging tool steel (material number: 1.2709, chemical notation: X3NiCoMoTi18-9-5); according to the EOS datasheets [18] its yields strength is $R_{p0.2}$ = 2010 MPa; ultimate tensile strength is R_m = 2080 MPa, and elongation at fracture is A = 4%. Material 316L is an austenitic stainless steel grade (material number: 1.4404, chemical notation: X2CrNiMo17-12-2); according to the EOS datasheets [18] its yields strength is $R_{p0.2}$ = 490-535 MPa; ultimate tensile strength is R_m = 590-650 MPa, and elongation at fracture is A = 35-45%.

As both materials are steel grades, elastic properties are highly similar: Young's modulus was about 190 GPa; Poisson's ratio was 0.3 and density was about 8000 kg/m^3 for both materials.

Flow curves for MS1 at room temperature in the simulations only scale to a plastic strain of ε_p = 0.007, while yield strength is in the range of about 1500-1900 MPa. Flow curves for 316L at room temperature in the simulations scale up to a plastic strain of ε_p = 0.2, while yield strength are in the range of about 600-700 MPa.

3.4. Inherent strains as parameters

The inherent strain distribution was considered as uniform in the build space, with the planar ε_{i-x} and ε_{i-y} values varied in {0; -0.003; -0.016; -0.030} while off-plane value of ε_{i-z} was set to -0.030 in all simulations. This parameter variation resulted in 16 simulations for one material, all together 32 simulations.

4. Results and Discussions

Stress-strain analysis reveals (details not shown here) that in the build parts the strains are the highest at the U-notch, and there is a slight plastic deformation occurs in the parts at the bottom near the supports, while the upper region of the specimen is less loaded internally. In the MS1 specimens, the maximum of yield stress is about 1870 MPa, and in the 316L specimens, the maximum yield strength is about 680 MPa.

Results for the planar displacements at the measurement points are shown in figure 3 and figure 4 for MS1 and 316L specimens, respectively.

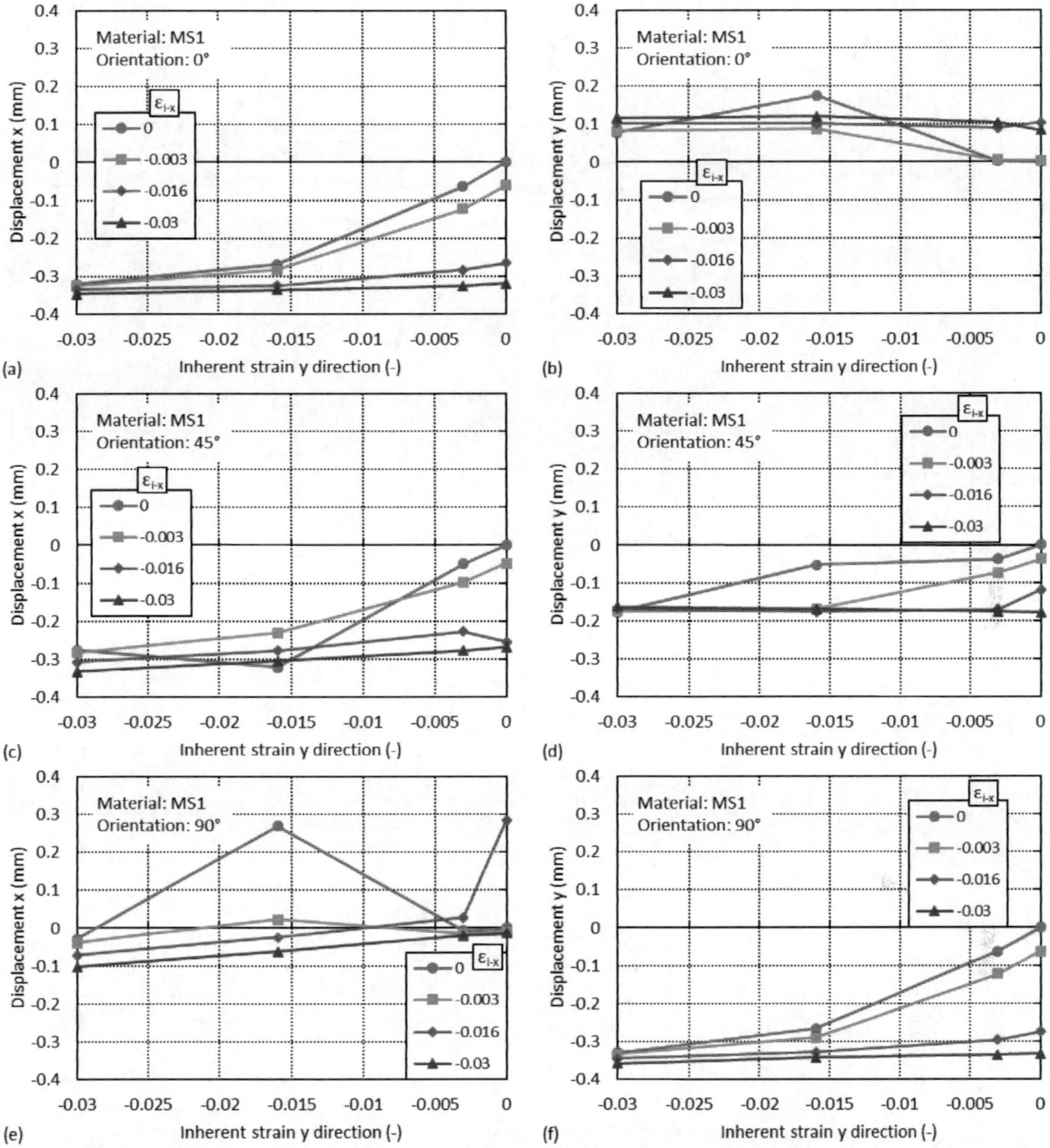

Figure 3(a)-(f). Planar displacements (left column: x; right column: y directions) as a function of planar inherent strains (horizontal axis: inherent strain in y direction, parameter for curves: inherent strain for x direction) for MS1 specimens with different orientations (from top to bottom: 0°, 45°, 90°).

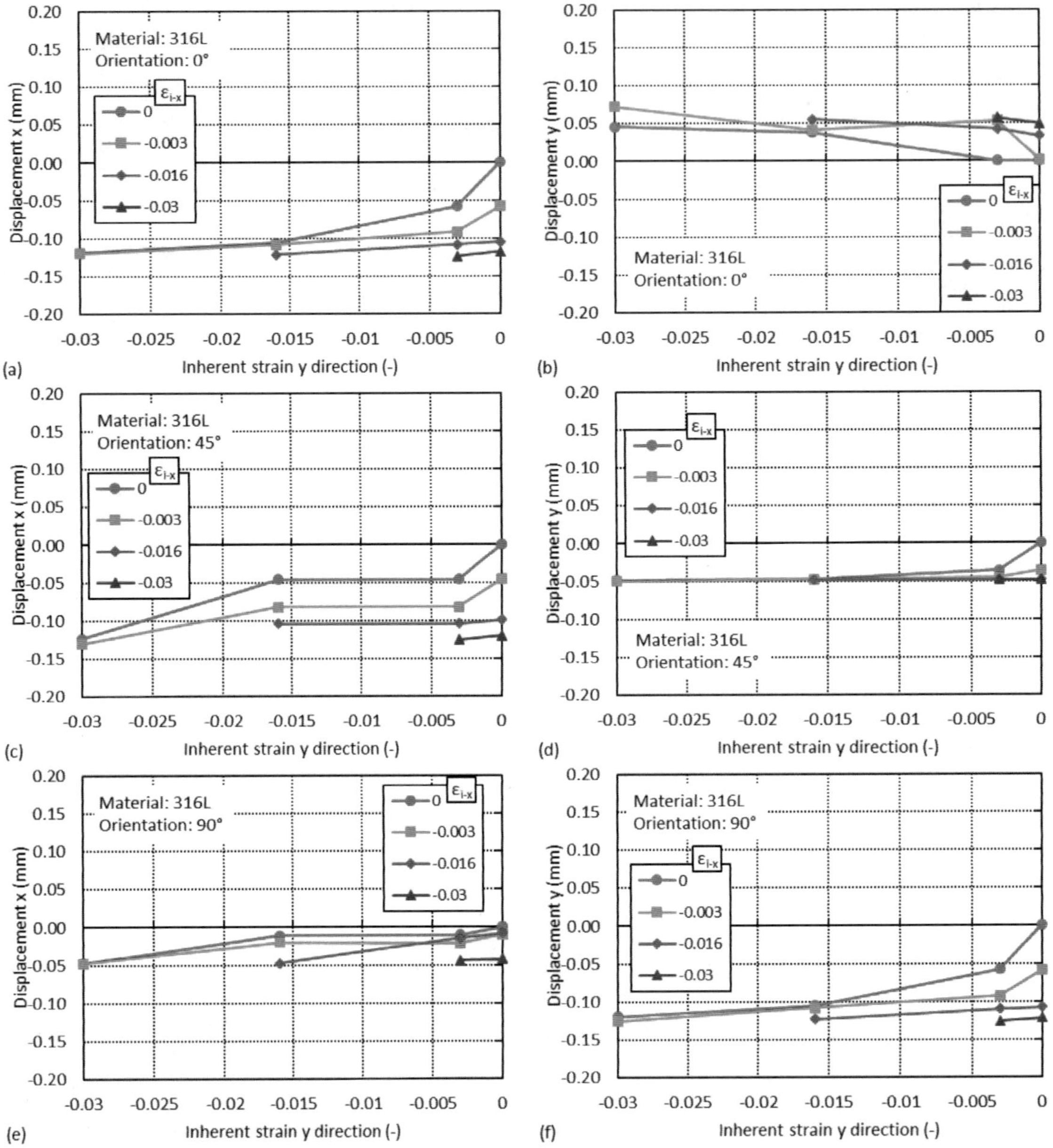

Figure 4(a)-(f). Planar displacements (left column: x; right column: y directions) as a function of planar inherent strains (horizontal axis: inherent strain in y direction, parameter for curves: inherent strain for x direction) for 316L specimens with different orientations (from top to bottom: 0°, 45°, 90°).

4.1. Comparison of planar displacements for MS1 specimens

By comparing the planar displacements with respect to the orientations of the specimens, the following can be found: for 0° orientation, the specimen's long axis is parallel to the x axis; with increasing the x and y inherent strains magnitude, displacement in x direction increases in negative direction as seen in figure 3(a). Figure 3(b) shows, that displacement in the y direction slightly increases, while inherent strain in y direction has small effect on the y displacement. Overall, strains are small in the transversal direction due to the short base length.

At 45° orientation, with increasing the x and y inherent strains magnitude, displacement in x direction increases in negative direction as shown in figure 3(c). Zero value for inherent strain makes calculations difficult, which results the deviating trend in the corresponding curve. Figure 3(d) shows, that displacement in the y direction slightly increases with the increasing inherent strain magnitude in x, while inherent strain in y direction has small effect on the y displacement.

At 90° orientation, the specimen's long axis is parallel to the y axis. Figure 3(e) shows, that displacement in the y direction slightly increases with the increasing inherent strain magnitude in x, while inherent strain in y direction has small effect on the y displacement. With increasing the x and y inherent strains magnitude, displacement in x direction increases in negative direction as shown in figure 3(f). Note that this figure corresponds well with figure 3(a).

4.2. Effect of orientation for MS1 specimens

At 0° orientation, both planar inherent strain component have the same effect on the x displacements, which converge to the same value. At 45°, the trend is similar, while uncertainty in the displacements are high. At 90°, the effect of the inherent strains are smaller on the x displacements, uncertainty is high at low x inherent strains. For y displacements, the same statements apply for the opposite orientation order.

4.3. Comparison of planar displacements for 316L specimens

By comparing the planar displacements with respect to the orientations of the specimens, the following can be found: for 0° orientation, the specimen's long axis is parallel to the x axis; with increasing the x and y inherent strains magnitude, displacement in x direction increases in negative direction as seen in figure 4(a). Figure 4(b) shows, that displacement in the y direction slightly increases, while inherent strain in y direction has small effect on the y displacement. Simulations failed to complete with high inherent strain pairs.

At 45° orientation, with increasing the x and y inherent strains magnitude, displacement in x direction increases in negative direction as shown in figure 4(c) and curves show a very clear trend. Figure 4(d) shows, that displacement in the y direction slightly increases with the increasing inherent strain magnitude in x with a very clear trend, while overall the displacement values are small.

At 90° orientation, the specimen's long axis is parallel to the y axis. Figure 4(e) shows, that displacement in the y direction slightly increases with the increasing inherent strain magnitude in x, while inherent strain in y direction has small effect on the y displacement. With increasing the x and y inherent strains magnitude, displacement in x direction increases in negative direction as shown in figure 4(f). Note that this figure corresponds well with figure 4(a).

4.4. Effect of orientation for 316L specimens

For x displacement, at 0° orientation both inherent strain components have strong effect, displacements converge to the same value. At 45° orientation, the trend is similar, but the effect is weaker. At 90° orientation, the sensitivity of displacements are smaller, at small inherent strains in x, the uncertainty is large. For y displacements, the same statements apply for the opposite orientation order.

4.5. Comparison of material grades

With respect to the magnitude of the deformations, for identical inherent strains, displacements are generally higher for MS1 than for 316L material. This difference is caused by the difference in the strength of the materials, and it is inversely proportional to the strength.

4.6. Effect of the geometry of the part

Obviously, larger displacement occurs at larger a dimensions, as shown of displacements in the long axial and short transversal dimensions.

5. Conclusions

Upon the simulation results of deformation of printed MS1 and 316L specimens with varying the planar inherent strain values, the following conclusions can be drawn: (1) A parameter study of inherent strains prior to the calibration reveals the deformation behaviour of the part. (2) The complex deformation behaviour of a part is difficult to predict in advance only upon the inherent strains, as it is affected by material and structural nonlinearities, and the gradual gain of new material during the build. (3) Simulations showed, that high strength MS1 material specimen deforms more than the less high strength 316L for the same set of inherent strains. Real inherent strain values may differ for the materials.

References

[1] Langi E, Zhao L G, Jamshidi P, Attallah M M, Silberschmidt V V, Willcock H and Vogt F 2021 Microstructural and Mechanical Characterization of Thin-Walled Tube Manufactured with Selective Laser Melting for Stent Application *J. Matls. Eng. and Performance* **30** 696
[2] Radhamani A V, Lau H C and Ramakrishna S 2021 316L Stainless Steel Microstructural, Mechanical, and Corrosion Behavior: A Comparison Between Spark Plasma Sintering, Laser Metal Deposition, and Cold Spray *J. Matls. Eng. and Performance* **30** 3492
[3] Bean G E, McLouth T D, Witkin D B, Sitzman S D, Adams P M and Zaldivar R J 2019 Build Orientation Effects on Texture and Mechanical Properties of Selective Laser Melting Inconel 718 *J. Matls. Eng. and Performance* **28** 1942
[4] Ueda Y and Yuan M G 1993 Prediction of Residual Stresses in Butt Welded Plates Using Inherent Strains *J. Eng. Matls. and Technology* **115** 417
[5] Setien I, Chiumenti M, van der Veen S, San Sebastian M, Garciandía F and Echeverría A 2019 Empirical methodology to determine inherent strains in additive manufacturing *Computers and Mathematics with Applications* **78** 2282
[6] Lu Q, Beauchesne E and Liszka T 2019 Enhancements to the Inherent Strain Method for Additive Manufacturing Analysis *Int. J. Multiscale Computational Eng.* **17** 65
[7] Liang X, Chen Q, Cheng L, Hayduke D and To A C 2019 Modified inherent strain method for efficient prediction of residual deformation in direct metal laser sintered components *Comput. Mech.* **64** 1719
[8] Simufact Additive finite element analysis software, Simufact Engineering GmbH, Germany
[9] Deng D, Murakawa H and Liang W 2007 Numerical simulation of welding distortion in large structures *Comput. Methods Appl. Mech. Eng.* **196** 4613
[10] Hill M R and Nelson D V 1995 The inherent strain method for residual stress determination and its application to a long welded joint *Proc. Asme PVP1995* **318** 343, New York, NY
[11] Kang S, Kim J, Jang Y and Lee K 2019 Welding Deformation Analysis, Using an Inherent Strain Method for Friction Stir Welded Electric Vehicle Aluminum Battery Housing Considering Productivity *Appl. Sci.* **9** 3848
[12] Kim T-J, Jang B-S and Kang S-W 2015 Welding deformation analysis based on improved equivalent strain method considering the effect of temperature gradients *Int. J. Nav. Archit. Ocean Eng.* **7** 157
[13] Lindgren L-E 2006 Numerical modelling of welding *Comput. Methods Appl. Mech. Eng.* **195** 6710

[14] Ma N, Nakacho K, Ohta T, Ogawa N, Maekawa A, Huang H and Murakawa H 2016 Inherent Strain Method for Residual Stress Measurement and Welding Distortion Prediction *Proc 35th Int. Conf. Ocean, Offshore and Arctic Eng.*, Busan, South Korea, June 2016

[15] Ma N and Huang H 2017 Efficient Simulation of Welding Distortion in Large Structures and Its Reduction by Jig Constraints *J. Matls. Eng. and Performance* **26** 5206

[16] Murakawa H, Deng D and Ma N 2010 Concept of inherent strain, inherent stress, inherent deformation and inherent force for prediction of welding distortion and residual stress *Transactions of JWRI* **39** 103

[17] Murakawa H, Deng D, Ma N and Wang J 2011 Applications of inherent strain and interface element to simulation of welding deformation in thin plate structures *Comput. Matls. Sci.* **51** 43

[18] Eos material datasheet, 2017, EOS GmbH, www.eos.info

IOP Conf. Series: Materials Science and Engineering 1246 (2022) 012011 doi:10.1088/1757-899X/1246/1/012011

Preparation and characterization of Fe-SiO_2 soft magnetic composites prepared by classic and modified Stöber Method

B Kocsis[1], L K Varga[2] and I Zsoldos[1]

[1] Széchenyi István University Department of Materials Science and Technology, 1. Egyetem tér, Győr 9026, Hungary

[2] Wigner Research Center for Physics Institution for Solid State Physics and Optics, 29-33 Konkoly-Thege út, Budapest 1121, Hungary

Abstract. In this study, gas atomized iron powder with the particle size of 10 - 100 μm was coated in two different ways. The coating was created by a chemical procedure, the powder was treated for different periods of time with the classical and modified Stöber process. Toroidal samples were then prepared using different pressures. These samples were heat treated in an argon shielding gas. The complex permeability spectra of the green and heat-treated samples were measured and the effect of pressure, chemical coating time and heat treatment was investigated. The formation of the coatings in each case was confirmed by scanning electron microscopic examinations. Samples with long-term chemical treatment and without heat treatment were appropriate in high-frequency applications ($\mu_{stat} = 50, f_{FMR} = 5 - 15\ MHz$). After the heat treatment, the relative initial permeability has increased significantly (160-180) and the resonance frequency has decreased (400-4000Hz). Furthermore, it can be stated that a more coherent, thicker coating can be obtained with the modified Stöber method.

1. Introduction

The development of modern magnetic materials has once again become the focus of development thanks to the growing market of energy-efficient electrical equipment. Due to their low coercivity and the low eddy current losses induced in them, soft magnetic composites can result in a more efficient, faster, more compact, high-permeability, high frequency device [1,2]. In the majority of cases, an iron-based powder was coated with an organic, or an inorganic insulator, or a mixture of the two. The powder is then sintered at high pressure and temperature to obtain the final geometry. The type of insulation phase is primarily determined by the application conditions of the given equipment (mechanical, thermal and radiation loads) [3,4]. The magnetic properties of the soft magnetic composites (SMCs) are strongly degraded by accumulated dislocations and lattice distortions, however their effects can be reduced by heat treatment processes. Organic insulators can be heat-treated at low temperatures (less than 400 °C), where lattice defects are not annihilated, so in these cases only inorganic insulators can be used [1]. The selection of the ferromagnetic element or alloy focuses on the electrical resistance of the powder particles and thus on the locally induced eddy current losses. For high-frequency equipment, it is practical to choose “high” resistance alloys (compared to pure iron), while keeping the price of the final products low. It is important to note that for soft magnetic composites, the type of electrically insulating material reduces magnetic permeability. The insulating layer interrupts the particle-to-particle current path (minimizes eddy current loss), while reducing the permeability by having a non-magnetic gap between the particles and (in proportion to the volume fraction of the matrix material) it is also reducing the saturation magnetization [5]. There are efforts to develop electrically insulating magnetic matrix materials that can further improve saturation induction and permeability. The proportion of ferromagnetic particles and the insulating material is given by the so-called filling factor. Using the filling factor, the coefficient of demagnetization and the magnetic permeabilities of the two components, we can mathematically estimate the macroscopic property of the heterogeneous structure using Equivalent Homogeneous Material (EHM) techniques [6,7]. In this case we wanted to estimate the magnetic permeability of the heterogeneous structure. This procedure can help to design iron core for end use. The effective permeability was estimated using the Maxwell-Garnett (MG) formula:

$$\tilde{\mu}_{MG} = \mu_1 + \frac{\xi \mu_1 (\mu_2 - \mu_1)}{\mu_1 + N(1 - \xi)(\mu_2 - \mu_1)} , \quad (1)$$

where ξ – is the filling factor, μ_1 – is the magnetic permeability of the insulator, μ_2 – is the magnetic permeability of the ferromagnetic powder, and N is the demagnetization coefficient: $N = \frac{1}{3}$ for spherical inclusions. Based on the [2] literature, we chose the filling factor to be 0.976 (thickness of SiO_2 coating was approx. 200 nm). We took the value of μ_1 as 1 and the value of μ_2 as 5000. The estimated effective permeability according to the MG formula was 120.

2. Materials and Methods(1)

2.1. Materials and reagents

The core material was high purity, atomized iron powder (grade ASC 100.29), supplied by Höganäs AB Sweden. The particle size was less than 100 µm in all instances. The classical and modified Stöber methods were used for the chemical coating process. Tetraethyl orthosilicate (TEOS), aqueous ammonia (NH_4OH), isopropyl alcohol, deionised water and (3-aminopropyl)triethoxysilane (APTES) were utilized. These chemicals were of analytical grade and were applied without further dilutions.

2.2. Coating Methods

Stöber-silica synthesis consists of two basic steps. The first step is hydrolysis of TEOS, the second one is a polycondensation. In general, the ratio of the velocities of two sub-steps determines whether chain-like (acidic) or isometric (basic) microphases are obtained. The coating was prepared in four separate portions of 25 grams. The surface of the elemental powder was neutralized, degreased and then mixed with the solution. The solution was stirred continuously at 30 Hz by a Union Process 1S Szegvari attritor. We controlled the temperature of the solution by a ceramic hotplate. For the classical Stöber method, 25 grams of iron powder was added to 800 ml of isopropyl and 160 ml of deionized water. After a few minutes, 20 ml of ammonia and 80 ml of TEOS were added [8]. We prepared 3 doses from this concentration and stirred them for 8, 16 and 24 hours. In case of the modified Stöber method, 25 grams of iron powder was stirred in 250 ml of isopropyl alcohol with continuous stirring at room temperature for 10 minutes. Then 1.5 gram of APTES and 5 ml of deionized water were added. The temperature of the solution in this step was raised to 50 °C and stirred at 30 Hz for 1 hour. In the next step, 7.5 ml of TEOS and 1 ml of ammonia were added to the solution. The temperature was raised up to 60 °C and the powder was left in the solution for an additional 23 hours with continuous stirring [9]. Finally, the powder samples were dried in a drying chamber after several wash down with ethanol at 60 °C for 24 hours.

2.3. Preparation of Fe/SiO_2 SMC toroidal samples

Rectangular toroidal samples were prepared from the coated powder for complex permeability spectrum measurements. The diameters of the toroids were D = 12 mm, d = 7 mm and the height was altered between h = 2 – 3.5 mm. The samples were subjected to a pressure of 500 – 1300 MPa for 5 minutes at room temperature using a hydraulic press. The toroids were than heat treated to reduce microstructural defect and residual stress. They were treated at 700 °C for 90 minutes under argon shielding gas.

2.4.Characterization of coated powders and Fe/SiO2 toroids
Microstructural and compositional investigations of the powder and compacted samples were performed on a HITACHI 3400 scanning electron microscope and its Bruker energy dispersive spectrometer (EDS). The permeability spectra of the pressed (green) and heat treated toroid samples were measured using a Hewlett-Packard 4274A, an Agilent 4285A, and an Agilent 4291B multi-frequency LCR meter from 100 Hz to 1 GHz. Static and dynamic permeability values were examined, from which the Snoek limit was determined as a basis for comparison. The samples were then heat treated to reduce microstructural anisotropy and their permeability spectra were remeasured.

3. Results and discussion
The experiments were started with scanning electron microscopic measurements. Powders treated with the classical and the modified Stöber method were also tested before compression. The presence of SiO_2 coating on the surface of some particles was detected by line analysis. Energy-dispersive point analyses were performed on the surface of the particles to quantify the precipitated silica. Line analysis, EDX spectra and composition ratios are shown in figure 1.

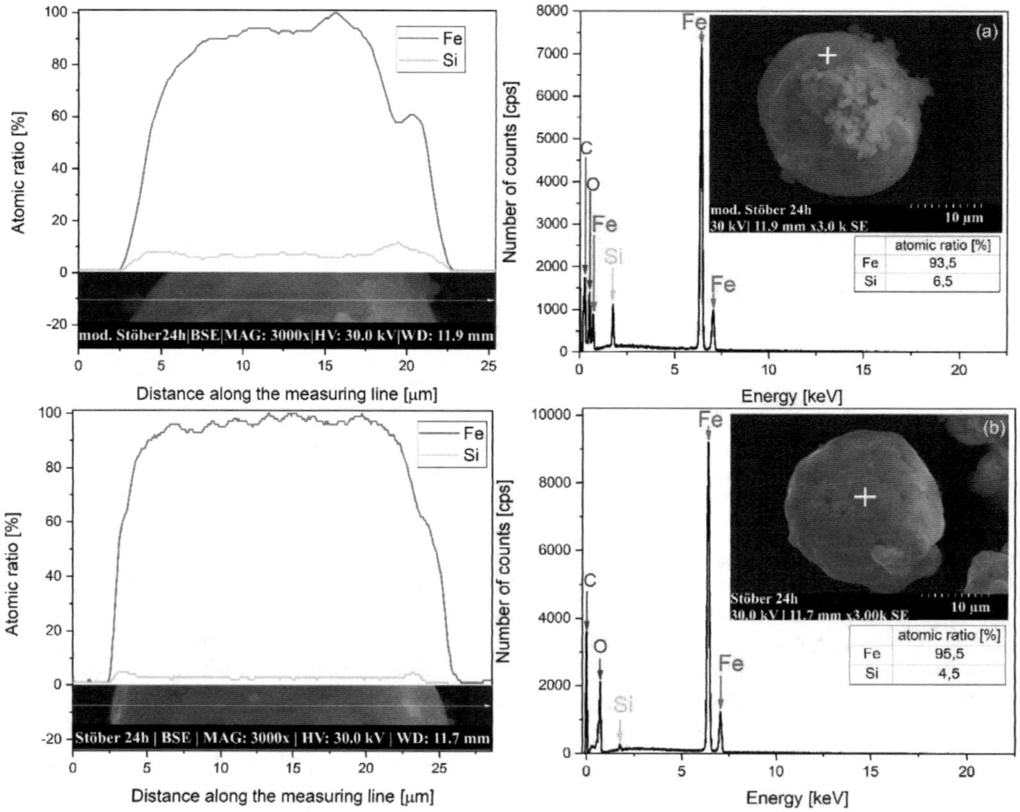

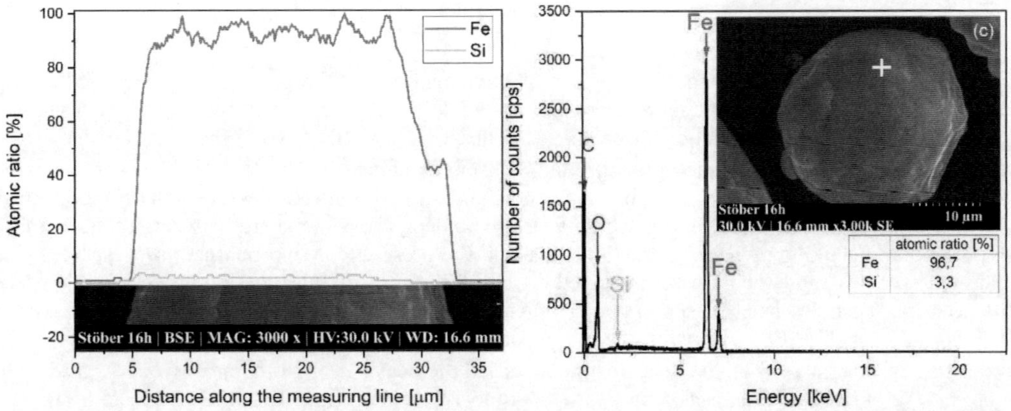

Figure 1. Line and point analyses and energy dispersive spectrum of powder particles treated with the modified Stöber method for 24 hours (A) and treated with the classical method for 24 (B) and 16 (C) hours.

Based on the line analyses, it can be stated that both Stöber methods create a kind of SiO_2 coating on the surface of powders. Microscopic images of their thickness and coherence could not be obtained. During grinding and polishing, the coating was either fragmented from the hot mounting resin or so thin that it could not be clearly detected by SEM. Transmission microscopic examinations or magnetic measurements should be performed to answer these questions. The permeability spectra shown in figure 2 can be used to further classify coatings created by different methods.

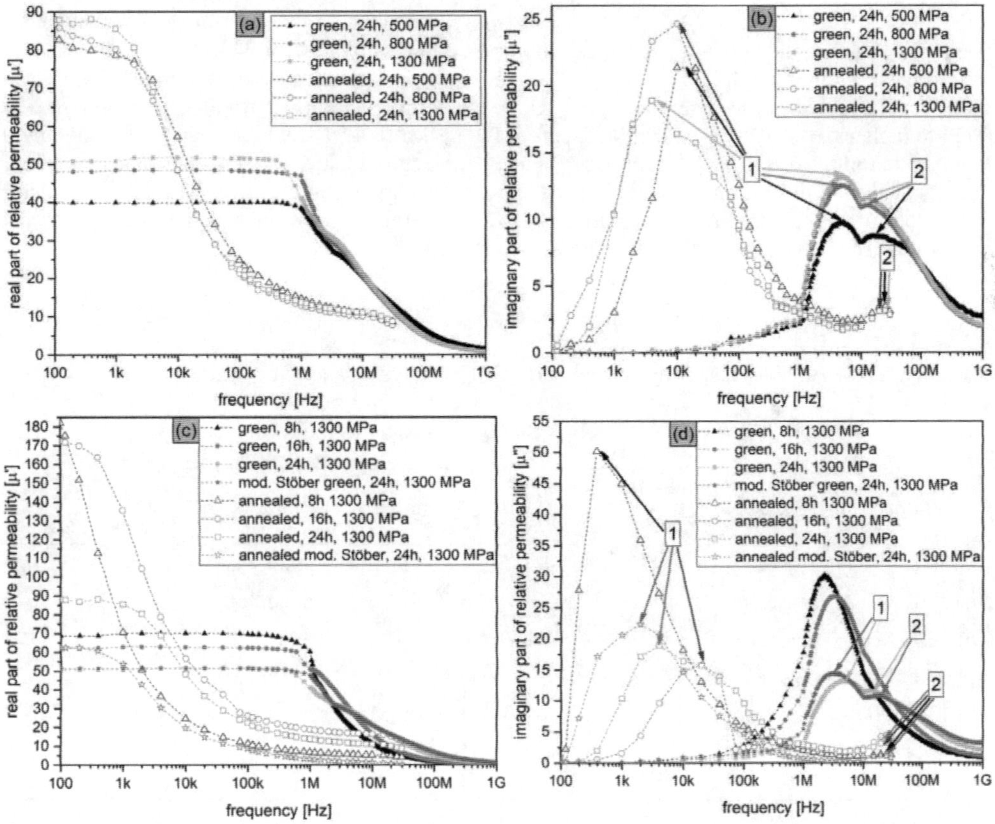

Figure 2. Investigation of the pressure dependence (A, B) and the duration of the chemical treatment (C, D) of the relative permeability spectra of toroidal samples prepared by the classical andthemodified Stöber methods.

Diagrams A and B in figure 2 show the effect of the magnitude of the pressure value applied during compression on the real and imaginary permeability values. It can be clearly seen that higher applied pressure resulted in higher static permeability values ($\mu_{stat} = 40 - 50$). This may be due to an increased number of interparticle contact, which can be explained by the failure of the brittle oxide coating. In this case, a direct proportionality can be drawn between the pressure values and the loss factor. In each case, this static, initial segment begins to decrease somewhere between 400 kHz and 1 MHz. Examining the imaginary permeability values, the presence of two phases was observed. The peaks (marked 1) denote the particles arranged in agglomerates during the coating process, while peaks 2 denote iron particles isolated separately. After heat treatment, the real permeability values almost doubled ($\mu_{stat} = 82 - 88$). However, the initial permeability begins to decrease drastically at 1-2 kHz. In the case of imaginary part, the heat treatment shifted the peaks 1 (agglomerates) from 4 - 5 MHz to approx. 4 – 20 kHz, while the peaks marked with number 2 went up from 15 – 17 MHz to 25 MHz. The unique particles were presumably relieved of lattice defects and internal stresses during heat treatment. Diagrams C and D in figure 2 show the effect of duration of the classical and the modified Stöber chemical coating process to the complex permeability spectra. It can be observed in diagram C that the powder treated for 8 hours has the highest initial static permeability due to inadequate electrical interparticular insulation. This is closest to the permeability spectra of iron powder toroid. It can also be observed that the Snoek limit of this sample is the lowest, as its initial permeability starts to decrease already at 500 kHz, its resonance frequency is approximately 2 MHz. In the instance of the samples treated for 16 and 24 hours, the static initial permeability changes only between 50-62 and starts to decrease significantly only at 600 – 1000 kHz, their resonance frequency is ~10 MHz. As a result of the heat treatment, the permeability of the samples treated for 8 and 16 hours increases almost threefold ($\mu_{stat} = 182$), which begins to decrease already at 100 Hz and their resonance frequency also deteriorates significantly. The static permeability of the samples treated for 24 hours was already much less improved. Examining the imaginary part of the relative permeability (see Diagram D), we reach a similar conclusion as in the case of diagram B. Peaks 1 shifted to lower frequencies, while peaks 2 shifted to higher frequencies. The displacement of the peaks marked 1 is presumably explained by the compression failure of the large grains arranged in the agglomerates and the increasing number of contacts, while the displacement of the peaks 2 can be explained by the decrease of microstructural defects due to the heat treatment relaxation. The Snoek-limit calculation was determined based on equation (2). The values for each sample are shown in Table 1.

$$Snoek\ limit = \mu_{stat} + f_{FMR}\ , \tag{2}$$

where μ_{stat} is the static permeability of real part, f_{FMR} is the resonance frequency.

Table 1. Snoek-limits of SiO_2 coated green and annealed toroidal samples.

Sample	Status	Snoek limit [MHz]		Sample	Status	Snoek limit [MHz]	
		1st peak	2th peak			1th peak	2th peak
24h; 500 MPa	green	97	374	24h; 500 MPa	annealed	4.1	1025
24h; 800 MPa	green	108	360	24h; 800 MPa	annealed	4.3	1075
24h; 1300 MPa	green	127.5	394.74	24h;1300 MPa	annealed	1.76	1100
8h; 1300 MPa	green	77.52	-	8h; 1300 MPa	annealed	0.034	2150
16h; 1300 MPa	green	111.6	-	16h;1300 MPa	annealed	18.2	2275
mod. Stöber 24h; 1300 MPa	green	80.25	520.38	mod. Stöber 24h; 1300MPa	annealed	6.2	775

Based on table 1, it can be concluded that the best iron core was the 24h, 1300 MPa, green toroid that was treated with the modified Stöber method, when it comes to discussing the two-phase curves. However, it is important to note that if only the second phase is examined, 2.275 GHz Snoek limit can be reached.

The microstructural condition after pressing was also examined with a scanning electron microscope. The images of this step are illustrated in figure 3.

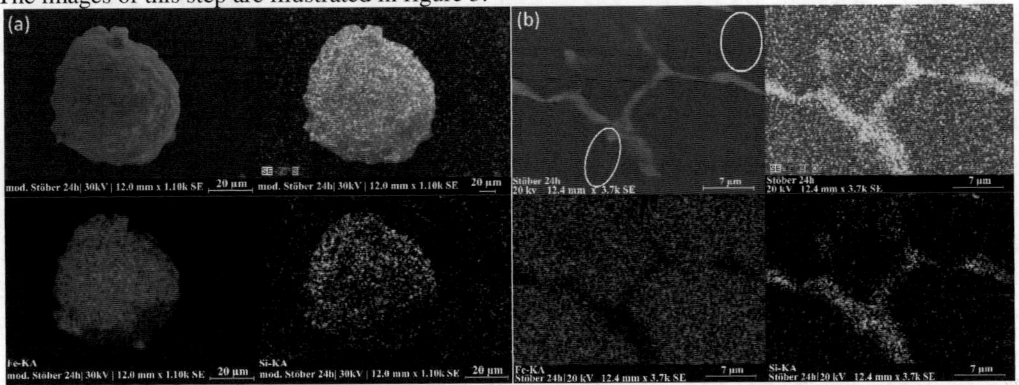

Figure 3. SEM images of powder (A) and toroidal sample (B) pressed with 1300 MPa, treated for 24 hours with the modified Stöber process.

It can be clearly seen that the 24-hour treatment resulted in a highly significant coating on the surface of the particles, which only partially exerts its insulating effect after compression. In the microscopic image of Figure 3B, the portions marked with white ellipsoids are thought to be contact surfaces formed by compression. In the pre-compression state, a uniform coating was observed on the surface of the powder particles.

4. Conclusion

In light of the examinations and results described above, we can conclude that, in principle, both types of chemical coating processes can provide an adequate coating quality. However, it is important to emphasize that the duration of the coating process can greatly affect the magnetic permeability, and it is advisable to dispense with the heat treatment step in high frequency applications. Overall, the following findings were made:

- Based on the Maxwell-Garnett formula and the preliminary literature data, the effective permeability of SMCs can be estimated with a very good approximation.
- The effect of increasing the pressure on the degradation of the insulating layer is becoming more and more significant, which was inferred from the increase in μ' and μ" values and SEM investigations.
- The duration of the chemical treatment has a significant effect on the quality and quantity of the coating formed.
- During the Stöber methods, it may be worth using a different stirring process that can further reduce the amount of agglomerates formed.
- It may be desirable to re-segment the powder after the coating process in order to eliminate large particles and agglomerates from the samples to reduce the biphasic nature.
- Based on the SEM examinations, with the modified Stöber method, at least thicker and more coherent coatings are deposited on the iron surfaces in 24 hours than during the classical process. Unfortunately, this does not appear significantly in the magnetic properties under the parameters described above.
- The pressing force must be kept to the minimum required level in order to maintain the best possible particle-to-particle insulation.

- The Stöber processes can be used to create soft magnetic composites with highly favourable filling factors.

5. Acknowledgements

The research presented in this paper was funded by the "Thematic Excellence Programme 2021 (TKP2021) – National Defence, National Security Subprogramme – Research on military systems resilience at Széchenyi István University (TKP2021-NVA-23) and NTP-NFTÖ-21. The project was also supported by the Hungarian Scientific Research Fund (OTKA K 128229).

6. References

1. Slovenský P, Kollár P, Mei N, Jakubčin M, Zeleňáková A, Halama M, et al. Mechanical surface smoothing of micron-sized iron powder for improved silica coating performance as soft magnetic composites. Applied Surface Science. 2020 Nov;531:147340.

2. Neamţu BV, Opriş A, Pszola P, Popa F, Marinca TF, Vlad N, et al. Preparation and characterisation of soft magnetic composites based on Fe fibres. J Mater Sci. 2020 Feb;55(4):1414–24.

3. Merta V, Beňo J, Obzina T, Radkovský F, Kroupová I, Lichý P, et al. Innovative Inorganic Binder Systems for the Production of Cores for Non-Ferrous Metal Alloys Reflecting the Product Quality Requirements. Metals. 2021 Apr 29;11(5):733.

4. Appino C, de la Barrière O, Fiorillo F, LoBue M, Mazaleyrat F, Ragusa C. Classical eddy current losses in soft magnetic composites. Journal of Applied Physics. 2013 May 7;113(17):17A322.

5. Périgo EA, Weidenfeller B, Kollár P, Füzer J. Past, present, and future of soft magnetic composites. Applied Physics Reviews. 2018 Sep;5(3):031301.

6. Timonen JVI, Ras RHA, Ikkala O, Oksanen M, Seppälä E, Chalapat K, et al. Magnetic Nanocomposites at Microwave Frequencies. In: Bârsan V, Aldea A, editors. Trends in Nanophysics [Internet]. Berlin, Heidelberg: Springer Berlin Heidelberg; 2010 [cited 2022 Jan 19]. p. 257–85. (Engineering Materials). Available from: http://link.springer.com/10.1007/978-3-642-12070-1_11

7. Sihvola A. Electromagnetic Mixing Formulas and Applications [Internet]. The Institution of Engineering and Technology, Michael Faraday House, Six Hills Way, Stevenage SG1 2AY, UK: IET; 1999 [cited 2022 Jan 19]. Available from: https://digital-library.theiet.org/content/books/ew/pbew047e

8. Stöber W, Fink A, Bohn E. Controlled growth of monodisperse silica spheres in the micron size range. Journal of Colloid and Interface Science. 1968 Jan;26(1):62–9.

9. Wang J, Fan X, Wu Z, Li G. Intergranular insulated Fe/SiO 2 soft magnetic composite for decreased core loss. Advanced Powder Technology. 2016 Jul;27(4):1189–94.

IOP Conf. Series: Materials Science and Engineering 1246 (2022) 012012 doi:10.1088/1757-899X/1246/1/012012

Numerical Investigation of Hot Roll Bonding of Multilayer Sheet Metal

D E Jemal[1], Dr. L Zsolt[1] and Dr. S Máté[2]

[1] Institute of Materials Science and Technology, Faculty of Mechanical Engineering and Informatics, University of Miskolc, H-3515 Miskolc, Egyetemváros, Hungary
[2] Institute of Physical Metallurgy, Metal forming and Nanotechnology, University of Miskolc H-3515 Miskolc-Egyetemváros, Hungary

Abstract. The roll bonding process of the sandwich sheet is a preferable production method to manufacture multilayer sheets and it needs clear investigation because some industrial parts cannot fulfill all of the requirements by single-layer material. In this study, a theoretical model of roll bonding has been reviewed and hot sandwich rolling processes with different plate thickness and flow stress were numerically investigated. The contact pressure-dependent bonding properties in the deformation zone were defined. Then, the liner thickness and reduction ratio are important parameters to determine the total deformation zone. When the thickness of the liner becomes large, unbonded length of deformation can be a serious problem and at the same time, the good bonding strength can be found at a higher reduction ratio of the thinner liner.

1. Introduction

Some industrial parts cannot fulfill all of the required properties by single-layer material. Therefore, Multilayer sheet metals are highly demanded in the automotive, aviation, and electrical industries [1]. Flat rolling is a suitable choice for manufacturing sandwich sheets because applying enough contact pressure between sheets is the best option to create bonding [2]. It is a more efficient and economical method [3]. Flat rolling is a common continuous metal forming process in which a piece of metal sheet is reduced by passing it between a pair of rotating cylindrical rollers [4].

Khan H A and Asim K [5] reviewed the key aspects of the roll bonding process such as process parameters, microstructure, and defects, and also reviewed the influence of those key aspects on the resultant properties in detail. Da Silva and Laurie [6] also reviewed the cold roll bonding of aluminum alloy with steel and the effect of the main process parameters of cold roll bonding and bond strength was discussed. The progress in cold roll bonding of metals was reviewed by Li Long, Kotobu Nagai, and Fuxing Yin [7]. Surface preparation-related process parameters that affect bonding have been discussed and described the possibilities that help to process roll bonding in the future. Szabó G [8] also researched an experimental study of hot roll bonding to optimize the bonding properties of the layer-clad aluminum sheets using a von roll mill. They were considered AlMn1Si0.8 and AlSi10 materials for core and liner respectively. Both the tensile shear test and T-peel test were performed to define the bonding strength. Pan S C and Huang [9] investigated the roll bonding of unbounded clad sheets under constant shear friction at both room and elevated temperatures. The authors develop a new analytical approach of asymmetrical roll bonding of clad sheets to find the various stress distributions easily and rapidly. Lee C H and Park J P [10] studied the warping behavior, which occurs due to the mismatch in longitudinal elongation of flat rolling of two layer-sheets based on differential analysis. Use FORGETM as a finite element code to support the investigation. Han J and Niu H [11] researched the influence of mechanical surface treatment on the bonding mechanism of the cold roll bonding process. The authors discussed the properties and bond strengthening mechanisms of cold-rolled Cu/Al clad sheets using both tensile shear and peeling tests.

In this study, the effect of the reduction ratio and liner thickness of the sandwich rolling process on the deformation zone, and the bonding strength of the sandwich roll bonding process have been numerically investigated.

2. Theoretical models of roll bonding

Some theoretical models have been proposed to predict the bond strength developed from roll bonding [12].

$$\eta = \frac{\sigma_B}{\sigma_0} = R_f(2 - R_f) \tag{1}$$

where η is bond efficiency, σ_B is the bond strength, σ_0 is the strength of the base metal, and $R_f = (l_f - l_o)/l_f$ is the final reduction at the final rolling pass, l_f is the length of the sheet after deformation and l_o is the length of the sheet before deformation. Subsequently, modified relationships have also been proposed by considering the length at the threshold deformation as equation (2) [12].

$$\eta = 1 - \left(\frac{(1-R_f)^2}{(1-R_t)^2}\right) \tag{2}$$

where $R_t = (l_t - l_o)/l_t$ is the reduction in threshold deformation, l_t is the length at the threshold deformation.

On the basis of the oxide and contaminant film mechanisms, another theoretical model showing the basic effects of both surface and normal pressures on the bond strength was proposed [13], and the equation is:

$$\frac{\sigma_B}{\sigma_0} = (1 - \psi^2)Y\frac{p - p_E}{\sigma_0} + \psi^2\frac{Y - Y'}{1 - Y'}\frac{p}{\sigma_0} \tag{3}$$

where p is normal pressure at base metal surfaces, p_E is extrusion pressure, Y is Surface exposure of weld interface surface, Y' is threshold surface exposure for contaminant film and ψ is the fraction of the film layer with respect to the total area.

On the other hand, in strip rolling the relationship between strain and deformation can be described by the following equation [14]:

$$\varepsilon = \frac{2}{\sqrt{3}}\ln\frac{1}{(1-R_f)} \tag{4}$$

From the above equation, a new model for estimating bond efficiency from peel test results for a 5754-aluminum alloy was proposed. In this model, the effect of work hardening of material was considered as equation (5) [14].

$$\sigma_0 = K\varepsilon^n \tag{5}$$

where K and n are the material plastic constants

Therefore, it can be obtained

$$\eta = \frac{\sigma_B}{\sigma_0} = \frac{\sigma_B}{K}\left[\ln\frac{1}{1-R_f}\right]^{-n} \tag{6}$$

However, the general mechanism of bonding during rolling is still unclear and almost all the above theoretical models were focused on the cold roll bonding process of symmetrical material properties and sizes at a high amount of reduction ratio. There are some difficulties in precisely determining the hot roll bonding process with a small amount of deformation ratio, particularly in the case of bonding different metals with different thicknesses.

3. Finite element modeling

The finite element (FE) model in this study was carried out using Marc Mentat commercial FE software. Given the assumption of isotropic material response, the FE model developed for this study comprised a two-dimensional axisymmetric full & Hermann formulation Quad element, with a finer mesh. The model is composed of one rigid solids roller and two deformable solid workpieces as shown in figure 1. The roller is completely fixed and only free to rotate counterclockwise direction. The simulation work plan is summarized in table 1. All contacts between the work and roller are treated as hard contacts and an Arctangent (shear) model with a 0.3 coefficient of friction was used and the coefficient of friction between sheets was assumed 1. The constant hot rolling parameters used in the simulation are listed as follow:

Roll speed: 2.272 rad/sec
Workpiece temperature: 400 °C
Roller temperature: 20 °C
Heat transfer coefficient: 20 kW/m^2.C°
Film coefficient (air, free convection): 25x10^{-3} W/(m^2.C°)

Table 1. Simulation work plan summary

Roller diameter, Do (mm)	Liner thickness, t_{liner} (mm)	Core thickness, t_{core} (mm)	Percentage reduction (%)
220	2.5 & 7.5	20	1.25, 2.5, 5, 10, 15 and 20

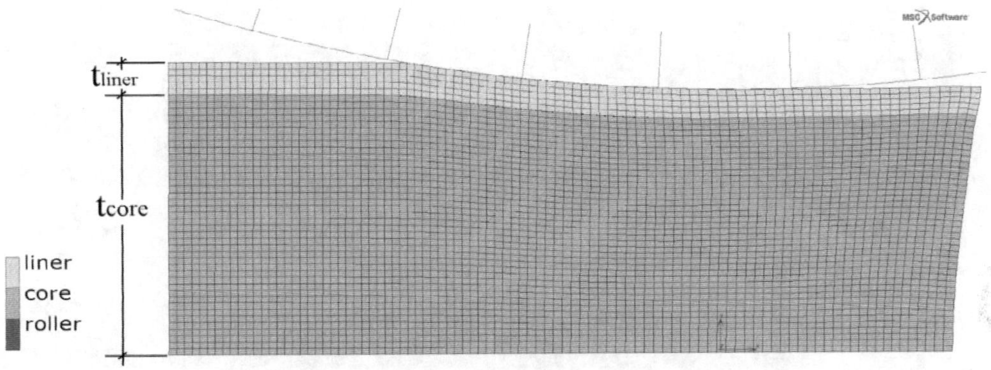

Figure 1. Geometrical model of the multilayer sheet.

3.1 Material model

In this simulation, the isotropic hardening model was used and yield stress is dependent on equivalent plastic strain, the inelastic strain rate, and temperature. The work material was Al3003 aluminum alloy and the mechanical properties are summarized in figure 2. The physical and thermal properties are summarized in table 2 [15].

Table 2. Physical and thermal properties of Al3003

Density (g/cm^3)	Elastic Modulus (Mpa)	Poisson's ratio	Thermal Conductivity (W/m.C)	Specific Heat Capacity (J/(kg·C)
2.73	68900	0.33	154	893

IOP Conf. Series: Materials Science and Engineering 1246 (2022) 012012 doi:10.1088/1757-899X/1246/1/012012

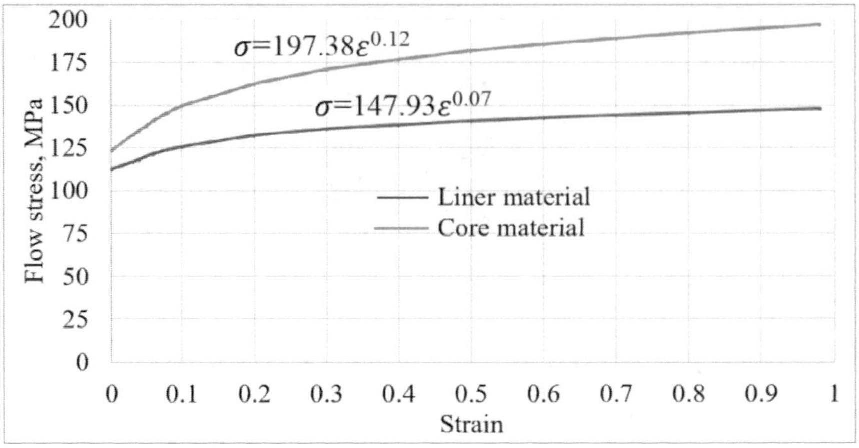

Figure 2. Flow curves of Al3003 aluminum alloy.

4. Simulation results and analysis

The temperature and contact pressure distribution from the FEM simulation is shown in figure 3 and 4 respectively.

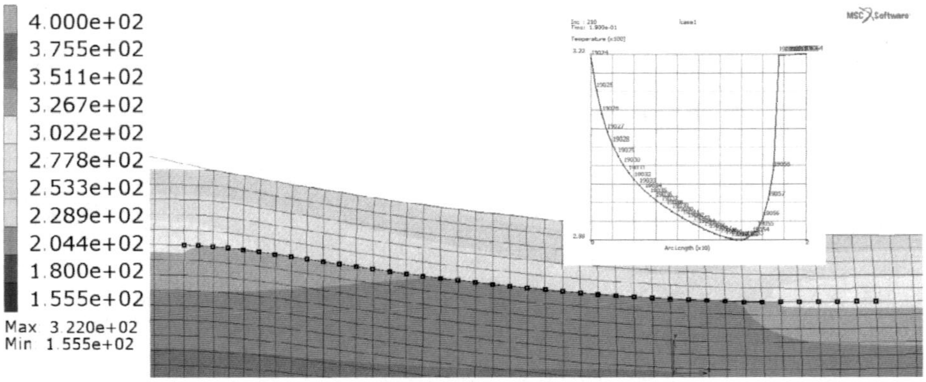

Figure 3. Temperature distribution.

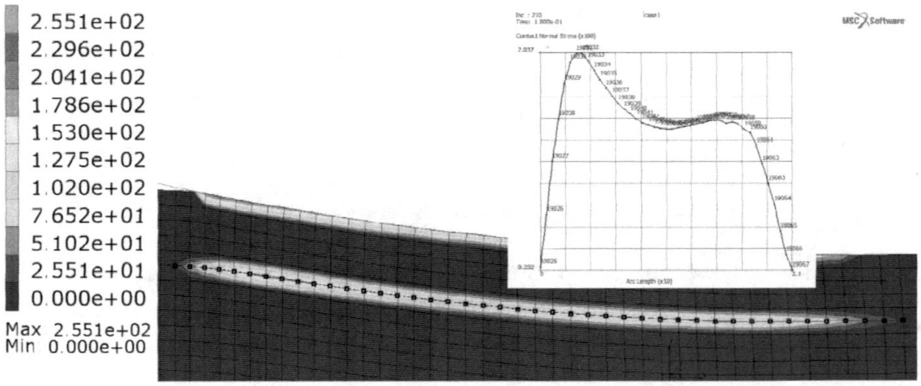

Figure 4. Norml contact pressure distribution.

In Figures 5 and 6, the effect of the liner thickness and percentage reduction on the normal contact pressure along the deformation length between sheets and between roller and liner sheet were graphically presented. The graph clearly explained that when the liner thickness decreases, the normal contact pressure in both interfaces becomes increases at a small percentage reduction. At the same time, the figure indicated there will be the possibility of un-bonding when decreasing the amount of percentage reduction at larger liner thickness.

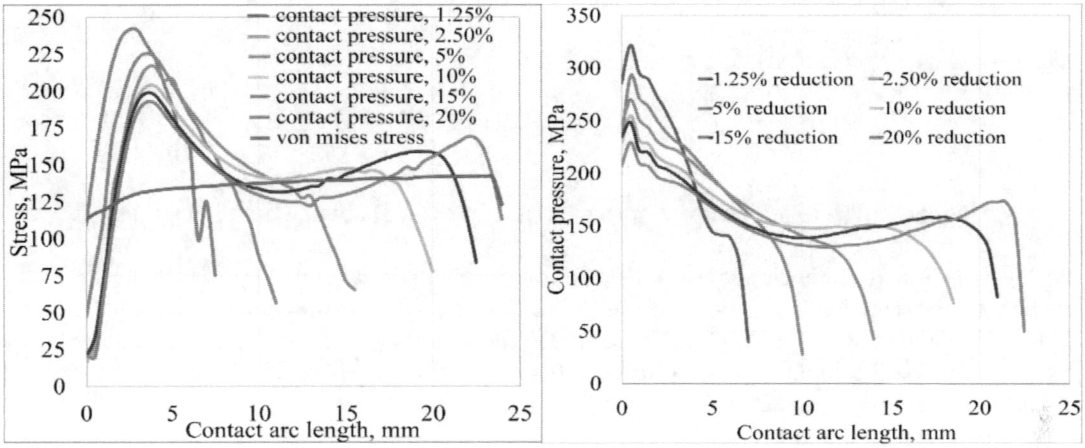

Figure 5. Normal contact pressure distribution along the deformation length at 2.5 mm liner thickness: (a) between sheets, (b) between roller and sheet.

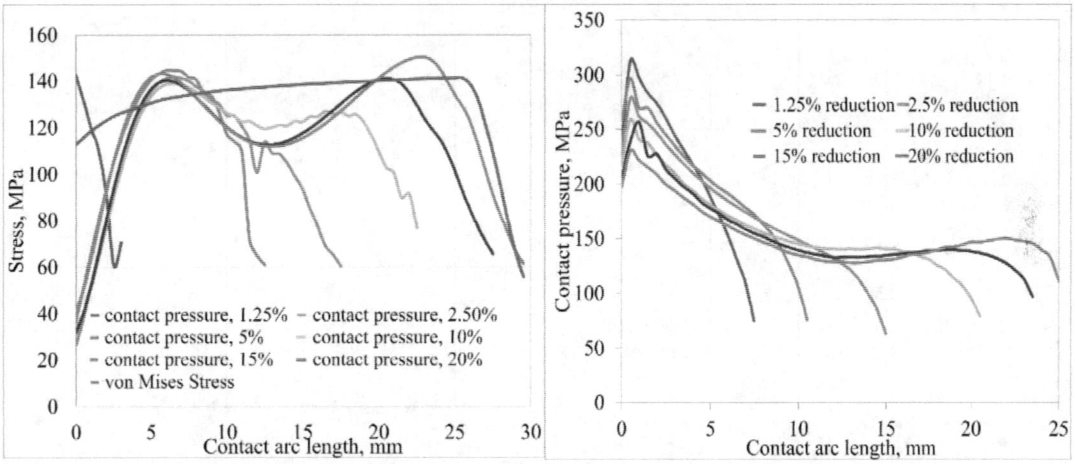

Figure 6. Normal contact pressure distribution along the deformation length at 7.5 mm liner thickness: (a) between sheets, (b) between roller and sheet.

The effect of liner thickness and percentage reduction on the contact pressure relationship between both interfaces has been shown in figure 7. The figure indicated the normal contact pressure between sheets becomes closer to the normal contact pressure required between roller and liner at a higher amount of reduction while decreasing liner thickness.

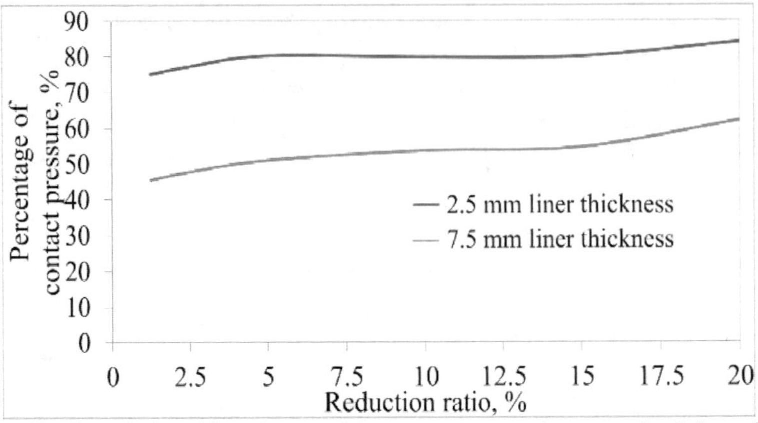

Figure 7. The relationship of normal contact pressure between both interfaces.

The deformation zone developed from the position of some critical contact pressures between sheets along the total deformation length is shown in figure 8. Point A was positioned at the pressure required to start yielding the hard sheet. This study considered the general assumption that the bonding between sheets started while the contact pressure reaches the yield strength of weaker material and position at point B. At the maximum contact pressure and the exit contact pressure required in the deformation zone were positioned points C and D respectivelly.

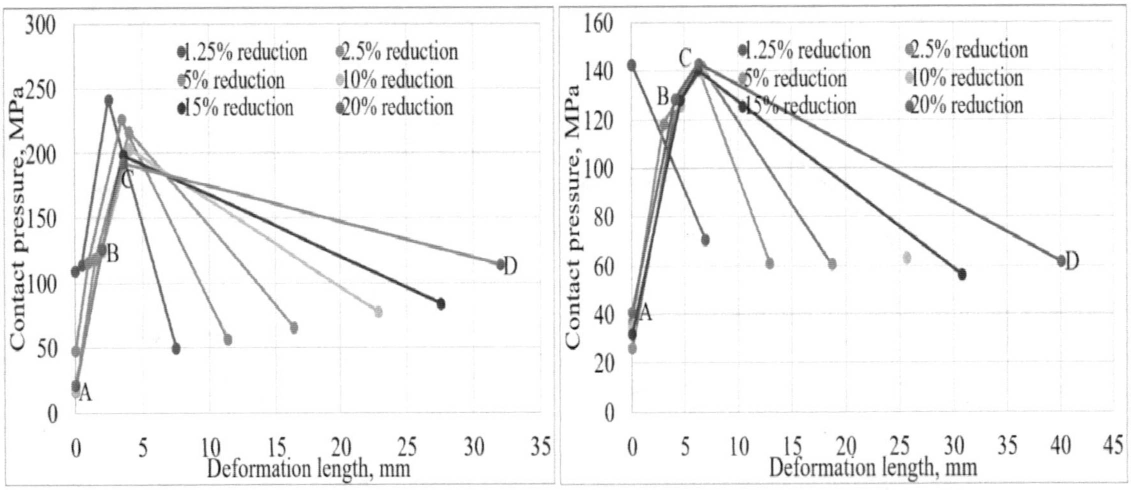

(a) (b)

Figure 8. Critical normal contact pressure in the deformation zone: (a) at 2.5 mm liner thickness, (b) at 7.5 mm liner thickness.

The effect of liner thickness on the strength of the bond which was assumed at maximum contact pressure required between sheets is presented in figure 9. As the figure indicated, while the liner thickness increases the strength of the bond becomes weak and decreases while increasing the reduction ratio because of the increasing threshold area in the contact interface.

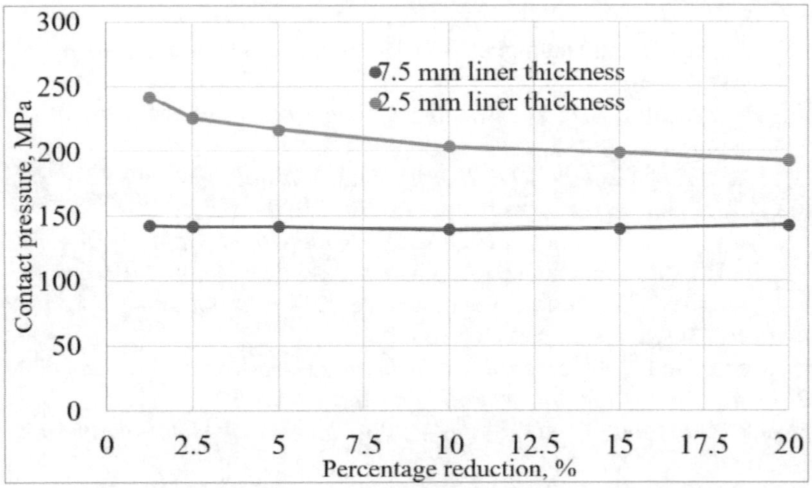

Figure 9. The effect of liner thickness on the bond efficiency.

The ratio of bonded length from the total deformation length of different liner thicknesses is shown in figure 10. The figure indicated a higher ratio of bonded length from the total deformation length can be found from the thinner liner and at a small reduction ratio of the thicker liner the unbonded sheet was obtained.

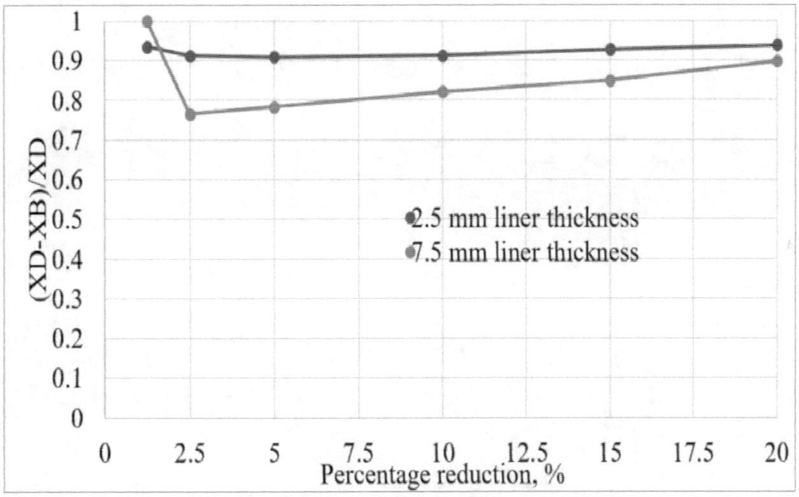

Figure 10. The ratio of bonded length from the total deformation length.

5. Conclusion

Numerical modeling and simulation are acceptable for understanding and predicting the deformation zone in hot roll bonding of multilayer sheets. Normal contact pressure required between roller and liner as well as between each sheet interface needs precise computation to analyze the deformation zones. A high percentage of contact pressure could be transferred from the roller-liner interface to each sheet interface while increasing the reduction ratio. The amount of liner thickness has a higher impact on the influence of the amount of contact pressure transfer. The ratio of the bonded length with the total deformation length increases at the high amount of reduction ratio while decreasing liner thickness. Finally, this investigation finds out the liner thickness and reduction ratio are important parameters to optimize a quantitative mathematical model related to the normal contact pressures required to start bonding.

References

[1] Mollapour Y, Afshari D and Haghighat H 2018 *Iranian Journal of Materials Forming* **5(2)** 36-53

[2] Haghighat H and Saadati P 2015 *Transactions of Nonferrous Metals Society of China* **25(5)** 1605-13

[3] Hwang YM, Hsu HH and Lee HJ 1996 *International Journal of Machine Tools and Manufacture* **36** 47-62

[4] Al-Maqdi S, Qudeiri JA and Ziout A 2020 *IEOM Society International (Dubai, UAE)* 2031-44

[5] Khan HA, Asim K, Akram F, Hameed A, Khan A and 2021 *Metals* **11(9)** 1344

[6] Da Silva L, El-Sharif M, Chisholm C and Laidlaw S 2014 *International Conference on Metallurgy and Materials (Metal)* (Brno, Czech Republic)

[7] Li L, Nagai K and Yin F 2008 *Science and technology of advanced materials* 2008 **9(2)** 023001

[8] Szabó G 2017 *Archives of Metallurgy and Materials* **62** 1205-8

[9] Pan SC, Huang MN, Tzou GY and Syu SW 2006 *Journal of Materials Processing Technology* **177** 114-20

[10] Lee CH, Park JP, Tyne CV and Moon YH 2015 *Journal of Mechanical Engineering Science* **229(17)** 3153-61

[11] Han J, Niu H, Li S, Ren Z, Jia Y, Wang T, Plokhikh AI and Huang Q 2020 *Chinese Journal of Mechanical Engineering* **33** 1-3

[12] Wright PK, Snow DA and Tay CK 1978 *Metals Technology* **5** 24-31

[13] Bay N 1983 *Welding j* **62(5)** 137

[14] Madaah-Hosseini HR and Kokabi AH 2002 *Materials Science and Engineering* **335** 186-90

[15] MatWeb Material Property Data, http://www.matweb.com, 14.10.2021

IOP Conf. Series: Materials Science and Engineering 1246 (2022) 012013 doi:10.1088/1757-899X/1246/1/012013

Impact of plasma treatment on solderability of printed circuit boards

E. Kocsis*, A. Lukács and I. Szalai

Institute of Mechatronics Engineering and Research, University of Pannonia, H-8900, Zalaegerszeg, M. Gasparich St. 18/A, Hungary

Email: kocsis.eszter@phd.uni-pannon.hu

Abstract. Flux is essential in the process of soldering. White flux residues are not only an aesthetic issue, but the inactivated components can cause severe problems. With miniaturization of products in the electronics industry, the short circuits caused by ion migration are more likely to occur during the lifetime of the product. Plasma treatment has been proved to be effective in improving the wettability in case of metal and polymer surfaces. The aim of this study is to investigate the possibility of promoting wettability of pads on printed circuit boards by plasma treatment. PCBs with three different types of surface finishes were investigated before and after plasma treatment. Wettability and solderability were examined by contact angle measurement and wetting balance tester. This method has the potential to reduce the usage of flux during soldering as an alternative surface treatment to create oxide-free metal surfaces.

1. Introduction

In lead-free soldering the usage of different fluxes is essential in order to create oxide-free, clean metal surfaces as well as to promote wetting of solder and the joining surfaces [1]. Although necessary, the application of flux has numerous and significant disadvantages, causing high risk for the electronic products. Flux residues contain ionic contaminants capable to ionic migration in certain conditions, which can cause short circuit. Flux residues are able to absorb water from the environment that is enough to initiate ionic migration in presence of voltage. In this process dendrites are growing which can cause short when connecting two conductive pads [2]– [4]. In consequence of the development and miniaturization of the electronic products in the automotive industry, these malfunctions could be fatal.

Atmospheric pressure plasma devices are in interest since more than 50 years for several industrial applications [5]. Plasma treatment is able to modify only the surface and not the bulk material, it is more gentle than chemical surface treatments. Various plasma jets and different applied gases make the method suitable for numerous purposes. Important benefits - regarding the manufacturing industry - are the possibilities of automatization and in-line installation [5]. Forming gas that contains 5 vol% hydrogen and 95 vol% nitrogen was used for plasma generation, utilizing the reducing properties of hydrogen to remove the oxide layer from the metal surfaces [6].

PCBs with three different types of surface finishes were investigated. These surface finishes are frequently used in the electronics industry because of their good solderability and reliability. However, all three types are susceptible to oxidation over time and in certain environmental circumstances [7], [8].

The aim of this study is the reduction of flux amount in the soldering process as well as introduction and evaluation of a new, alternative surface treatment method.

IOP Conf. Series: Materials Science and Engineering 1246 (2022) 012013 doi:10.1088/1757-899X/1246/1/012013

2. Materials and Methods

For the plasma surface treatment, a Plasmatreat FG5001 atmospheric plasma generator with RD1004 rotating plasma nozzle was used. Forming gas - containing 5 vol% hydrogen in nitrogen - was applied. Distance between the rotating plasma nozzle and PCB surface was set to 10 mm.

In this study three different surface finishes were investigated: Electroless Nickel/Immersion Gold (ENIG), Immersion Silver (ImAg) and Immersion Tin (ImSn).

Contact angle measurements were conducted with a Krüss DSA30 Drop Shape Analyzer, „Advance" software and Sessile Drop Method. The machine is equipped with an automated sampler and a high-speed camera. Standard solutions of water (polar component of SFE) and diiodo-methane (dispersive component of SFE) were used to determine the contact angles of drops on the PCB pads. 1 μl of solution was placed on the surface with an automated sampler and a manually controlled stage. No flux was used during this investigation.

An important factor influencing the efficiency of plasma treatment for industrial applicability is the storage time in air after treatment. To investigate the durability of the treatment effect, contact angles were measured before, immediately after the plasma treatment and 4, 12, 24 and 48 hours later respectively. From the measured contact angles the surface free energy was calculated using the Owens/Wendt Theory. Mathematically, the theory is based on two fundamental equations (Young and Good equations) which describe interactions between solid surfaces and liquids [9].

$$\frac{\sigma_L(\cos\theta+1)}{2(\sigma_{L,D})^{1/2}} = (\sigma_{S,P})^{1/2}\frac{(\sigma_{L,P})^{1/2}}{(\sigma_{L,D})^{1/2}} + (\sigma_{S,D})^{1/2} \quad (1)$$

Where σ_L is overall surface tension of the wetting liquid, $\sigma_{L,D}$ is dispersive component of the surface tension of the wetting liquid, $\sigma_{L,P}$ is polar component of the surface tension of the wetting liquid, σ_S is overall surface energy of the solid, $\sigma_{S,D}$ is dispersive component of the surface energy of the solid, $\sigma_{S,P}$ is polar component of the surface energy of the solid, and θ is the contact angle between the liquid and the solid surfaces. The unit of surface tension and surface energy is [N/m] and the unit of the contact angle is [°] degree in equation (1).

Wetting Balance Must 3 machine was used for the wetting tests. The wetting balance test is a quantitative method for solderability determination through measurement of the wetting speed and wetting force. The specimen is fixed to a sensitive balance and immersed into molten solder. The resultant vertical forces of buoyancy and surface tension on the immersed specimen are detected by a transducer and converted into an electrical signal. The machine used in this study is also equipped with a camera to registrate the testing process.

SAC305 lead-free solder was used in this analysis. The solder was heated to 250°C. Immersion speed was 6 mm/s and immersion depth was 3 mm, the test lasted 30 seconds. Interflux 2005c type flux was dosed to the samples. In order to quantify the effectiveness of plasma treatment a series of diluted flux solutions were prepared. The original flux was diluted to 1/2, 1/4 and 1/8 concentration with isopropyl alcohol. One drop of flux solution was dosed to the samples immediately after plasma treatment. Original samples, without surface modification were tested for reference.

In order to investigate the qualitative and/or topographic changes of the samples, scanning electron microscopy examination was carried out using Tescan Mira 3 machine.

3. Results and Discussion

3.1. Contact angle measurement

All three types of PCB pad samples were investigated by contact angle measurement before and after plasma treatment (PT) immediately and 4, 12, 24, 48 hours later. Results of the tests are shown in tables 1-3 and figure 1.

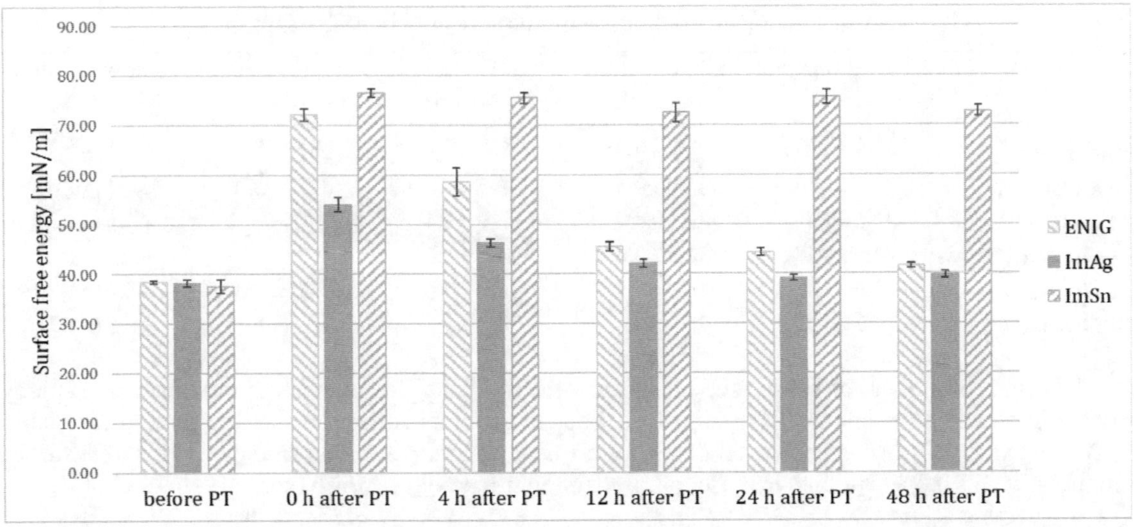

Figure 1. Surface free energy in correlation with time elapsed after plasma treatment (PT).

Table 1. Contact Angle measurement results in case of ENIG

	SFE total [mN/m]	SFE disperse [mN/m]	SFE polar [mN/m]	water - Mean CA [°]	diiodo-methane - Mean CA [°]
before PT	38.4 ±0.4	36.0 ±0.3	2.4 ±0.1	85.7 (±0.3)	46.9 (±0.5)
0 h after PT	72.1 ±1.2	41.8 ±0.6	30.3 ±0.7	24.6 (±1.5)	35.5 (±1.2)
4 h after PT	58.6 ±2.8	39.5 ±1.1	19.1 ±1.7	48.4 (±2.7)	40.2 (±2.2)
12 h after PT	45.5 ±1.0	38.1 ±0.4	7.4 ±0.5	71.0 (±1.1)	42.9 (±0.8)
24 h after PT	44.3 ±0.7	36.9 ±0.3	7.5 ±0.4	71.7 (±0.9)	45.3 (±0.6)
48 h after PT	41.6 ±0.5	37.2 ±0.3	4.3 ±0.2	79.0 (±0.5)	44.6 (±0.5)

The results show that the contact angles dropped significantly due to PT mostly in case of water, increasing the polar component of the SFE and at the same time the total SFE value jumped up. 4 hours after surface modification SFE was still higher than the original however started to decrease. The measurement of the 12-, 24- and 48-hour samples indicates that the effect of plasma treatment starts to fade, approaching the initial state after 48 hours.

Table 2. Contact Angle measurement results in case of ImAg

	SFE total [mN/m]	SFE disperse [mN/m]	SFE polar [mN/m]	water - Mean CA [°]	diiodo-methane - Mean CA [°]
before PT	38.2 ±0.7	35.6 ±0.4	2.6 ±0.3	85.4 (±0.9)	47.6 (±0.8)
0 h after PT	54.0 ±1.4	44.5 ±0.4	9.5 ±1.0	62.6 (±1.9)	29.3 (±1.1)
4 h after PT	46.23 ±0.9	42.6 ±0.5	3.7 ±0.4	77.7 (±1.2)	33.8 (±1.1)
12 h after PT	42.1 ±0.9	40.1 ±0.6	2.0 ±0.2	84.9 (±0.7)	39.0 (±1.3)
24 h after PT	39.1 ±0.6	37.4 ±0.5	1.7 ±0.1	87.5 (±0.5)	44.3 (±0.9)
48 h after PT	39.8 ±0.6	37.9 ±0.4	1.8 ±0.2	86.8 (±0.8)	43.3 (±0.8)

The results of immersion silver demonstrate similarities to results of ENIG surface finish. After plasma treatment the contact angles decreased, and surface free energy increased. The effect of surface modification starts to decline, approaching the initial state after 48 hours of storage in air.

Table 3. Contact Angle measurement results in case of ImSn

	SFE total [mN/m]	SFE disperse [mN/m]	SFE polar [mN/m]	water - Mean CA [°]	diiodo-methane - Mean CA [°]
before PT	37.6 ±1.3	28.9 ±0.7	8.7 ±0.6	74.5 (±1.0)	59.5 (±1.3)
0 h after PT	76.5 ±0.8	41.0 ±0.5	35.5 ±0.4	8.2 (±1.3)	37.2 (±1.0)
4 h after PT	75.5 ±1.1	39.6 ±0.7	35.9 ±0.4	10.9 (±0.7)	40.1 (±1.3)
12 h after PT	72.5 ±1.9	37.8 ±0.6	34.6 ±1.3	19.3 (±3.7)	43.4 (±1.2)
24 h after PT	75.6 ±1.4	38.7 ±0.9	36.9 ±0.6	8.1 (±0.3)	41.7 (±1.7)
48 h after PT	72.7 ±1.1	35.7 ±0.6	37.0 ±0.5	16.3 (±1.2)	47.5 (±1.0)

The results of immersion tin surface finish show the most significant increase in surface free energy. Contact angles of water on the PCB pad surface dropped extremely after plasma treatment. In correlation with storage time no decreasing tendency of SFE was observed. This indicates that the surface modification is more durable in case of immersion tin surface finish contrary to the other two investigated metal surfaces.

Figure 2. illustrates contact angle measurement: 1 μl water droplets on ENIG surface. The most significant difference is demonstrated on the images below (before and immediately after plasma treatment). Contact angles dropped immensely from 85° to 24°. Continuous contact angle increase can be observed on the images presenting the samples 4-, 12-, 24- and 48-hours storage after plasma treatment.

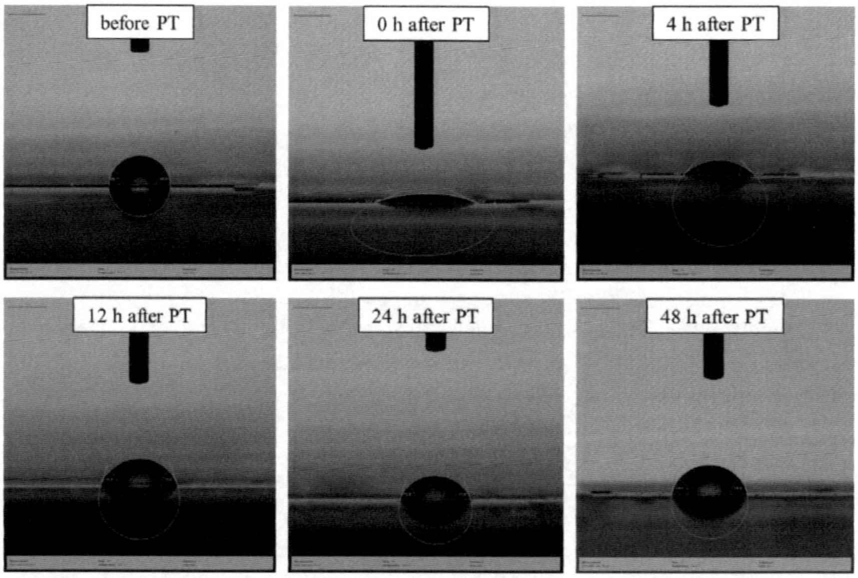

Figure 2. Contact angle measurement of water droplets on ENIG surface.

3.2. Wetting Balance test

Wetting balance test was conducted with different diluted solutions of Interflux 2005C flux after plasma treatment. Untreated samples were used for reference. Measured values were corrected with the buoyancy, so the effect of geometric differences is eliminated in the comparison. The test results are shown in figures 3-5.

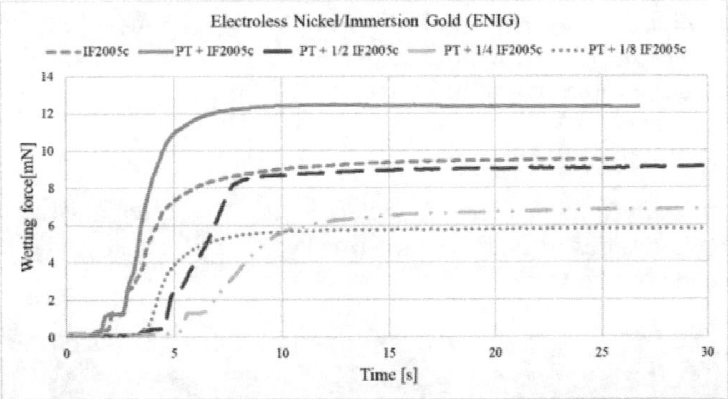

Figure 3. Wetting Balance test results for ENIG, where PT means Plasma Treatment and IF2005c is the name of the flux used in different ratio solution (eg. 1/2 IF2005c=half IF2005c and half IPA).

Results demonstrate the effectiveness of plasma treatment via the difference produced between the original sample and the surface treated sample. Wetting force is higher in case of the sample after plasma treatment than without treatment. Similarity can be observed between the untreated sample with concentrated flux and after plasma treatment with 1/2 diluted flux. This correlation indicates that the amount of flux can be reduced to half if plasma treatment is applied before soldering.

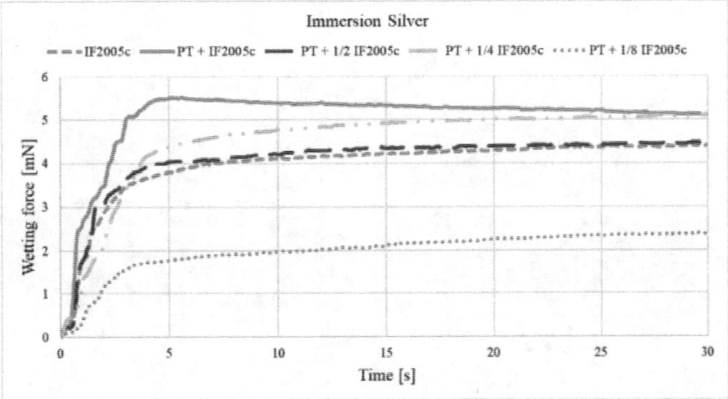

Figure 4. Wetting Balance test results for ImAg, where PT means Plasma Treatment and IF2005c is the name of the flux used in different ratio solution (eg. 1/2 IF2005c=half IF2005c and half IPA).

The ImAg results show similar efficiency of plasma treatment like in case of ENIG. Wetting force increased after treatment of surface with forming gas plasma.

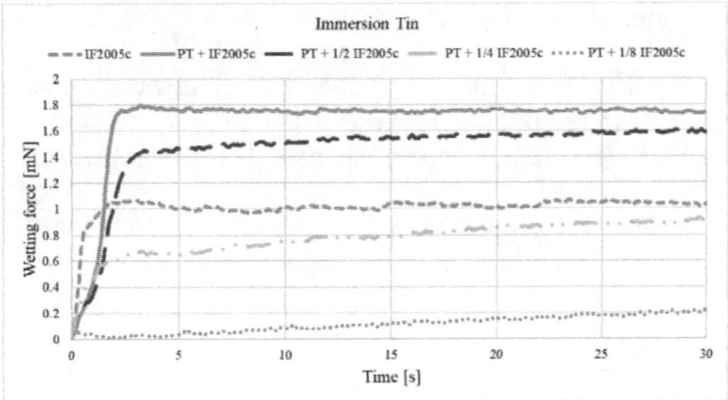

Figure 5. Wetting Balance test results for ImSn, where PT means Plasma Treatment and IF2005c is the name of the flux used in different ratio solution (eg. 1/2 IF2005c=half IF2005c and half IPA).

Finally, the wetting balance test results of ImSn present similar behaviour to the other investigated surface finishes. After plasma treatment both concentrated and 1/2 diluted flux solutions produced higher wetting forces than the reference samples (without PT).

3.3. Scanning electron microscopy investigation

Scanning electron microscopy images were collected before and after plasma treatment on the same samples of each investigated types of PCB surface finishes. No flux was used during this examination. Images of the samples can be seen in figure 6.

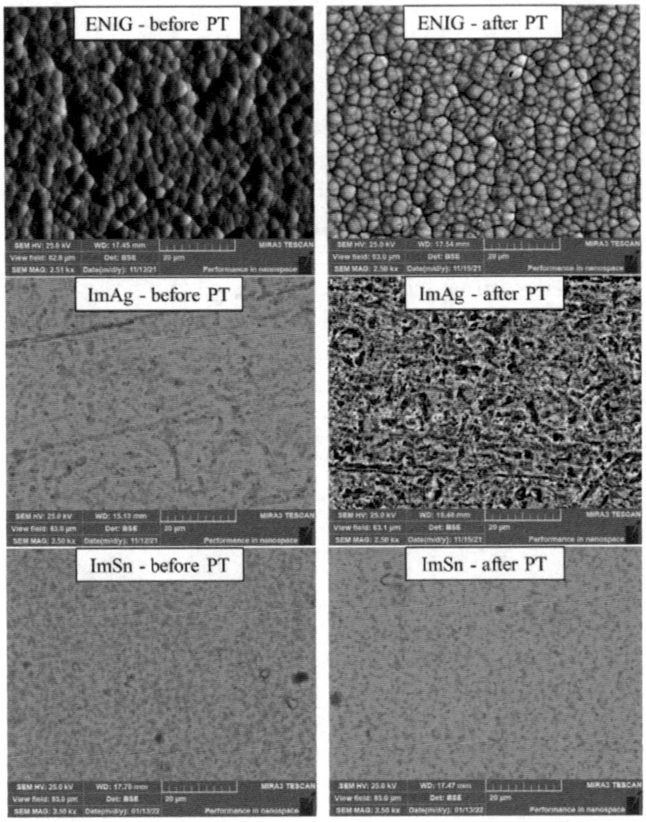

Figure 6. SEM images of the investigated surfaces before and after plasma treatment.

The collected BSE SEM images show the surface of the samples. Based on the images after plasma treatment the surfaces are presumably rougher in nano scale which promote better wetting between the metal surface and molten solder.

The composition of the surfaces was investigated by energy-dispersive X-ray spectroscopy (EDS). The amount of oxygen has decreased (table 4.) in case of all three investigated surface finish after plasma treatment. The same samples were studied before and after surface modification.

Table 4. EDS results in Weight% of Oxygen before and after plasma treatment (PT)

Surface finish	Before PT	After PT
ImSn	2.6	2.2
ImAg	2.7	0.7
ENIG	0.9	0.8

Figure 7. demonstrates the spectras of ImSn before and after plasma treatment, the change in the intensity of the oxygen peak can be observed.

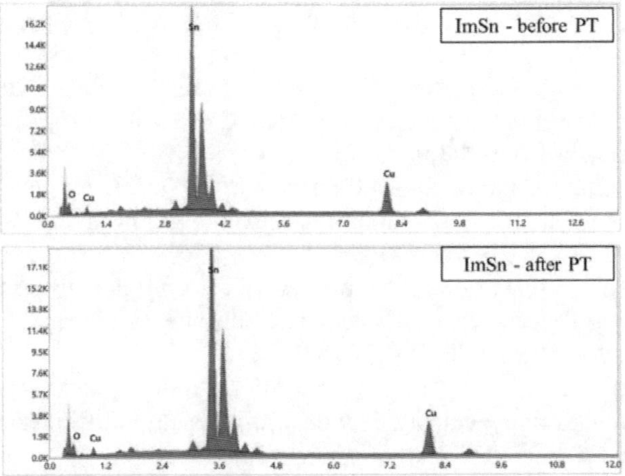

Figure 7. EDS spectras of ImSn surface before and after plasma treatment (PT).

4. Summary

In this study the effect of plasma treatment on three different PCB surface finishes (Electroless Nickel/Immersion Gold, Immersion Silver, Immersion Tin) were investigated. The aim of PCB pad surface treatment was to remove oxides and promote wettability allowing minimization of the necessary flux amount. The samples were investigated and evaluated by contact angle measurement, wetting balance test and scanning electron microscopy.

- The measured contact angles decreased in case of all three surface finishes, the calculated surface free energy values increased as result of plasma treatment. Significant change can be observed, especially in case of immersion tin surface finish. As result of atmospheric storage for 48 hours (after PT), the samples were approaching the original state of the surfaces in the case of immersion silver and ENIG. On the other hand, measurements on immersion tin surfaces suggest more durable effect.
- Wetting balance test results confirm the efficiency of plasma treatment. Wetting forces increased in case of all three types of samples, the most significant change can be observed in case of immersion silver surface finish. As revealed by the series of experiments with diluted flux solutions, the amount of flux can be reduced by plasma treatment. For instance, the results of ENIG surface finish properly demonstrate that half of flux amount can be replaced with plasma treatment.
- Scanning Electron Microscopy images appear to be presumably rougher and energy-dispersive X-ray spectras indicate that the amount of oxygen on the surfaces is decreased as a result of plasma treatment.

Based on these experiments the beneficial effects of plasma treatment are evident. Hypothetically the hydrogen radicals from the forming gas plasma contribute removing metal oxides from the surfaces, and mechanically modifying it on a sub-microscopic level at the same time. This modification might move or remove the particles of the top of the surface while also creating a more even protective layer.

5. References

[1] J. F. Shipley, Influence of Flux, Substrate and Solder Composition on Solder Wetting, Weld. J. (Miami, Fla), vol. 54, no. 10, 1975.

[2] H. Huang, Z. Pan, Y. Qiu, and X. Guo, Electrochemical corrosion behaviour of copper under periodic wet-dry cycle condition, Microelectron. Reliab., vol. 53, no. 8, pp. 1149–1158, 2013, doi: 10.1016/j.microrel.2013.05.002.

[3] K. K. Ding, X. G. Li, K. Xiao, C. F. Dong, K. Zhang, and R. T. Zhao, Electrochemical migration behavior and mechanism of PCB-ImAg and PCB-HASL under adsorbed thin liquid films, Trans. Nonferrous Met. Soc. China (English Ed., vol. 25, no. 7, pp. 2446–2457, 2015, doi: 10.1016/S1003-6326(15)63861-4.

[4] P. Yi, K. Xiao, K. Ding, C. Dong, and X. Li, Surface failure mechanism of PCB-ENIG in typical outdoor atmospheric environments, Mater. Res. Bull., vol. 91, pp. 179–188, 2017, doi: 10.1016/j.materresbull.2017.03.017.

[5] J. Winter, R. Brandenburg, and K. D. Weltmann, Atmospheric pressure plasma jets: An overview of devices and new directions, Plasma Sources Sci. Technol., vol. 24, no. 6, p. 64001, 2015, doi: 10.1088/0963-0252/24/6/064001.

[6] M. Godard, D. Drouin, M. Darnon, S. Martel, and C. Fortin, Plasma Treatment for Fluxless Flip-Chip Chip-Joining Process, Proc. - Electron. Components Technol. Conf., vol. 2018-May, pp. 419–424, 2018, doi: 10.1109/ECTC.2018.00069.

[7] H. Huanga, X. Guoa, F. Bua, G. Huang, Corrosion behavior of immersion silver printed circuit board copper under a thin electrolyte layer, Engineering Failure Analysis, Volume 117, November 2020, 104807

[8] P. Yia, K. Xiaoa, K. Dingb, C. Donga, X. Li, Surface failure mechanism of PCB-ENIG in typical outdoor atmospheric environments, Materials Research Bulletin, Vol. 91, 2017, pp. 179-188

[9] D. K. Owens and R. C. Wendt, Estimation of the surface free energy of polymers, J. Appl. Polym. Sci., vol. 13, no. 8, pp. 1741–1747, 1969, doi: 10.1002/app.1969.070130815.

Acknowledgements

The authors wish to thank Dr. László Tóth, who supported and motivated the research, for his help and expertise on the present paper.

Project no. 2020-1.1.2-PIACI-KFI-2020-00141 has been implemented with the support provided from the National Research, Development and Innovation Fund of Hungary, financed under the 2020-1.1.2-PIACI KFI funding scheme.

Prepared with the professional support of the Doctoral Student Scholarship Program of the Co-operative Doctoral Program of the Ministry of Innovation and Technology financed from the National Research, Development and Innovation Fund."

Production of syntactic metal foams reinforced in the matrix material

D Károly[1,2], J E Maróti[2], J Lolbert-Szabó[2], A Kemény[1,2] and I N Orbulov[1,2]

[1] MTA-BME Lendület Composite Metal Foams Research Group, Bertalan L. u. 7., Budapest H-1111, Hungary

[2] Budapest University of Technology and Economics, Faculty of Mechanical Engineering, Department of Materials Science and Engineering, Műegyetem rkp. 3., H-1111, Hungary

E-mail: karoly.dora@gpk.bme.hu

Abstract. Metal foams are cellular materials where the matrix material is metallic, and gaseous cells are present in this enclosing material. Due to these cells, metal foams are low-density materials and have good energy-absorbing properties to their structure. Reinforcing materials can highly increase the load capacity of the structure. Our goal is to produce a low-cost, reinforced syntactic metal foam with an easy production process. The mechanical properties of A356 (AlSi7Mg) aluminium matrix syntactic foams are investigated, produced with lightweight expanded clay aggregate particles (LECAPs) as filler material and reinforced with aluminum-oxide (Al_2O_3) with a diameter of 1 mm and 0.5 mm, and silicon carbide (SiC) with a diameter of 0.4 mm in the matrix. 20 wt.% of the reinforcement was used in each case. The foams were produced by low-pressure infiltration, and the specimens, which were cut by conventional machining methods, were subjected to heat treatment. The grain structures of the materials were studied via optical microscopy, their mechanical properties were tested with compression tests. Results show that the production is successful, and reinforcement can be achieved with these materials.

Keywords: syntactic metal foam, reinforcement, LECAP, mechanical property

1. Introduction

There are many materials in our environment whose structure contains hollow cells. Such materials are called cellular materials. Cellular materials found in nature include, but are not limited to, trees, sea corals, sponges, and, for example, the skull and femur in the human body. These naturally occurring materials inspired the development of the first artificial cellular materials. By the term „metal foam" a cellular material is meant, in which the matrix material is made of metal and inside, there are porous, gaseous cells. There are open- and closed cell foams, as well as metal matrix syntactic foams. Between the open- and closed cell metal foams, the difference is that while in the open cell case, the particular material deficient areas (in terms of the matrix material) are connected, in the case of closed cell material, these cavities are completely separated [1]. Metal matrix syntactic foams belong to the latter one, the group of closed cell materials, where the porous structure is created with the addition of filler material [2–4].

Compared to solid materials, the metal foams' density is lower, and their specific energy absorption capacity is higher. Due to these properties, they are more and more widespread in research and industrial fields, including automotive components and insulation materials in construction. They also have applications in medical technology; furthermore, due to their high energy absorption capacity, warfare applications are more and more common [2,3].

Another major area of effort is to maximize the load capacity and minimize the weight of composite materials. This is accomplished by embedding a relatively small amount of heavy-duty reinforcing material in a lower density matrix material. This creates a compromise, as the weight is reduced with unchanged enclosure dimensions, but the structure can only withstand loads in specific directions. Reinforcement is grouped according to material and direction. The latter category is formed by the one-, two- and three-dimensional division based on the direction(s) in which the given form of reinforcement plays a reinforcing role.

In metal-based composites, there are three types of reinforcement based on the shape of the reinforcing material, long fiber, short fiber, and particle reinforcement. In the case of metal matrix composites, the matrix material is mostly aluminum, but titanium, magnesium, nickel, copper and cobalt matrix composites have also been used [5–8].

Particle reinforcement is meant the amplification of the properties by adding particles with a diameter of 0.1–10 μm. The most commonly used are Al_2O_3 and SiC, but several other substances have experimented with SiB_2, B_4C, TiC, WC, BN, ZrO_2, etc. In general, reinforcement can be used to achieve better strength, lower density and higher hardness. During the production of a particle-reinforced composite, unfavorable chemical processes can occur in the case of some matrix-reinforcement pairings. An example is the reaction of SiC particles with Al matrix is: $4\,Al + 3\,SiC \rightarrow Al_4C_3 + 3\,Si$. The resulting Al_4C_3 precipitates degrade the mechanical properties of the composite. Aluminum oxide is generally less reactive than SiC, so it is also more advantageous to use an aluminum matrix [5,9–12]

Sajjadi et al. produced an A356 (AlSi7Mg) aluminum matrix Al_2O_3 micro- and nano-sized particle reinforced composite by a mixed casting process. The research examined the effect of the different manufacturing parameters such as mixing speed, casting process or heat treatment of the reinforcing material, and particle size and ratio to the matrix material regarding the properties of the composite. The wetting properties of the ceramic have been improved by pre-heat treating. The low tendency to wet means that the melt will poorly wet the surface of the ceramic particles, so it will float to the top of the melt where it can form an oxide film. The contact angle for wetting is above 90° in such cases. Heat treatment can remove absorbed gases from the surface of the reinforcement particles and improve wetting by casting under an inert gas. Inhomogeneity of the composite can also be a problem when using ceramic particle reinforcement. Due to the density difference between the Al_2O_3 particles and the aluminum matrix, the reinforcing materials were grouped and not evenly distributed in the melt. Because of that, different mixing processes were experimented with during the manufacturing process. The best results were obtained at 300 rad/sec. It has also been observed that as the amount of reinforcing particles in the melt increases and the size decreases, the porosity and wetting properties deteriorate, however, in these cases, the hardness of the composite increases [13].

Tao et al. produced an aluminum matrix syntactic metal foam reinforced with aluminum particles by pressure infiltration. The authors use Al 6082 (AlSi1MgMn) aluminum alloy for matrix material with 0.5–1 mm aluminum powder Al 6082 reinforcement and 75–125 μm ceramic microspheres. In the study, five metal foams with different compositions were produced. In the case where only Al matrix and ceramic hollow spheres were used, the volume ratio of aluminum was 37%. In the other case where Al particles were added, the proportion of aluminum was 43%, 50%, 60% and 70%. The foam which did not contain added Al particles, the ceramic hollow spheres show a homogeneous distribution. In the other specimens that already contained added aluminum particles, these aluminum particles did not melt during infiltration [14].

In the course of the research, we sought to answer two questions as to whether it is possible to produce a syntactic metal foam reinforced in a matrix material by low-pressure infiltration in one step, and what effect the larger reinforcing particles have on the mechanical properties of the produced metal foam.

IOP Conf. Series: Materials Science and Engineering 1246 (2022) 012014 doi:10.1088/1757-899X/1246/1/012014

2. Materials and methods

2.1. Materials

In this study, A356 (AlSi7Mg) was used as matrix material. Aluminum is an excellent matrix material because it has low density, good casting properties, and low energy demand for melting. In addition, it also has good resistance to corrosion and fire, which is also the advantage of the subsequent performance of the metal foams.

Lightweight expanded clay aggregate particles (LECAPs) were used as filler material, obtained from Liapor GmbH & Co. KG. The diameter of the particles was Ø3–3.5 mm, with stereomicroscopic measurements the average diameter was found to be 3.24±0.65 mm. From this, the calculated density of the filler was 0.76±0.2 g/cm^3. The chemical composition of the used LECAPs (according to the manufacturer) is 60±5 wt.% SiO_2, 17±3 wt.% Al_2O_3, 14±2 wt.% Fe_2O_3 and ~9 wt.% other oxides (CaO, MgO, Na_2O, K_2O).

As reinforcing material, three different particles were used: two different aluminum oxide, Al_2O_3 (1 and 0.5 mm) and silicon carbide, SiC (0.4 mm), shown in Figure 1. The nominal and measured sizes can be seen in Table 1. For each specimen 20% of reinforcing particles were used, which is the amount of reinforcing material relative to the matrix material's volume.

Table 1. Marking of test specimens and sizing of particles.

	Filler material	Diameter of filler (mm)	Reinforcing material	Nominal size of particles (mm)	Measured size of particles (mm)
0	LECAP	3–3.5	-	-	-
AL1			Al_2O_3	1	1.20±0.22
AL0.5			Al_2O_3	0.5	0.63±0.10
SI0.4			SiC	0.4	0.44±0.06

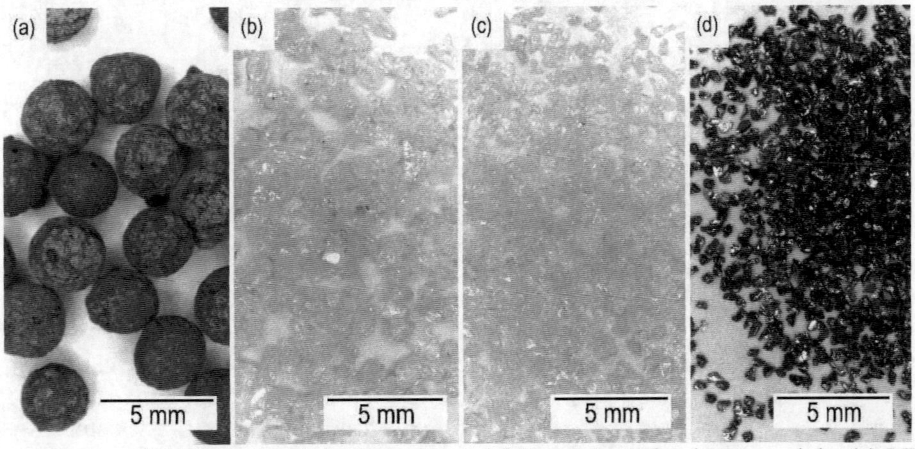

Figure 1. Geometrical features of the investigated filler and reinforcing materials, (a) LECAPs Ø3–3.5mm, (b) Al_2O_3 1 mm, (c) Al_2O_3 0.5 mm, (d) SiC 0.4 mm.

2.2. Production

The specimens were produced by low pressure infiltration. The LECAPs and the Al_2O_3 particles were mixed until the mixture was homogenous in the production process. A stainless-steel mesh was inserted atop the mixture, which was wedged against the walls of the mold to prevent the movement of the filler material. After preheating the prepared molds for 30 min at 600 °C, the aluminum melt at minimum 840 °C was poured onto the top of the mixture and then pressed between the particles using argon gas at a pressure of 500 kPa. The infiltration time was 5 sec. The air remaining under the melt exited the mold through a condensed tube at the bottom.

Heat treatment was applied to the specimens to reach the T6 treatment state. The first step is heating at 300°C/h rate to 535°C, holding for 4 hours then quenching in water. The second step was heating at 200°C/h rate to 150°C, holding for 15 hours, then again quenching in water.

2.3. Methods

Microstructural examination was done on the manufactured metal foams by Olympus PMG 3 optical microscope and an Olympus SZX16 stereomicroscope. Material composition was measured by a Zeiss MA10 scanning electron microscope (SEM) with EDAX energy-dispersive X-ray spectroscopy.

Compression tests were carried out on a total of 12 specimens with an MTS810 universal hydraulic testing machine between two flat steel platens. The device was equipped with a 250 kN load cell. 0.3 mm thick Kolofol Teflon foil was used for lubrication, placed on the contact surfaces between the specimens and the platens. Each specimen was compressed with a 4 mm/min cross-head speed to at least 50% engineering strain value for comparability, based on ISO13314:2011 [15].

The following characteristic property values were evaluated from the results: compressive strength (σ_c - compressive stress at the first peak after the elastic deformation) and energy absorption ($W_{50\%}$ - area under the engineering stress-engineering strain curve up to 50% strain).

3. Results and discussion

The images of the microstructure (Figure 2.) shows that the aluminum matrix completely filled the space between the LECAPs and the reinforcing materials. Based on these, the production method was suitable, and the mixture was homogenous.

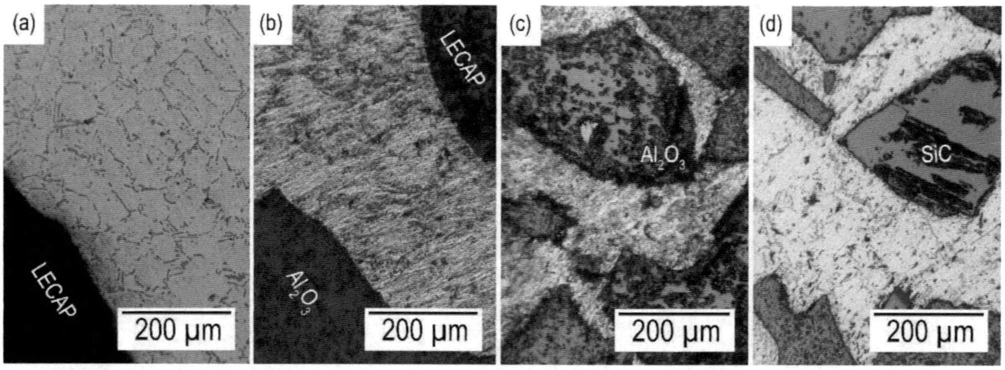

Figure 2. Microstructure of the produced metal foams, (a) specimen 0 with no reinforcement, (b) specimen AL1 with 1 mm Al_2O_3, (c) specimen AL1 with 0.5 mm Al_2O_3, (d) specimen SI0.4 with 0.4 mm SiC.

The results of the quasi-static compression tests are shown in Figure 3 showing the engineering stress and energy absorption curves in the function of the engineering strain. Each curve is the average of three compression tests. It can be seen that reinforcing results in a significant compressive strength peak at ~3.5% strain but adding ceramic particles to the aluminum matrix make the foams more brittle compared to the only LECAP filled foam as the stress decrease occurs earlier. The combination of these two effects results in similar energy absorption values between the reinforced and not reinforced foams.

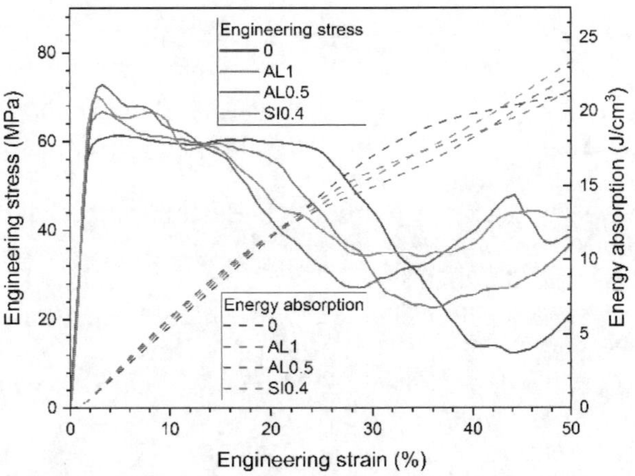

Figure 3. Engineering stress-strain diagram and energy absorption of the specimens.

Table 2 shows the numerical results of the compression tests. Reinforcing the LECAP filled syntactic metal foams with Al_2O_3 and SiC particles increase their density with 7–8% in average. However, it also increases the average compressive strength with 12–20%. The energy absorption values show a slight increase with the decrease of the size of the reinforcement particles.

Table 2. Compressive characteristic properties.

	Density (g/cm³)	σ_c (MPa)	$W_{50\%}$ (J/cm³)
0	1.58±0.01	59.3±0.16	20.9±0.30
AL1	1.71±0.05	67.1±2.58	21.4±1.25
AL0.5	1.69±0.11	73.4±8.79	22.7±1.45
SI0.4	1.71±0.06	70.5±6.19	23.8±2.43

4. Conclusion

The manufacturing of syntactic metal foams reinforced with Al_2O_3 (1 mm and 0.5 mm) and SiC (0.4 mm) particles in the matrix material A356 (AlSi7Mg) was successful with low pressure infiltration.

- It can be seen from the microstructural images that the material of the matrix completely surrounds the reinforcing particles, no porosity defects was observed in the arrays.
- From the mechanical tests, it can be stated that the compressive strength of the metal foam could be increased by reinforcing the matrix. The most significant increase was observed for Al_2O_3 with a particle size of 0.5 mm, with an increase of ~ 20% compared to the foam without reinforcement.
- The energy absorption values were nearly equal with each specimen, despite the increased compressive strength since reinforcing promotes the brittleness of the syntactic metal foams.

5. Acknowledgements

The research reported in this paper is part of project no. BME-NVA-02, implemented with the support provided by the Ministry of Innovation and Technology of Hungary from the National Research, Development and Innovation Fund, financed under the TKP2021 funding scheme, and the NRDI Fund (TKP2020 NC, Grant No. BME-NCS) based on the charter of bolster issued by the NRDI Office under the auspices of the Ministry for Innovation and Technology. This work was supported in part by the Hungarian Scientific Research Fund under Grant agreement OTKA FK_21 138505.

6. References

[1] Naeem MA, Gábora A and Mankovits T 2020 *Period. Polytech. Mech. Eng.* **64** 172–178
[2] Ashby MF, Evans AG, Fleck NA, Gibson LJ, Hutchinson JW and Wadley HNG 2000 *Metal Foams* (Oxford:Butterworth-Heinemann) p 6–23
[3] Ashby MF, Evans AG, Fleck NA, Gibson LJ, Hutchinson JW and Wadley HNG 2000 *Metal Foams* (Oxford:Butterworth-Heinemann) p 40–54
[4] Rahmani M and Petrudi AM 2020 *Period. Polytech. Mech. Eng.* **64** 224–232
[5] Clyne TW 2018 *Comprehensive Composite Materials II Vol 4* (Amsterdam: Elsevier) p 1–21
[6] Shirvanimoghaddam K, Hamim SU, Karbalaei Akbari M, Fakhrhoseini SM, Khayyam H, Pakseresht AH, Ghasali E, Zabet M, Munir KS, Jia S, Davim JP and Naebe M. 2017 *Compos. Part A Appl. Sci. Manuf.* **92** 70–96
[7] Kubelka P, Kádár C and Jost N 2021 *Mater. Lett.* **287** 129293
[8] Gubicza J, Jenei P, Han G, Hung PT, Song Y, Park D, Szabó Á, Kádár C, Kim JH and Choe H 2021 *Materials (Basel)* **14** 2691
[9] Nan CW and Clarke DR. 1996 *Acta Mater.* **44** 3801–3811
[10] Sritharan T, Chan LS, Tan LK and Hung NP 2001 *Mater. Charact.* **47** 75–77
[11] Ye T, Xu Y and Ren J 2019 *Mater. Sci. Eng. A* **753** 146–155
[12] Zaiemyekeh Z, Liaghat GH, Ahmadi H, Khan MK and Razmkhah O.2019 *Mater. Sci. Eng. A* **753** 276–284
[13] Sajjadi SA, Ezatpour HR and Beygi H 2011 *Mater. Sci. Eng. A* **528** 8765–8771
[14] Tao XF and Zhao YY 2009 *Scr. Mater.* **61** 461–464
[15] ISO13314:2011 - Mechanical testing of metals - Ductility testing - Compression test for porous and cellular metals

IOP Conf. Series: Materials Science and Engineering 1246 (2022) 012015 doi:10.1088/1757-899X/1246/1/012015

Ionic contamination reduction with dry ice cleaning

Zs Tóth, A Lukács and I Szalai

Institute of Mechatronics Engineering and Research, University of Pannonia, Gasparich Márk utca 18/A, Zalaegerszeg, H-8900, Hungary

E-mail: toth.zsolt@phd.uni-pannon.hu

Abstract. Compared with routine methods, dry ice cleaning is an innovative, inexpensive and green technology, which efficiently cleans printed circuit boards (PCBs) and decreases levels of ionic contamination. Due to its outstanding performance, dry ice cleaning is suitable for removing flux residues. In this study, boards that have been artificially contaminated by flux were blasted by dry ice with different parameters. The influence of dry ice cleaning on different surface finishes was comprehensively investigated. It was shown that the efficiency of dry ice cleaning is comparable with other methods and is quantifiable. Furthermore, dry ice cleaning can effectively remove pollution on PCB surfaces, yielding a noticeable and ideal cleaning effect without decreasing the solderability of different surface finishes.

1. Introduction

Nowadays, lead-free soldering is the preferred interconnecting method between components and printed circuit board (PCB). This process results in satisfactory metallurgical joints and good electrical connections. The quality of solder joints is critical to ensure the reliability and lifetime of products. For the purpose of producing acceptable intermetallic compounds, solderable surfaces are often treated with flux.

This is necessary because the organic or inorganic acids in the fluxes can chemically remove the oxide layer from the metal surface, thereby rendering it more wettable for the molten solder. The solution evaporates as a result of soldering and a good solder joint develops between the etched metal surfaces. The remaining flux residues can reduce the lifetime and reliability of the product. The unused acids contain dissolved ions on the surface so are able to induce corrosion or short circuits [1].

Electrochemical migration is the movement and dissolution of metal ions in the presence of an electric voltage, which occasionally results in the growth of dendritic structures on the PCB surface. This phenomenon may reduce the resistance between the metallic areas or cause insulating issues. The occurrence of electrochemical migration can be a frequent issue in the case of high-voltage electrical assemblies [2].

The highly efficient removal of ionic contaminants is key to the long-term reliability of a product. One such exciting approach may be to minimize the amount of flux produced during soldering as obviously less residue will be present. Numerous research and industrial developments have addressed this issue. During the post-soldering process, water or a solvent is usually used to remove the residue, however, this can be costly as well as environmentally unfriendly. Many flux manufacturers have developed so-called "no-clean" mixtures that theoretically do not require post-solder cleaning. They leave less residue on the PCB surface, moreover, are less prone to corrosion and dendritic formation over the lifetime of the product due to encapsulated ionic species. On the other hand, "no-clean" flux residues can be risky especially across short insulating distances or in terms of high-voltage assemblies. In this work, preliminary studies of a new approach are presented concerning dry ice cleaning of PCBs and the reduction in flux residue.

Dry ice blasting to remove surface contaminants is a relatively new method that has only been used in industry since the mid-1980s [3].

The process whereby particles are accelerated with a compressed airstream towards the polluted surface to clean it is comparable with sandblasting. Unlike classical abrasive agents, the effect of dry ice blasting is based on three different phenomena. Due to its mechanical effect, high-speed, even supersonic, particles collide with the surface and knock the dirt off it. After these collisions, the upper layer of the surface cools down and becomes brittle due to the thermodynamic effect: The resulting sublimation results in an immediate expansion in volume by approximately 600-800 times[4].

As a result of the collisions, an extreme increase in pressure can also occur and liquid CO_2 can appear on the surface as well. Liquid carbon dioxide can be a good solvent for many organic impurities, e.g. non-polar hydrocarbons [5]. The carbon dioxide solidifies at approximately -78.5 °C under normal conditions [6].

Dry ice cleaning has been used successfully in many fields to clean contaminated surfaces [7] or prepare components before an adhesive coating is applied [8]. The method is promising in the food industry, resulting in low recontamination risks [9]. Dry ice blasting is also environmentally friendly as no harmful substances are involved.

2. Methods and experiments

The experimental platform in this study consists of a CPU-6000-2 CO_2 Cleaning Station, a compressed air system and a heated test stand. The predefined pressure was provided by the compressed air system. The dry ice mass flow rate and nozzle speed were controlled by the dry ice cleaner, while the heating system controlled the temperature of the sample. The operational parameters of the cleaning system are shown in table 1:

Table 1. Operational parameters of a dry ice blasting machine.

Parameter	Value	Unit
CO_2 pressure	55	bar
CO_2 jet diameter	5	mm
Operating pressure of the compressed air	6	bar
Nozzle-to-surface distance	47	mm
CO_2 jet/surface angle	90	°
CO_2 mass flow rate (per nozzle)	0.9	kg/h

The thermal conductivity and response of the material to be cleaned with regard to the reduction in temperature must also be carefully considered. Materials with high thermal conductivities have a tendency to quickly condense atmospheric moisture on their surfaces if the temperature drops below the dew point. Condensation can be moderate with the use of a dry cover gas [10]. The test surface was heated from the bottom to minimize the appearance of dew (see figure 1).

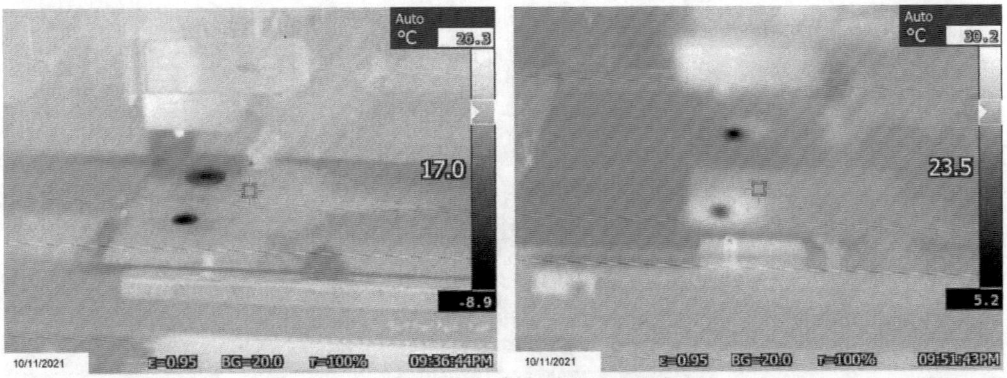

Figure 1. Appearance of dew minimized by controlled surface heating. Condensation without (left) and while (right) controlling the surface temperature.

Some typical PCBs with different surface finishes, coated in immersion tin (ImSn) and treated with electroless nickel immersion gold (ENIG) (as objects to be cleaned) were used in this experiment to study the effects of dry ice cleaning under different treatment parameters. Contaminations resulting from a no-clean flux (IF 2005C) were analyzed. This type of flux was chosen since it is widely used while soldering. Overall, the study was comprised of 30 experiments. 16 experiments were conducted to compare the CO_2 snow cleaning of immersion tin surface finishes at different scanning frequencies (1 and 2 cycles) and an untreated reference.

A further 5 experiments performed on PCBs with an ENIG surface finish focused on the influence of the CO_2 snow-cleaning efficiency. The methods used for the comparison were Fourier-transform infrared spectroscopy (FTIR) quasi-quantitative and optical analyses. The ENIG surface treatments were performed over 3 cycles (scanning frequency).

An additional 6 experiments were performed on PCBs coated in ImSn and with a ENIG surface finish. These experiments only focused on the solderability of pads after the CO_2 snow treatment. The raw ImSn and ENIG surfaces were treated over 3 cycles.

The levels of ionic contamination were measured by a C3/CI system (Foresite), the optical analysis was performed using a digital microscope (Leica DVM6-A) and the FTIR spectra were collected by an FTIR microscope in reflectance mode (Thermo Scientific Nicolet iN10). The surface-wettability tests were performed by the sessile drop method using a Drop Shape Analyzer (Krüss DSA30).

3. Cleaning efficiency analysis

3.1. Electrochemical measurements

As far as ionic migration is concerned, the presence of ions in flux residues is the most critical. To determine the CO_2 cleaning efficiency, the conductivity of an extract from the affected metal surfaces was measured. The local ionic contamination of the surface can be easily specified by the Corrosivity Index (CI) [11].

The samples were ImSn pads artificially contaminated by flux. It is evident from the measurement results that the uncleaned samples are on average more contaminated than the treated ones. The cleanliness can be further improved by increasing the scanning frequency or duration of the treatment. Although trends can be observed in the measurements, the initial amount of flux and the amount remaining after the treatment in the same area cannot be determined precisely due to the principles of the measurements. Therefore, the measurements are characterized by a large standard deviation even at the same settings. Measurements were also performed on ENIG boards with a scanning frequency of 3 cycles. Post-treatment CI values were close to the values of the flux-free pads (see figure 2).

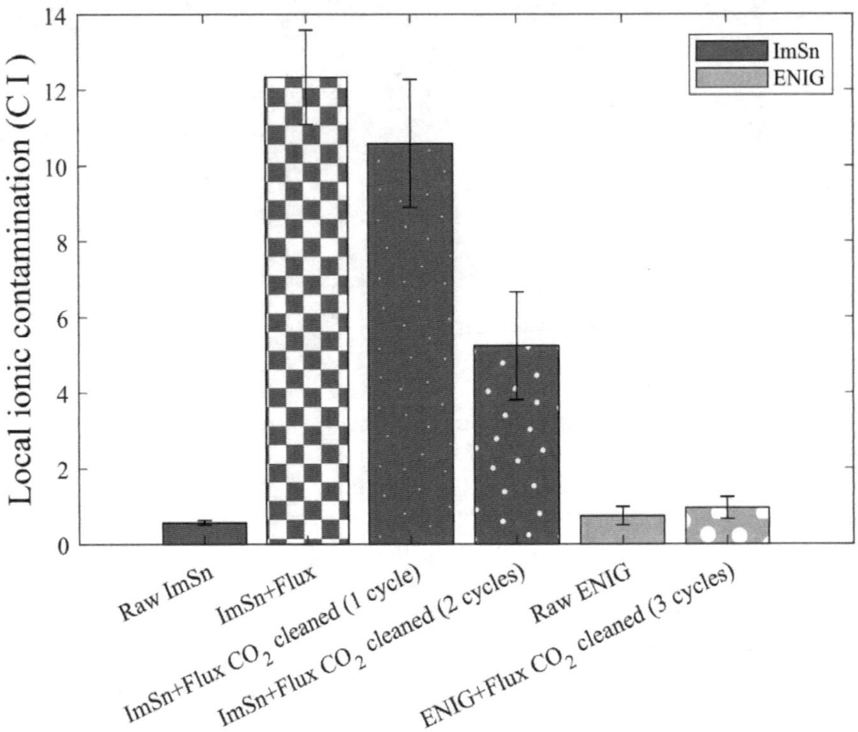

Figure 2. CO_2 cleaning reduces the level of ionic contamination.

3.2. Optical investigations

In the experiments, a crystalline residue was investigated on the surface of the ENIG finish after the flux had been dried. Contamination was easily detectable with polarized light under an optical microscope. By analysing the images of the examined areas with an image processing algorithm based on the Otsu's thresholding method [12] implemented in the MATLAB environment, the contaminated areas could be quantified.

By comparing the effect of the treatment according to the following context, a kind of residue rate was determined. The residue rate of flux residues on the specimen was employed as a smaller-the-better quality characteristic. The residue rate R was formulated as [13]:

$$R = \frac{covered\ area\ of\ contaminants\ after\ cleaning}{covered\ area\ of\ contaminants\ before\ cleaning} \times 100\% \quad (1)$$

3.3. Fourier Transform Infrared Spectroscopy (FTIR) analysis

Spectrometric studies were also performed to determine the cleaning efficiency. The absorption spectra of a dried ENIG surface contaminated by flux- and one cleaned by CO_2 were collected in the reflectance mode (see figure 3). Based on our own measurements and Huang's study [14], weak organic acids could be deduced from the peak surrounding 1690 cm^{-1}. Assuming a linear relationship, concentrations and residues could be inferred. Therefore, the determined cleaning efficiency was over 95%.

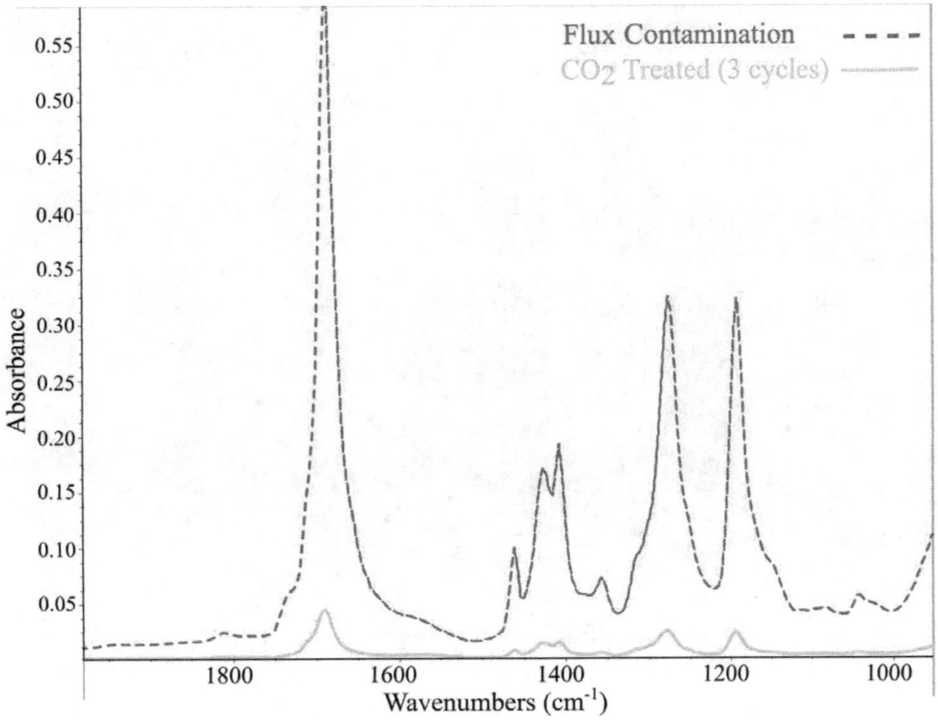

Figure 3. The characteristic peak of the C = O group (WOA) is significantly weaker after cleaning. FTIR spectra of surfaces contaminated by flux and treated with CO_2.

Based on the electrochemical, spectrometric and optical studies performed during the analysis, the results show that CO_2 snow cleaning removed more than 95% of the flux contamination from our test samples (see table 2).

Table 2. Efficiency of CO_2 snow cleaning on ENIG surfaces contaminated by flux.

Sample/Method	Electrochemical	Spectrometric	Optical
Cleaning efficiency	97%	96%	95%

4. Analysis of the wettability on CO_2 snow cleaned surface finishes

In addition to the cleaning efficiency, the effect of dry ice cleaning on the solderable surfaces of the PCB could also be an important issue. The effect of CO_2 snow cleaning on the wettability was investigated on ImSn and ENIG pads. During the experiment, the surface free energy (SFE) of raw pads treated with flux or CO_2 was determined. Untreated ImSn and ENIG pads clearly present worse levels of wettability compared to flux-treated ones. Our studies have also shown that the polar component of SFE changed significantly. After the dry ice cleaning of the original raw pads, no significant change in the SFE values was observed compared to under untreated conditions. It can be established that CO_2 snow cleaning alone does not inhibit solderability (see figure 4).

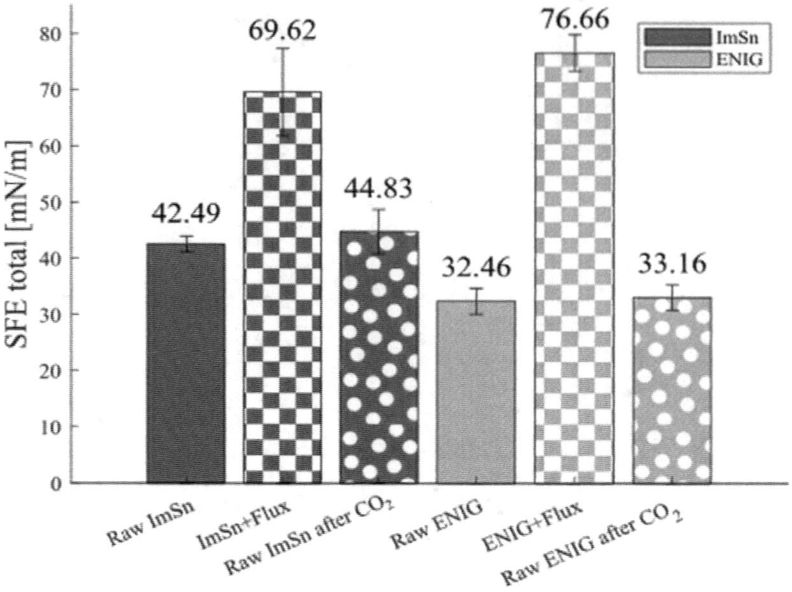

Figure 4. Measured surface free energies. The ImSn and ENIG samples exhibit similar behaviour. Dry ice cleaning has no effect on the wettability of the raw pads.

5. Conclusions

- Dry ice cleaning can be effectively used to remove flux residues, weak organic acids (WOAs) and other ionic contaminants commonly used in soldering.
- By controlling the temperature of the sample and keeping it above the dew point, condensation can be reduced so that the ionic residues do not contaminate the surface.
- The results of electrochemical, spectrometric and optical studies performed during the analysis show that CO_2 snow cleaning removed more than 95% of the flux contamination from our test samples (table 2).
- CO_2 snow cleaning alone (in the absence of flux) does not affect the wettability of the metal finishes.

Acknowledgments

The authors would like to thank László Tóth for his comments as well as the reviewers for their insightful suggestions and careful reading of the manuscript.

Project no. 2020-1.1.2-PIACI-KFI-2020-00141 has been implemented with the support provided from the National Research, Development and Innovation Fund of Hungary, financed under the 2020-1.1.2-PIACI KFI funding scheme.

PREPARED WITH THE PROFESSIONAL SUPPORT OF THE DOCTORAL STUDENT SCHOLARSHIP PROGRAM OF THE CO-OPERATIVE DOCTORAL PROGRAM OF THE MINISTRY OF INNOVATION AND TECHNOLOGY FINANCED FROM THE NATIONAL RESEARCH, DEVELOPMENT AND INNOVATION FUND.

References

[1] Klein Wassink R J 1989 Soldering in Electronics, Electrochemical Publications Limited

[2] Steppan J J, Roth J A, Hall L C, Jeannotte D A and Carbone S P 1987 A Review of Corrosion Failure Mechanisms during Accelerated Tests: Electrolytic Metal Migration *J. Electrochem. Soc.* **134** 175–90

[3] Dzido A, Krawczyk P, Badyda K and Chondrokostas P 2021 Operational parameters impact on the performance of dry-ice blasting nozzle *Energy* **214** 118847

[4] Uhlmann E and El Mernissi A 2008 Pre-treatment by dry ice blasting for adhesive bonding *Prod. Eng.* **2** 133–8

IOP Conf. Series: Materials Science and Engineering 1246 (2022) 012015 doi:10.1088/1757-899X/1246/1/012015

[5] Sherman R 2007 Carbon dioxide snow cleaning *Part. Sci. Technol.* **25** 37–57
[6] Mikołajczak A, Wołowicz M, Kurkus-Gruszecka M, Badyda K and Krawczyk P 2018 Improving the efficiency of Liquid Air Energy Storage by organic rankine cycle module application *2018 International Interdisciplinary PhD Workshop (IIPhDW)* (IEEE) pp 99–102
[7] Jantzen S, Decarreaux T, Stein M, Kniel K and Dietzel A 2018 CO2 snow cleaning of miniaturized parts *Precis. Eng.* **52** 122–9
[8] Elbing F, Anagreh N, Dorn L and Uhlmann E 2003 Dry ice blasting as pretreatment of aluminum surfaces to improve the adhesive strength of aluminum bonding joints *Int. J. Adhes. Adhes.* **23** 69–79
[9] Witte A K, Bobal M, David R, Blaettler B, Schoder D and Rossmanith P 2017 Investigation of the potential of dry ice blasting for cleaning and disinfection in the food production environment *LWT* **75** 735–41
[10] Cano F 2000 Carbon Dioxide Dry Ice Snow Cleaning *Handbook for Critical Cleaning* (CRC Press) pp 353–60
[11] Isaacs P and Munson T 2019 Cleanliness Requirements: A Moving Target *2019 Pan Pacific Microelectronics Symposium (Pan Pacific)* (IEEE) pp 1–10
[12] Otsu N 1979 A threshold selection method from gray-level histograms *IEEE Trans. Syst. Man. Cybern.* **9** 62–6
[13] Yang S-C, Huang K-S and Lin Y-C 2007 Optimization of a pulsed carbon dioxide snow jet for cleaning CMOS image sensors by using the Taguchi method *Sensors Actuators A Phys.* **139** 265–71
[14] Huang C, Li M, Ku C, Hsieh H and Li K 2009 Chemical characterization of failures and process materials for microelectronics assembly *Microelectron. Int.*

Characterisation of defects in die cast aluminium parts

R. Horváth [1*], M. Réger[1], F. Oláh[2],

[1]Obuda University, Donát Bánki Faculty of Mechanical and Safety Engineering, H1081, Népszínház u. 8., Hungary

[2]Doctoral School on Materials Sciences and Technologies, Óbuda University, Budapest

*horvath.richard@uni-obuda.hu

Abstract. In this paper, the results of computed tomography and destructive testing of a die cast, gas tight aluminium part are compared and analysed with the aim of refining the porosity map. The mapping of the continuity defects detected by the two methods makes it possible to estimate the number, shape and distribution of continuity defects on and in the immediate vicinity of the machined surface, and to compare the characteristics of continuity defects detected by non-destructive and sequential fine machining. Based on the results of the CT scan of a given casting, we also performed destructive testing to find defects in a selected part of the casting by sequential fine machining. The basic criterion for the choice of the cutting parameters was to minimize the plastic deformation (smearing) on the machined surface, i.e. to clearly identify porosity. We then investigated the position, shape and other characteristics of porosity on the surfaces. The results and differences resulting from the two different methods are also analysed.

1. Introduction

The outer, fast solidifying surface of die castings is generally free of porosity when appropriate die casting technology is used. However, for a variety of reasons and in various distributions, porous parts can form inside the casting, some of which may be exposed during later machining and finishing. The location of individual porosity defects can be determined non-destructively by computed tomography (CT).

Ensuring gas tightness is one of the most important quality requirements for aluminium-silicon die cast compressor components, mainly used in the automotive industry, but also in other areas. Achieving a high level of gas tightness in such alloys can be difficult because:

- solidification is accompanied by a change in specific volume (shrinkage) of about 7%,
- during casting, the melt surface may be partially oxidised and the resulting aluminium oxide may cause continuity defects,
- dissolved gases may be released from the melt during solidification.

Several methods are used to seal the surface porosity or porosity that extends to the surface of castings, one of the most important being vacuum impregnation. It has been shown that even after the impregnation of carefully prepared quality castings, leakage can occur for the following reasons:

- a through cavity system (natural and/or artificial) of critical size and/or critical shape for impregnation is present in the wall of the casting,
- the gas leakage is mediated by the oxide film in the casting wall, which runs out to the surface.

Even when special die casting technologies are used to ensure gas tightness (e.g. HPDC, vacuum assisted HPDC, melt protection and filtration etc.), the casting can contain complex-shaped hollow systems with small cross-sections and volumes, in about 0.5–3% [1]. During machining, the shape of the cavity system in the casting does not change, but some of its elements may be exposed during subsequent machining, i.e. the proportion of open cavities may increase. Accordingly, impregnation should be carried out after the finishing (final machining) of the casting, to close any continuity gaps that may have been exposed during machining.

Laboratory impregnation experiments on sintered metal alloys have shown that all pores below 100 microns in diameter are typically filled, while between 100 and 500 microns, about a third of the pores are not closed. The experiments showed that above a pore size of 500 microns, impregnation essentially failed to ensure the gas tightness of sintered samples [2]. This phenomenon is due to the fact that closing the pores requires a low-viscosity resin which can leak into the small pores. Larger cavities allow some or all of the resin to escape before it is cured. In die cast aluminium parts, leakage is usually not through a single pore, but through a leakage path formed by an interconnection of pores. Therefore, the key issue is not the size of the pore that can be eliminated by impregnation, but whether the leakage path can be completely sealed.

There are many references in the literature, but very few actual test results concerning the effect of the oxide film on gas tightness. An oxide film is formed on the part of the molten aluminium alloy which comes into contact with air, and its thickness depends on how long it is in contact with air. Experiments have shown that the oxide layer is formed on the surface of the melt in a few tenths of a second, with its thickness in the nanometre range. To remove the oxide film, ceramic filters are used immediately before the molten metal is shot into the mould, but the film can also re-form during mould filling. The thin oxide film may change shape during flow, it can turn on itself and form a so-called bifilm. The geometry, location and final shape of this bifilm (even several layers can stick together) depends on the flow and solidification conditions and the gas content and gas dissolving capacity of the melt. Even if the gas content of the alloy is low, the inside of the oxide film may still be a potential leakage pathway, since the oxide film has some degree of gas permeability [3, 4].

Discontinuities in the casting also impair the gas tightness and mechanical properties of the casting. Large-cycle fatigue experiments have demonstrated that pore size and the proximity of pores to the surface significantly affect crack initiation generated by repetitive stress [5]. Reducing the size of shrink defects increases the fatigue resistance of the part [6]. The prediction of the formation, exact location and shape of shrink cavity and material continuity defects is fraught with many uncertainties. Thus, the knowledge of the size and distribution of continuity defects, i.e. the porosity map, can provide important information in the process of manufacturing design. Shrink cavities can be identified by computed tomography scanning, but metallographic studies are also useful to explore the pores in detail and to determine their geometric characteristics [7].

2. Materials and methods

The tests were carried out on a mass-produced, nearly axisymmetric, eutectic Al-Si die cast part, the complex geometry of which is shown in figure 1. The non-destructive tests were performed with a ZEISS Metrotom 1500 computed tomography (CT) machine. In addition, a sample of the tested part was cut out as shown in figure 1. For sequential fine machining, the cut-out part was embedded in resin; this made the digital recording of the machined surface features possible.

IOP Conf. Series: Materials Science and Engineering 1246 (2022) 012016 doi:10.1088/1757-899X/1246/1/012016

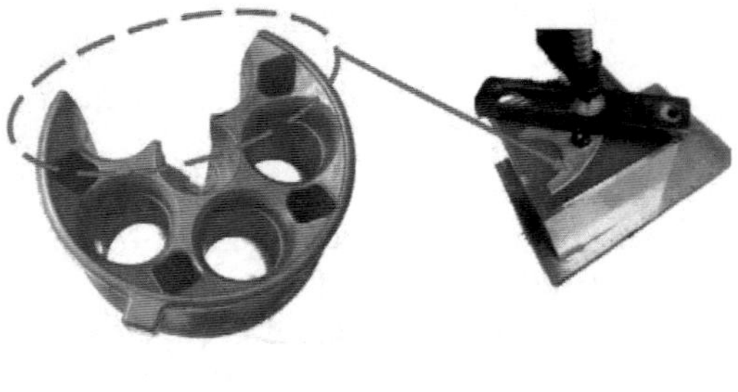

Figure 1. The tested part and the cut-out part embedded in resin for slicing by sequential cutting for the examination of shrink cavities.

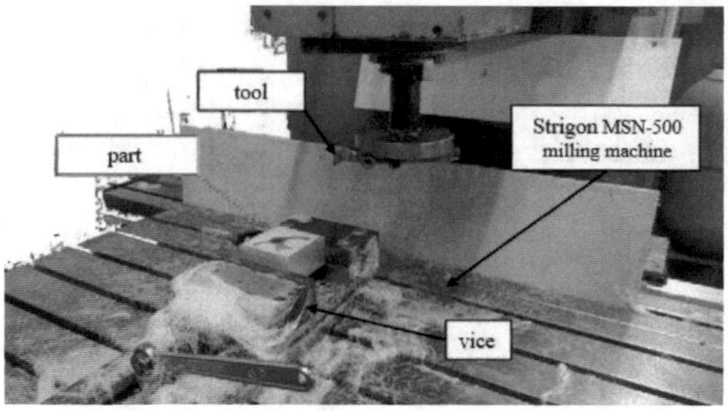

Figure 2. Sequential machining test layout.

The cutting experiments were carried out with a single-edge tool. For good surface finish, we used a D-shaped diamond-edged insert (DPGW 11T304FST). The diameter of the single-edge tool is 180 mm, due to the setting of the tool holder. The layout of the cutting experiments is described in figure 2. The machining parameters were: speed, n = 1400 rpm; feed, v_f= 30 mm/min; depth of cut, a_p= 0.05 mm (distance of slices).

At each cutting step, a high-resolution image (3600 dpi) of the explored surface (slicing plane) was taken, showing the contours of the shrink cavities in the current *x-y* plane at the appropriate magnification.

3. Results

3.1. Overall results of non-destructive testing

The CT scan identified 595 continuity defects, which are shown in fig. 3. a). Projecting these defects on the x-y plane shows the distribution of defects even better (fig. 3. b). The figure shows that most shrink cavities are in the centre of the piece, where wall thickness is greater.

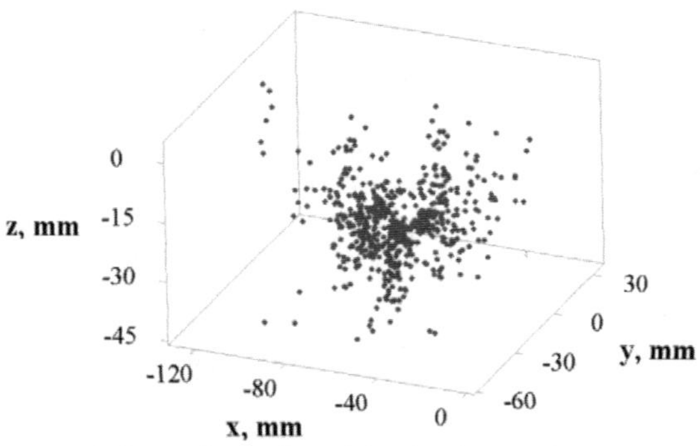

a) Location of cavities in the study area.

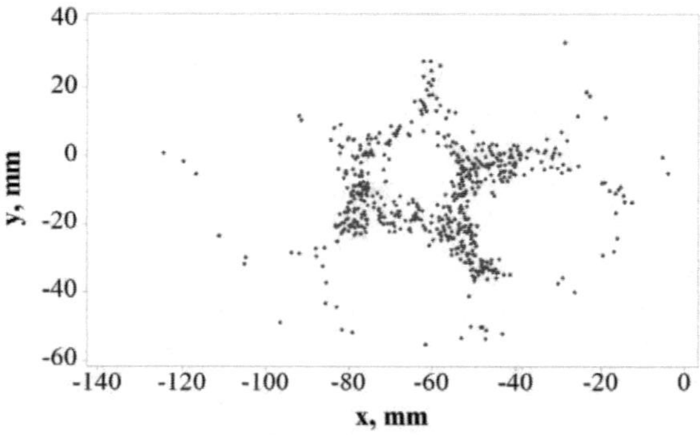

b) Distribution of cavities in the *x-y* plane.

Figure 3. Shrink cavities in the workpiece revealed by non-destructive testing.

The frequency plot in figure 4(a) shows that the volume of the shrink cavities is has a nearly exponential distribution. figure 4(b) also plots the size of the shrink cavities in the *x-y* plane. In the figure, each absorption cavity is represented by a circle with its diameter proportional to the volume of the cavity. It can be observed that the largest volume lunkers (cavities) are more frequent where wall thickness in the casting is greater.

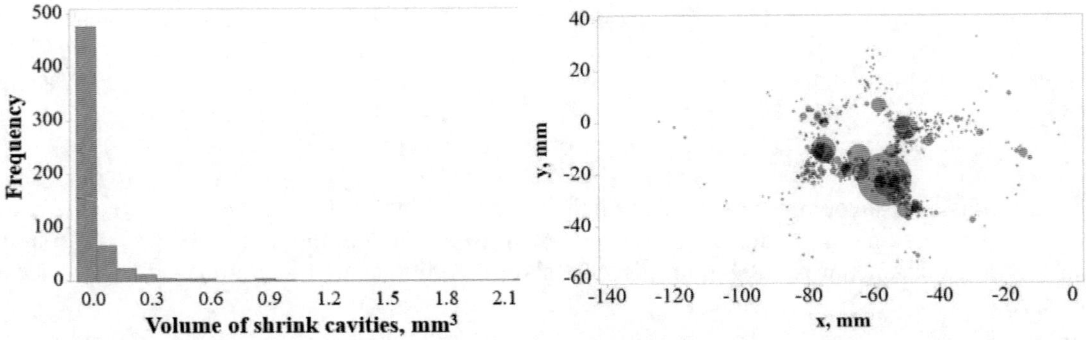

a) Volume distribution of cavities.

b) Plot of cavities versus volume in the *x*-*y* plane.

Figure 4. Volume analysis of porosity.

3.2. Frequency of continuity defects as a function of distance from the moulded surface

Solidification rates on the outer surface of castings, especially in metal moulds and in the case of die casting, are significantly higher than in the inner parts of the casting. In the crystallization front zone (mushy region), both the temperature gradient and the cooling rate decrease as we move away from the surface. As a result, the surface of castings is characterised by finer than average grain structure and a cleaner than average melt composition. This explains the fact that the near-surface region of the casting is less likely to exhibit continuity defects and porosity.

The distribution of continuity defects in the casting is mainly relevant for machining. If during the machining of the casting, internal defects get to the surface or the immediate vicinity of the surface, they can cause roughness and roughening of the machined surface, which can affect the operation of the machined surfaces, wear resistance, lubricant absorption, etc. A further problem is that the opening of a suction cavity system can even result in a "hole" through the part, which is unacceptable when gas tightness is required. The gas-tight sealing of such defects depends on the reliability of impregnation.

The mapping of the distribution of continuity defects was carried out in a section of the casting. Given that most of the defects detected are in the central region of the casting (figure 5), we performed the analysis on the central part of the material, which has a diameter of 70 mm and consists of five identical segments.

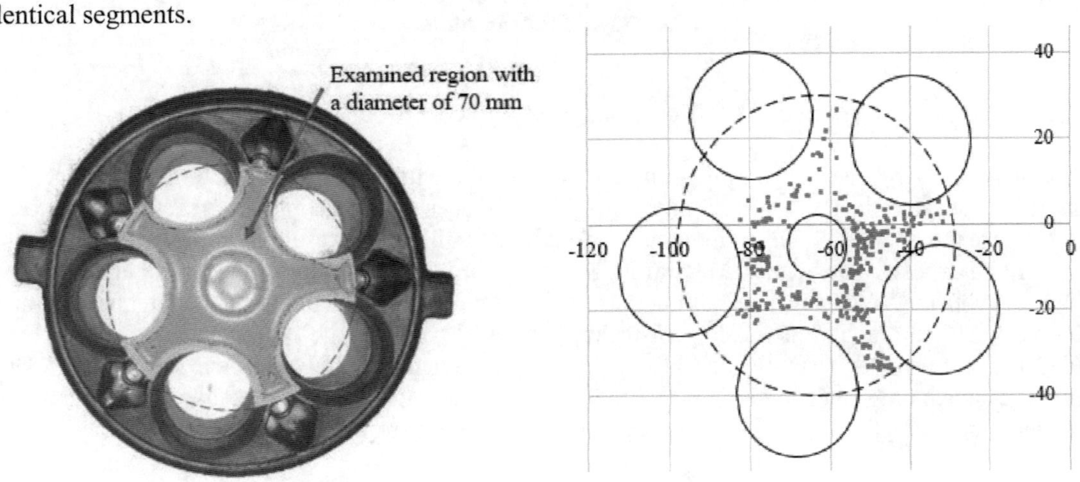

a) Cross-section of the part of the casting selected for the analysis of defect distribution in the *x*-*y* plane.

b) Continuity defects projected onto the x-y plane in the examined region.

Figure 5. Part of the casting selected for defect distribution analysis.

A total of 315 discontinuities were found in this volume during the non-destructive test. The total volume of the defects is 18.26 mm^3 and the average volume of defects is 0.058 mm^3. The largest defect volume is 2.21 mm^3 and the smallest is 0.001 mm^3. The continuity defects are characterised by a slightly elongated ellipsoidal shape with an average sphericity of 0.68. figure 5(b) shows the positions of the 315 defects in the *x-y* plane, and the contours of the examined part.

To estimate the amount of defects that might be exposed by machining, we calculated the distance between the location of each defect and the nearest surface. The minimum distance was 0.575 mm and the maximum distance was 8.258 mm. Figure 6 shows the frequency function of defect distance from the surface.

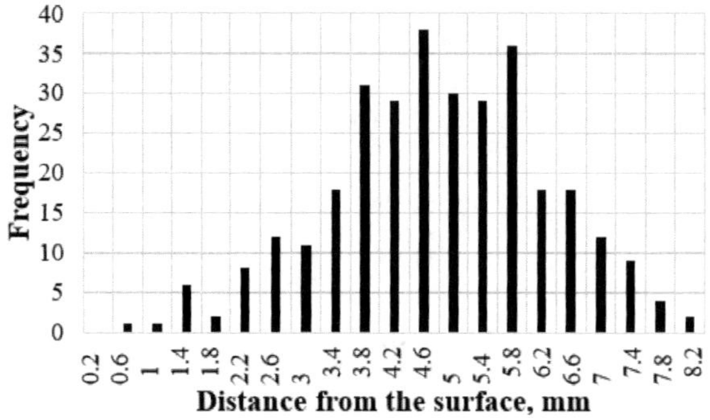

Figure 6. Frequency function of the distance of continuity defects from the surface.

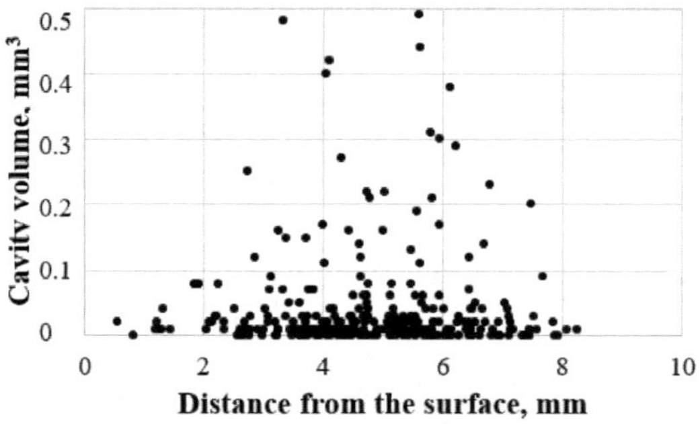

Figure 7. Volume of continuity errors as a function of distance from the surface.

The distribution results can be evaluated as a function of the thickness of the material machined from the surface. For example, if 0.4 mm is machined off the surfaces of the casting, there is a high probability that no internal defects will get to the machined surface at all. In the case of 1 mm, only one internal defect will be exposed statistically. Machining 3 mm or more off, however, will expose a larger number of defects. From the frequency distribution, the amount of defects getting to the surface can be estimated, given the machining requirements of each surface.

Figure 7 confirms that small internal continuity defects can be expected near the surface of the casting. For machining steps involving thicker parts of the casting (e.g. holes, grooves), larger amounts and sizes of continuity defects will be exposed on the machined surface.

IOP Conf. Series: Materials Science and Engineering 1246 (2022) 012016 doi:10.1088/1757-899X/1246/1/012016

3.3. Results of sequential fine machining tests and their comparison with CT analysis of shrink cavities

In order to check and refine the geometry of defects determined by the non-destructive method, we compared cavity characteristics obtained by the two methods (destructive and non-destructive) in a few cases. The destructive method is based on the visual evaluation of the surfaces revealed by cutting. As an example, figure 8(a) shows a fine-cut surface with a continuity defect in the centre of the area marked by the red circle. Figure 8(b) shows an enlargement of this area. This way, the slicing and imaging method can identify the contours of porosity cavities in the given plane.

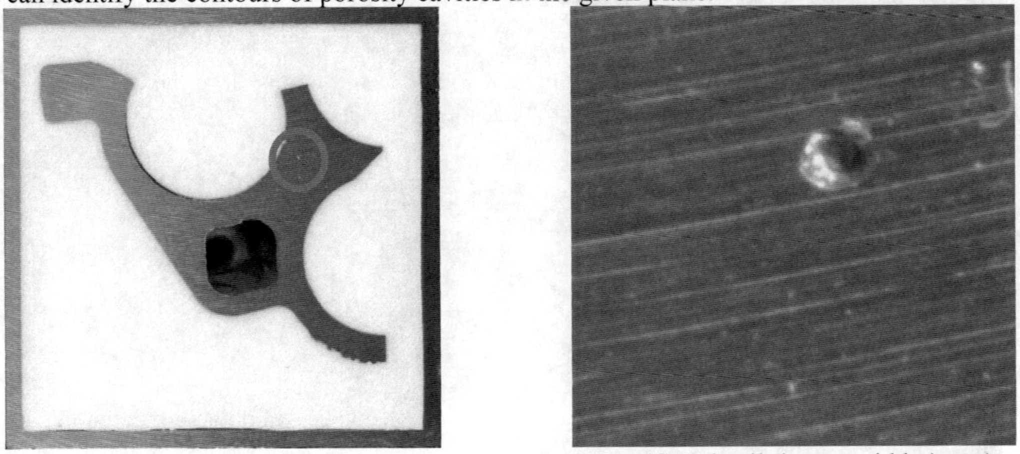

a.) Original image (image width 80 mm). **b.)** Magnified detail (image width 4 mm).

Figure 8. Photograph of a fine-cut section and an enlarged part of the area marked in the original image.

From the part examined by fine machining, we selected two cavities with relatively larger volumes. Table 1 shows their characteristics determined by CT: the diameter of the circumscribed sphere – d_{ct}, and the volume of the cavity – V_{ct}.

Table 1. CT results of the selected cavities.

Sign of the cavity	Diameter of the circumscribed sphere d_{ct}, mm	Cavity volume, V_{ct}, mm^3
a	1.69	0.96
b	1.08	0.29

The *x-y* contours of the cavities at a given *z* value were plotted on the surfaces explored by sequential chipping. From this set of contours, the shape of the cavity can be determined as a function of the steps in the z-direction. From the contour lines in the sectional planes, the 3D shape of the cavity can be formed and then, in a suitable model space, we determined the surface of the cavities and its characteristics, in our case the d_m diameter and V_m volume of the enclosing sphere. The main steps of the procedure are shown in figure 10.

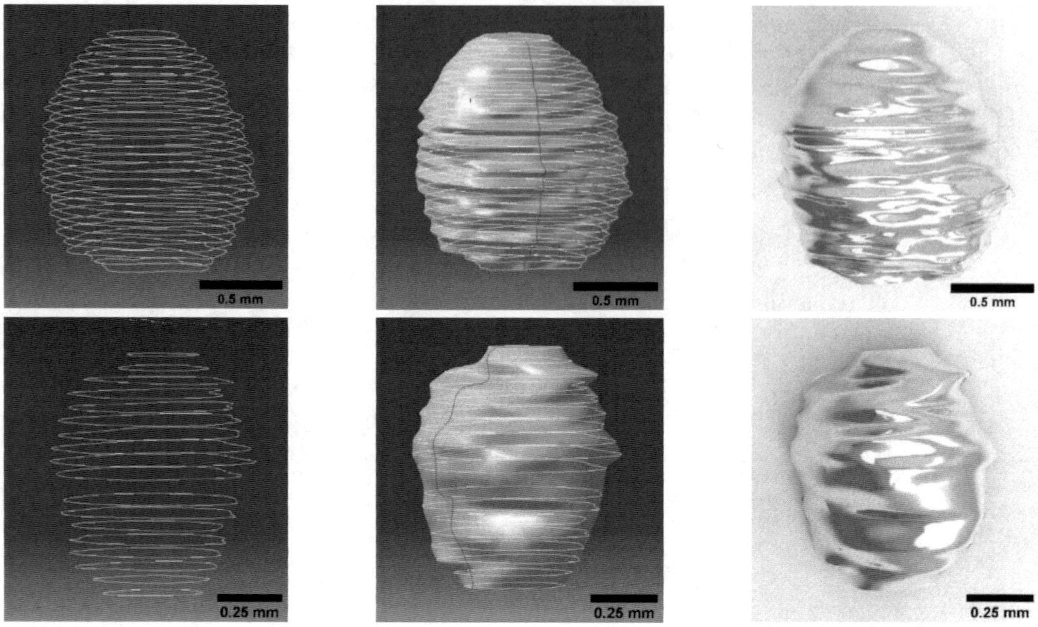

a) contouring and z-directional alignment (for cavities *a* and *b*).

b) modelling of the presumed geometry of the cavities from the recorded contours.

c) photo-rendered image of the cavities modelled back from contour layers (for cavities *a* and *b*).

Figure 9. The process of reproducing shrinkage cavities after sequential cutting and image processing.

Table 2 shows the results of the modelling of the three cavities and their comparison with the results of the non-destructive test.

Table 2. Results of the modelling of the selected cavities after sequential cutting.

Sign of cavity	Diameter of the circumscribed sphere, d_m, mm	Δd, %	Cavity volume, V_m, mm ^{3}	ΔV, %
a	1.4	−17.16	1.07	10.28
b	0.9	−16.6	0.33	12.12

The modelling results show that the shape of shrink cavities can be fairly accurately determined with an appropriate slicing technique. A comparison of the modelled cavity characteristics and the non-destructive test results shows that the volumes of both the circumscribed spheres and the modelled cavities are typically slightly larger when explored with the sequential cutting method (the diameter of the circumscribed sphere is on average 22% larger and modelled volumes are on average 14.5% larger). This discrepancy may be due to measurement inaccuracies in the two techniques, the method of image processing, the definition of contours by slicing, or the methodology of the non-destructive test. Further investigation is needed to identify the exact causes of the size differences.

4. Conclusion

In the present study, we investigated the porosity of a die cast eutectic Al-Si casting with gas tightness requirement. Two techniques were used: computed tomography and the sequential machining of a selected part of the sample. We analysed the size distribution of cavities in the whole volume of the sample, the frequency of cavities as a function of distance from the surface in a selected volume section of the casting, and the individual cavity shapes determined by non-destructive and destructive testing. Based on the tests performed, we drew the following conclusions:

- most shrink cavities were formed in the middle of the piece, in the thicker walls;
- the volume distribution of shrink cavities is near-exponential;
- larger cavities are formed in parts of the piece with greater wall thickness;
- the distribution of continuity defects in the central region of the casting is not uniform, even though the wall thicknesses in the segments are the same (this phenomenon is probably due to casting technology);
- it is likely that no internal defects will be exposed on the surfaces of the casting when 0.4 mm is machined off;
- If more than 3 mm is cut off, a larger number of internal defects can be expected to appear on the surface;
- the frequency distribution can be used to estimate the amount of defects appearing on the surface, given the machining requirements of each surface;
- with the modelling presented, the shape of shrink cavities can be determined with good approximation;
- Cavity parameters obtained by sequential fine cutting and modelling were about 10–17% greater than the results obtained by the non-destructive test.

Acknowledgements
This research was funded by the 2019-1.1.1-PIACI-KFI-2019-00462 project.
The authors would like to thank the staff of the Carl Zeiss Industrielle Messtechnik Austria GmbH Hungary Branch for the tests performed on the ZEISS Metrotom 1500.

Bibliography
[1] Weidt, M., Hardin. R. A., Garb, C., Rosc J., Brunner, R., Beckermann, C. (2018). Prediction of porosity characteristics of aluminium castings based on X-ray CT measurements *International Journal of Cast Metals Research*, *31* (5), 289-307. https://doi.org/10.1080/13640461.2018.1467105
[2] Lloyd, G. (2012). Finding a Solution to the Eternal Problem of Porosity in Casting *The Casting Engineer* 40.
[3] Campbell, J. (2003). *Castings*. Elsevier.
[4] Campbell, J. (2015). *Complete casting handbook: metal casting processes, metallurgy, techniques and design*. Butterworth-Heinemann.
[5] Gao, Y. X., Yi, J. Z., Lee, P. D., Lindley, T. C. (2004). The effect of porosity on the fatigue life of cast aluminum-silicon alloys. *Fatigue & Fracture of Engineering Materials & Structures*, *27* (7), 559-570. https://doi.org/10.1111/j.1460-2695.2004.00780.x
[6] Couper, M. J., Neeson, A. E., Griffiths, J. R. (1990). Casting defects and the fatigue behaviour of an aluminium casting alloy. Fatigue & Fracture of Engineering Materials & Structures 13 (3), 213-227. https://doi.org/10.1111/j.1460-2695.1990.tb00594.x
[7] Nicoletto, G., Anzelotti, G., Konečná, R. (2010). X-ray computed tomography vs. metallography for pore sizing and fatigue of cast Al-alloys *Procedia Engineering*, *2* (1), 547-554. https://doi.org/10.1016/j.proeng.2010.03.059

IOP Conf. Series: Materials Science and Engineering 1246 (2022) 012017 doi:10.1088/1757-899X/1246/1/012017

Finite element simulation of hardness testing with different material models

F. Oláh[1], M. Réger[2], V. Gonda[2], R. Horváth[2*]

[1]Obuda University, Doctoral School of Materials Sciences and Technologies, Hungary, Budapest
[2]Obuda University, Donát Bánki Faculty of Mechanical and Safety Engineering, H1081, Népszínház u. 8., Hungary

*horvath.richard@uni-obuda.hu

Abstract. Finite element modelling is a modern and efficient way to perform stress-strain analyses on both simple components and complex structures. If the material properties are properly set, the software can also model the deformation of parts. A very important criterion of accurate finite element calculations is an appropriate material model. The stress–strain function recorded during the tensile testing of standard specimens shows that stress is not constant during deformation and so it can be specified with a function in finite element software. Often, in finite element modelling, a bilinear model is used to simplify this function. The aim of this paper is to show the difference between the bilinear model, the multilinear model (the model split into several sections the stress-strain curve) and the real function in the analysis of deformation in hardness testing. We also analyse the minimum density of meshing which still provides acceptable accuracy in hardness testing calculations.

1. Introduction

Instrumented indentation testing has been widely used as one of the most versatile and practical methods of measuring mechanical properties of materials. It is particularly suitable for those applications where it is difficult to experimentally determine the mechanical properties by tensile testing of coupon specimens. Such applications include material processing and manufacturing of small and large engineering components and structures. Finite element simulation of hardness testing has been researched for decades. Experiments were simulated with the finite element method using conditions of frictionless and completely adhesive contact by Bhattacharya and Nix [1] as far back as 1988. According to Murakami and Matsuda [2], the accurate analysis of the Vickers hardness requires a special formulation of the three-dimensional finite element method based on the strain incremental plastic theory. They analysed the Vickers hardness of 0.46 percent carbon steel and brass with the proposed method. The calculated values showed good correlation with the experimentally measured hardness. They also found that at the maximum load, there was no tensile stress in the vicinity of the indenter. However, after they removed of the load, great residual tensile stresses were induced at the diagonal edge beneath the indentation. Beghini et al presented a numerical analysis based on a finite element model of the instrumented indentation test [3]. They concluded that the shape of the crater depends on the hardening behaviour of the material. Piling-up develops on low strain hardening materials, whereas sinking-in was observed in high strain hardening materials. Phenomena due to the rigid properties of the material (e.g. cracking) can also be modelled with the finite element method. Silicon nitride ceramics are state-of-the-art materials in the bearing industry. The Vickers indentation of this material is typically accompanied by the formation of cracks, which give much information about the material. Kadin et al developed two finite element models for crack geometry and indent size [4]. Their first model was based on fracture mechanics and the second on the cleavage stress criterion. Their calculations of crack geometry and indent size correlated well with the measurement results.

Also an important research field is measuring the hardness of multilayered or surface-treated metals. Essentially, the result of surface treatment and the in-depth hardness distribution can be determined by measuring the hardness of a section perpendicular to the treated surface, and by metallographic examination. The result of heat treatment can also be checked rapidly by surface hardness testing. However, surface hardness carries only indirect information regarding case depth and hardness distribution. Surface and cross-sectional hardness can be modelled mathematically taking into account the plastic zone developing in the indentation process. Réger et al. [5] shows an example of this; they applied a mathematical model to convert the surface hardness function into the in-depth hardness function and vice-versa. A finite element study of thin films was produced by Fabes et al [6]. They used two equations to determine the hardness of thin films from depth-sensing indentation data. Lichinchi et al. [7] used the finite element technique for the very complex stress-strain field of thin hard coatings subjected to nanoindentation. They simulated the Berkovich indentation experiments. The investigated system was titanium nitride on high-speed steel, as an example of a hard film on a softer substrate. Their numerical analysis allowed the plastic deformation history during indentation to be followed. Even with thinner layers, the nanoindentation test is the only viable approach to assess the properties of very thin coatings. Chen and Bull [8] presented a comprehensive investigation of the factors influencing the critical relative indentation depth using finite element simulation. They showed that the elastic property mismatch between coating and substrate also has an important effect on the measured hardness. Alaboodi and Hussain [9] presented the details of a nanoindentation experiment and its finite element analysis, focusing on thin films.

In this paper, the objective is to investigate the adequacy of finite element simulations for different material models for hardness measurement. Simulations were carried out for three different material models (real tensile curve, simplified multilinear and bilinear models) for S275JO steel.

2. Materials and methods

2.1. *Material*

S275JO steel was used for the hardness tests, which was chosen because it is a commonly available structural steel supporting the primary objective to perform a general modelling. Before the hardness test and finite element modelling, a tensile test was performed to characterize the hardening behaviour of the material. Subsequently, we used Brinell hardness testing, due to its geometrical simplicity, as it can be modelled axisymmetrically, unlike Vickers, where the indenter is a pyramid shape. Hardness is defined as resistance of a material towards penetration of an indenter. In this case, the penetration was applied by a spherical indenter with a diameter of 2.5 mm, and the load of 187.5 kg (ISO 6506-1:2014). The indentations were examined with an Olympus DSX 1000 digital microscope. This equipment allowed us to examine the depth and diameter of the indentations simultaneously. A top view and a 3D image of the indentation is shown in Figure 1.

IOP Conf. Series: Materials Science and Engineering 1246 (2022) 012017 doi:10.1088/1757-899X/1246/1/012017

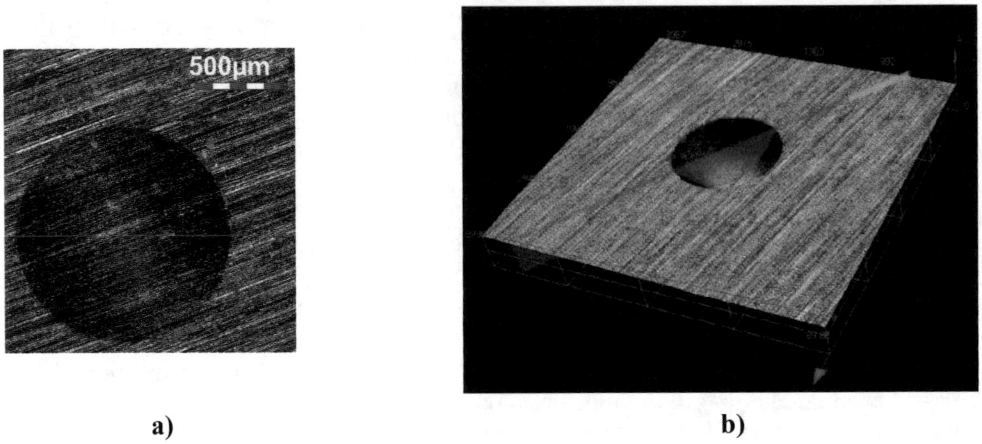

Figure 1. Optical microscopic images of an indentation a) top view, b) reconstructed 3D computer image of the intentation.

2.2. Material models

Figure 2 shows the measured tensile curves and its approximations in the analyses. The first section belongs to elastic deformation determined by Hooke's law in every case. Case a) represents a real tensile curve, where there is a transient yieldining followed by the hardening. Case b) represents a multilinear hardening approximation of the real tensile curce, where case c) represents a linear hardening model. Each of the modelled cases, true stress-true strain curves were calculated for the input of the finite element software plasticity material model data.

The section of the yield point is taken with a force increase of 100 N because the program cannot handle the oscillating section. Figure 2. (a) shows the real tensile curve up to *Fm* (maximum force of tensile test curve). Figure 2. (b) divides the section between the yield strength and F_m into three parts (multilinear model). The third, bilinear (Figure 2.c) model consists of two sections. The second line is up to the value of F_m.

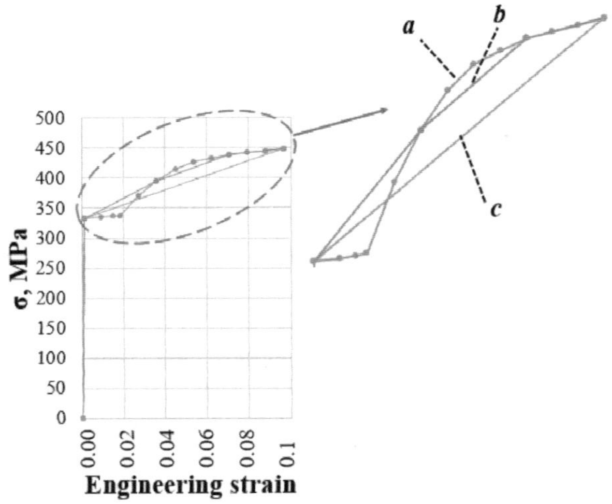

Figure 2. The models of the material.

2.3. Finite element modeling

Ansys workbench software was used for the finite element tests, as it is widely used in the industry and allows accurate calculations. In finite element modelling, the necessary boundary conditions must be defined. These boundary conditions effectively bound the degrees of freedom in the computational matrices. In practice, these determine the amount and direction of the rotation and the displacement at each node. Many conditions can be set, but in our case, we only needed a few.

First, we had to fix the bottom plane of the workpiece, which was set to fixed support, so every node in the bottom plane is locked against rotation and displacement in all directions. The penetrating body first had to have its displacement in the *x* and *y* directions fixed at zero and the appropriate constraints were set against rotation as well. The displacement in the z direction was given with a remote point, the value of which was iterated until the real measured depth was returned after unloading. The last necessary adjustment is symmetry, which is present in the *xz* and *yz* planes, which means that in practice only vertical displacement can be present along that plane.

Contact between the two bodies was set to frictional, but its value is zero, as several studies have shown that in this case, this setting has no influence on the calculation. For example, Antunes et al. [10] investigated this effect and found that it has no effective significance in the calculations. They examined the coefficient of friction at three different levels (0.08, 0.16, 0.24), but the load and unload curves cannot be distinguished, so they are independent of the effect of friction.

We used several different mesh densities for the model and the stress state in the participating bodies was continuously monitored. The appropriate mesh was selected when we did not see any distortions in the stress ramp-up. Figure 3 shows three different states of the mesh with the ramp-ups of each. In Figure 3(c) it is clearly visible that there is no need for a more sophisticated mesh.

The intender meshing setting were as follows:

- In all cases, the indenter was prepared with a tetrahedral mesh, since it is not necessary to measure the stresses generated in it accurately; in every case, mesh density was set to 0.05 mm within a 1 mm radius area.

For the workpiece, a hexahedral mesh was used under all settings, as it has the highest computational accuracy for the same number of cells. The settings of workpiece meshing were as follows:

- First, as shown in Figure 3(a), we set mesh size to 0.04 mm on the substrate within a radius of 1 mm. That gave a nonsatisfactory result, so further adjustments were necessary.
- In the second iteration, we kept the previous value of the mesh within a radius of 1 mm, but set a mesh size of 0.025 mm within a radius of 0.2 mm. This still did not give an adequate stress ramp-up value.
- In the third step, we set the mesh size to 0.025mm within the full one mm radius, which resulted in a much longer calculation time, but it is clear that there are no unnecessary stress peaks, so the setting is considered adequate.

The settings of the three types of workpiece meshing and the resulting stress ramp-up values are shown in Figure 3.

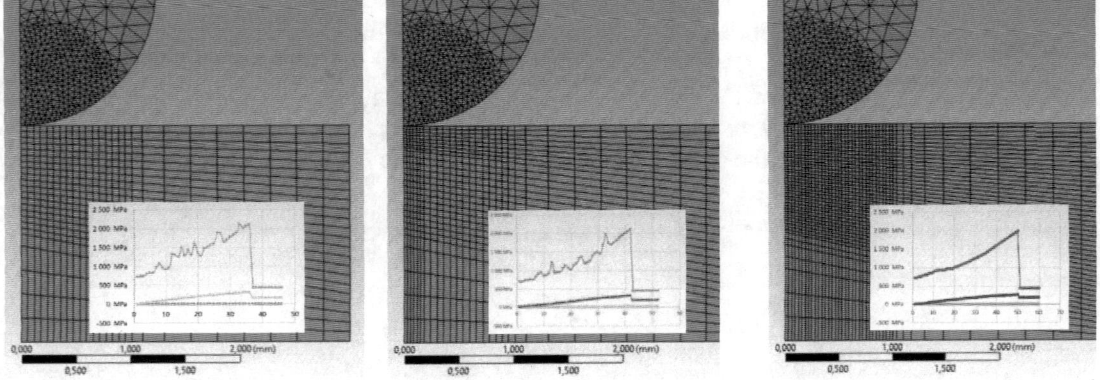

a) First mesh setting. **b)** Second mesh setting. **c)** Third mesh setting.

Figure 3. The setting of workpiece meshing.

3. Results

We performed the finite element calculations (Figure 2) with all the material models using the meshing in 2.3 (Figure 3 c). The force increases during penetration were very similar, which means that all three material models are usable. Figure 4 shows these force increases for all the material models used.

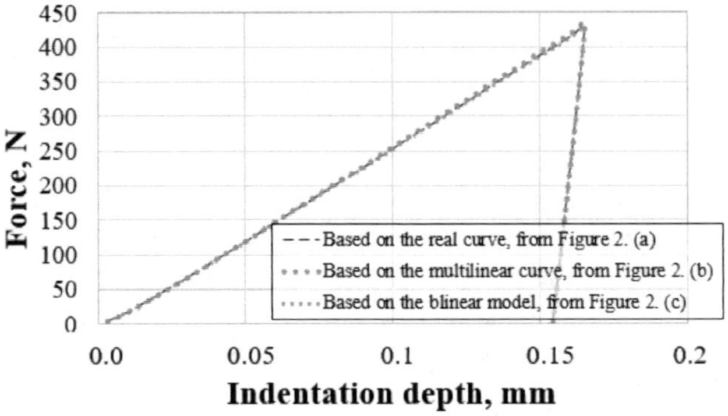

Figure 4. Force path diagrams of the 3 different material models.

Since the depth was iterated during the setup, the diameter of indentation is the same for all tests, but there were some differences in the force values.

Fig. 5 shows the differences in maximum forces between the material models used in the finite element model. Each run produced a lower maximum force than the theoretical force (the load of 187.5 kg). The deviation from the theoretical loading force is between –5.8% and –7.17%.

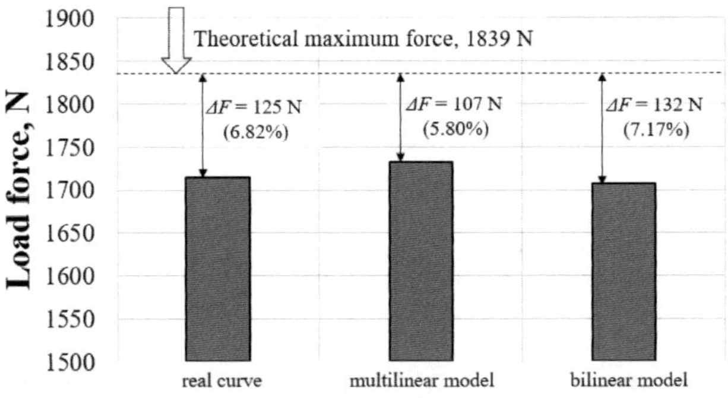

Figure 5. Force path diagrams of the 3 different material models.

The following images show the maximal stresses generated in the workpiece. The colours of stress values were set the same in all the images. This shows that stress distribution is almost identical in each case. It is also possible to see the geometrical extent to which the maximum stress values resulting from the hardness test load the test environment in the workpiece.

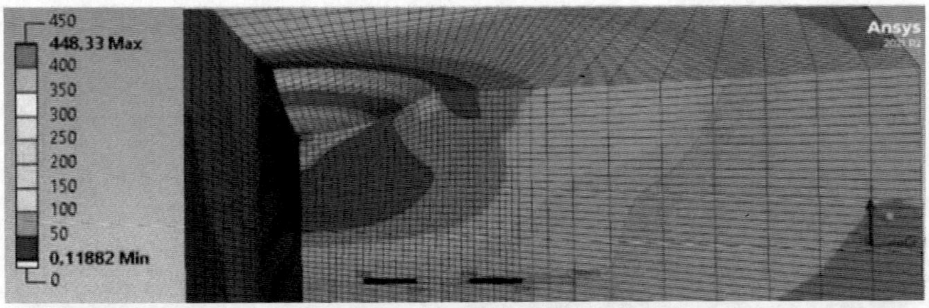

a) Real curve model.

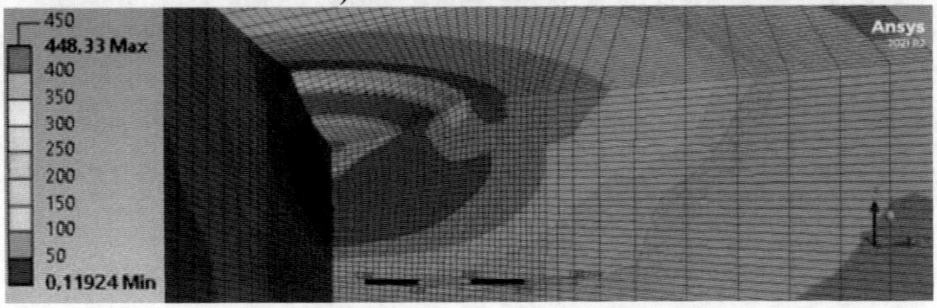

b) Multilinear material model.

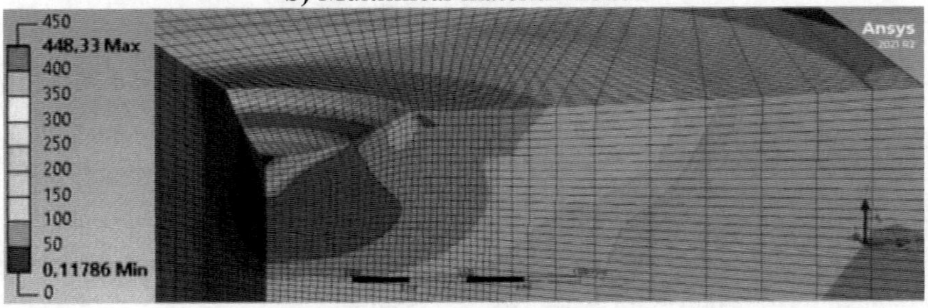

c) Biliniear material model.

Figure 6. Maximum values of the equivalent (von-Mises) stress, MPa.

As can be seen from the resulting forces and the images of stress distribution, the largest difference is between the multilinear and bilinear model, but it is still negligibly small. Fig. 7 shows such a difference between the finite element results of the two material models. There is a small difference in the vicinity of the Brinell indentation. The absolute values of the differences between the two material models in the same notes are 8.49, 23.68 and 17.29 MPa.

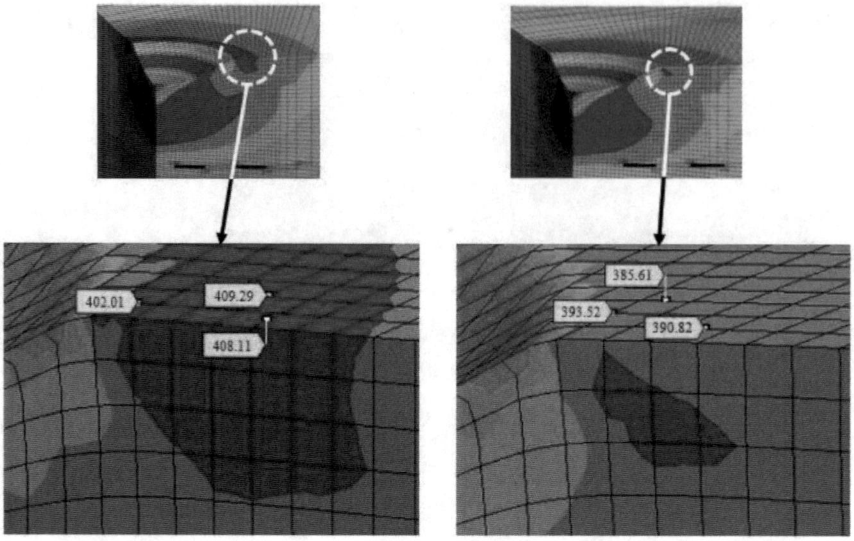

Figure 7. The difference between the multilinear and the bilinear material model.

4. Conclusion

In this paper, we performed finite element simulations of the Brinell hardness tests of S275JO steel used in mass production. The simulations were performed with three different material models: based on a real tensile curve, a multilinear and a bilinear model. The following conclusions can be drawn from the results:

- The distribution of the maximum stress values is similar for the models used.
- When different material models are used, the force-displacement diagrams for hardness tests are almost identical.
- In all cases, the maximum force values obtained from the modelled hardness measurements were smaller than the theoretical force value from loading.
- Our results indicate that the results of the runs based on the bilinear model are almost identical to those based on the real multi-point flow curve.
- For the determination of hardness and for further finite element simulations, the bilinear simplified model is suitable. This is because in this case, the tensile curve can be defined by specifying only three points in the finite element analysis and it gives adequately accurate results.

References

[1] Bhattacharya, A. K., & Nix, W. D. (1988). Finite element simulation of indentation experiments. *International Journal of Solids and Structures*, *24*(9), 881-891. https://doi.org/10.1016/0020-7683(88)90039-X

[2] Murakami, Y., & Matsuda, K. (1994). Analysis of Vickers hardness by the finite element method. https://doi.org/10.1115/1.2901563

[3] Beghini, M., Bertini, L., & Fontanari, V. (2001). Evaluation of the flow curve of metallic materials by means of spherical indentation. *WIT Transactions on Engineering Sciences*, *32*. https://doi.org/10.2495/CON010221

[4] Kadin, Y., Mazaheri, M., Zolotarevskiy, V., Vieillard, C., & Hadfield, M. (2019). Finite elements based approaches for the modelling of radial crack formation upon Vickers indentation in silicon nitride ceramics. *Journal of the European Ceramic Society*, *39*(14), 4011-4022. https://doi.org/10.1016/j.jeurceramsoc.2019.05.058

[5] Réger, M., Horváth, R., Széll, A., Réti, T., Gonda, V., & Felde, I. (2021). The Relationship between Surface and In-Depth Hardness for the Nitrocarburizing Treatment

Process. *Metals*, *11*(5), 812. https://doi.org/10.3390/met11050812

[6] Fabes, B. D., Oliver, W. C., McKee, R. A., & Walker, F. J. (1992). The determination of film hardness from the composite response of film and substrate to nanometer scale indentations. *Journal of Materials Research*, *7*(11), 3056-3064. https://doi.org/10.1557/JMR.1992.3056

[7] Lichinchi, M., Lenardi, C., Haupt, J., & Vitali, R. (1998). Simulation of Berkovich nanoindentation experiments on thin films using finite element method. *Thin solid films*, *312*(1-2), 240-248. https://doi.org/10.1016/S0040-6090(97)00739-6

[8] Chen, J., & Bull, S. J. (2009). On the factors affecting the critical indenter penetration for measurement of coating hardness. *Vacuum*, *83*(6), 911-920. https://doi.org/10.1016/j.vacuum.2008.11.007

[9] Alaboodi, A. S., & Hussain, Z. (2019). Finite element modeling of nano-indentation technique to characterize thin film coatings. *Journal of King Saud University-Engineering Sciences*, *31*(1), 61-69. https://doi.org/10.1016/j.jksues.2017.02.001

[10] Antunes JM, Menezes LF, Fernandes JV (2007) Influence of Vickers tip imperfection on depth sensing indentation tests. Int J Solids Struct 44(9):2732–2747

Analysis of inhomogeneous deformation occurring during post-necking phase of tensile test

D Harangozó[1], I Kozma[1], I Czinege[1], Sz Szalai[2]

1 Department of Material Science and Technology, Széchenyi University, 1 Egyetem tér, Győr H-9026 Hungary
2 Department of Vehicle Manufacturing, Széchenyi University 1 Egyetem tér, Győr H-9026 Hungary

Abstract. The aluminium-magnesium sheets used in vehicle body building are suitable for structural elements and internal panels based on their excellent strength and corrosion resistance. Due to their importance, particular attention should be paid to their strength and formability characteristics, including their behaviour during the tensile test at the stage of diffuse and local necking. With the help of the digital image correlation and 3D imaging techniques, the local deformations can be measured and the following characteristics can be calculated from them, namely the onset of diffuse and local necking, the change in geometry before fracture and the extrapolation of the flow curve from the measured axial stresses and the true cross-sectional area. AlMg3 sheets are used to measure the listed characteristics during tensile test.

Keywords
Digital image correlation, diffuse and local necking, flow stress curve, yield criteria

1. Introduction

The analysis of the post-necking phase of tensile tests is a priority area of research for the formability of sheets. The literature of this topic is rich, including the measurement results and finite element calculations of the necking process. Recent researches focused on the Digital Image Correlation technique (DIC), which enables to derive 3D geometric model of the specimen from measured local surface strains. The development of this method is presented in publication [1] from the discovery of the technique in 1980 to the recent past. Several authors published measurement results on inhomogeneous deformations during the tensile test, for example the Portevin-Le Chatelier (PLC) bands during uniform deformation, or the phenomenon of diffuse and local necking after reaching the maximum force [2-8]. But the 3D reconstruction of the necking is covered by relatively few publications, and more detailed analyses examine only the cross-section of the specimen after break, for example Bacha [9] and Zhang [10].

Changes in the geometric dimensions of the specimen at the necking stage lead to the problem of extrapolation of the flow stress curve. It should be noted that tensile tests enable identifying the hardening behaviour only up to the point of diffuse necking, and beyond this point it is necessary to have more measuring points in the necking range that refine the extrapolation. For cylindrical specimens, the Bridgman correction offers analytical solution for the calculation of effective stress from axial tensile stress, and later these formulae have been extended to flat specimens [11]. The use of finite element methods to quantify strain and stress distributions in the cross section of specimens was a significant contribution to the characterisation of stresses and strains during the necking phase before break, both for cylindrical and for flat specimens [12, 13].

Analyses carried out during the necking phase of the tensile test can basically be divided into two phases. In the range of diffuse necking, strain localization is moderate, so it is still acceptable that plane stress condition is valid. This significantly simplifies the application of theoretical models, as it can be found in publications [4,14-16]. Approximate calculations differ in whether the authors use yield criteria for homogeneous, isotropic or anisotropic material. The published results were systematised by the authors of this paper [17], short summary of the main findings is presented in the followings, derived from the non-quadratic yield criterion of Hill'79. Advanced anisotropic yield criteria – for example Banabic-Balan-Comsa (BBC) and others are not analysed in this paper.

Be a specimen loaded with uniaxial tension whose axial x coordinate coincides with the rolling direction, the direction y perpendicular to the axis shall be parallel to the plane of the sheet and perpendicular to the rolling direction. The ratios between stresses and strains are indicated by the following coefficients:

$$\alpha = \frac{\sigma_y}{\sigma_x};\ \beta = \frac{d\varepsilon_y}{d\varepsilon_x};\ \kappa = \frac{\sigma_e}{\sigma_x};\ \eta = \frac{d\varepsilon_e}{d\varepsilon_x}; \tag{1}$$

where ε strains always mean plastic logarithmic strain, σ_e and ε_e mean effective stress and strain respectively. The Hill'79 yield function can be described for plain stress in the following form:

$$f^a(\sigma_x,\sigma_y) = F\sigma_x^a + G\sigma_y^a + H(\sigma_x-\sigma_y)^a + A(2\sigma_x-\sigma_y)^a + B(2\sigma_y-\sigma_x)^a + C(\sigma_x-\sigma_y)^a \tag{2a}$$

Assuming that the ε_x, ε_y, σ_x, σ_y and anisotropy coefficients r_0 and r_{90} are available from measurements, in the equation (2a) A,B,C parameters equal to zero, the equation takes the following form:

$$f^a(\sigma_x,\sigma_y) = \frac{1}{1+r_0}\sigma_x^a + \frac{r_0}{r_{90}}\frac{1}{1+r_0}\sigma_y^a + \frac{r_0}{1+r_0}(\sigma_x-\sigma_y)^a \tag{2b}$$

In equations (2a) and (2b), the "*a*" stress exponent is an even number of 2 or more, so the Hill'79 yield criterion for a>2 is the same as the Hosford yield criterion. According to Hosford, for BCC metals a=6, for FCC-s a=8. If a=2, equation (2b) can be traced back to the Hill'48 yield criterion. Additional simplifications can be introduced for $r_0=r_{90}=r$, or, if r=1, in which case the equation takes the form of the Mises yield criterion [16].

The relationship between the yield function and the flow stress curve is

$$f^a(\sigma_x,\sigma_y) - \sigma_e(\varepsilon) = 0 \tag{3}$$

where $\sigma_e(\varepsilon)$ is a true stress-true strain function measured in the rolling direction. The flow stress curve can be calculated using measured stress-strain points from 0,2% engineering strain to the limit of uniform elongation. The relationship between stress components and deformations can be established on the basis of the flow rules (4a) and the equality of the plastic work (4b):

$$d\varepsilon_x = d\varepsilon_e\frac{\partial f^a(\sigma_x,\sigma_y)}{\partial\sigma_x};\ d\varepsilon_y = d\varepsilon_e\frac{\partial f^a(\sigma_x,\sigma_y)}{\partial\sigma_y} \tag{4a}$$

$$\sigma_e d\varepsilon_e = \sigma_x d\varepsilon_x + \sigma_y d\varepsilon_y \tag{4b}$$

Thus, starting from equation (2b) and forming the strain increments according to equation (4a), after replacing them in equation (4b), the coefficients in equations (1) can be calculated (see table 1).

Table 1. Calculation of α, η and κ from β

	Conditions	$\alpha=\sigma_y/\sigma_x$	$\eta=d\varepsilon_e/d\varepsilon_x$	$\kappa=\sigma_e/\sigma_x$
MIS	a=2; r=1	$\frac{2\beta+1}{\beta+2}$	$\frac{2}{\sqrt{3}}\sqrt{1+\beta+\beta^2}$	$\frac{\sqrt{3(1+\beta+\beta^2}}{\beta+2}$
H48	a=2; $r_0=r_{90}=r$	$\frac{(1+r)\beta+r}{1+r+r\beta}$	(5a)	(5b)
H79a	a=2; r_0; r_{90}	$\frac{r_{90}}{r_0}\frac{\beta+r_0+r_0\beta}{1+r_{90}+r_{90}\beta}$	(6a)	(6b)
H79b	a; r_0; r_{90}	(7a)	(7b)	(7c)

Abbreviations: MIS: Mises; H48: Hill'48; H79a és H79b: Hill'79 yield criteria.

Equations indicated in table 1:

$$\eta = \sqrt{(1+r)/(1+2r)\,[(1+r)(1+\beta^2)+2r\beta]} \tag{5a}$$

$$\kappa = \frac{1+2r}{1+r}\frac{1}{1+r+r\beta}\eta \tag{5b}$$

$$\eta = (1+\alpha\beta)^{\frac{1}{2}}\left\{\frac{1+r_0}{1+r_0(1-\alpha)}\right\}^{\frac{1}{2}} \tag{6a}$$

$$\kappa = (1+\alpha\beta)^{\frac{1}{2}}\left\{\frac{1+r_0(1-\alpha)}{1+r_0}\right\}^{\frac{1}{2}} \tag{6b}$$

$$\beta r_{90} + (1+\beta)r_0 r_{90}(1-\alpha)^{a-1} - r_0(\alpha)^{a-1} = 0 \tag{7a}$$

$$\eta = (1+\alpha\beta)^{\frac{a-1}{a}}\left\{\frac{1+r_0}{1+r_0(1-\alpha)^{a-1}}\right\}^{\frac{1}{a}} \tag{7b}$$

$$\kappa = (1+\alpha\beta)^{\frac{1}{a}}\left\{\frac{1+r_0(1-\alpha)^{a-1}}{1+r_0}\right\}^{\frac{1}{a}} \tag{7c}$$

From these equations, the coefficient α from the known coefficient β, furthermore η and κ can be determined. Thus, in the range of diffuse necking σ_e-ε_e values can be calculated. Table 1 shows the equations for each discussed yield criterion. It is also clear that knowing β, α values can be calculated using those formulas, except in case of Hill'79, when a>2. Here the implicit equation (7a) contains α, which can be determined by numerical method in the knowledge of β, a, r_0 and r_{90}.

3. Materials and methods

Two AlMg3 aluminium sheets were used for tests, both of them in annealed (O) and strain hardened (H22) state. Table 2 shows the main alloying elements, strength properties and tensile strains.

Table 2. Chemical composition, strength and ductility properties of sheets

	Si%	Fe%	Mn%	Mg%	$R_{p0,2}$, MPa	R_m, MPa	A_g, %	A_{80}, %
A1-O	0,17	0,23	0,18	3,41	80	217	21,0	25,3
A1-H22					147	234	15,6	20,7
A2-O	0,05	0,11	0,33	3,32	122	237	19,1	24,1
A2-H22					141	235	18,9	22,9

The tensile tests were carried out by Instron 5582 machine using flat specimens with a gage section of 80x20 mm and thickness of 2,5 mm. The DIC technique used is the GOM-ARAMIS® hardware-software system, which is suitable for determining local strains and illustrates the information obtained in the form of colour-coded strain maps and digitized data. The experimental results include strains in three principal directions as well as effective strain and distribution of effective stress.

The 3D measurements were made with YXLON CT MODULAR computer tomography equipment and GOM ATOS® 3D optical scanner. In both cases, the measurement results were processed and visualized by a 3D CAD system. The detailed description of DIC and 3D measuring technique together with the evaluation of results is described in the [18] publication. The same publication includes the validation of measurement results.

4. Test results

As an example, figure 1 shows the strain maps of a tensile specimen in the range of diffuse and local necking. Based on visual analysis the upper row contains the strain maps of diffuse necking and lower row belongs to local necking. Figure 2 shows effective strain curves in the diffuse and local necking phase, the lower curves characterise diffuse necking while the five upper curves can be connected to local necking. Strains measured on longitudinal and transverse directions can be used to determine the dimensional changes of specimen in length, width and thickness direction, which can be used for evaluation of the actual cross-section area. Knowing the acting forces, axial stresses can be calculated, which can be used for extrapolating flow-stress curves.

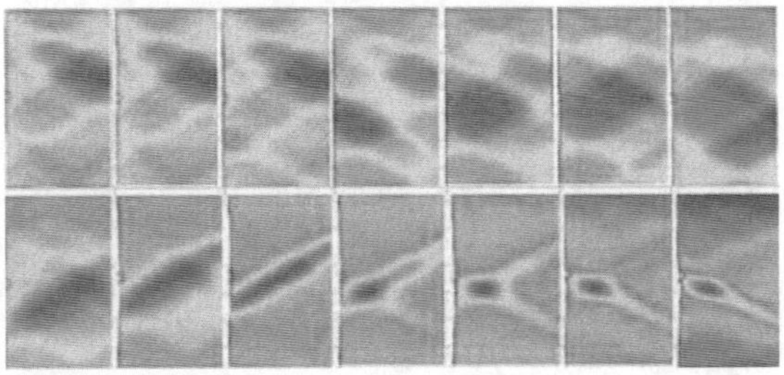

Figure 1. Longitudinal strain maps

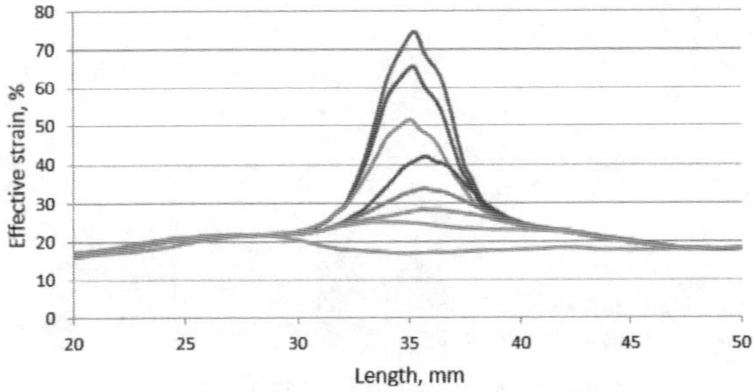

Figure 2. Effective strain evolution

4.1. Numerical analysis of diffuse and local necking

A number of time-dependent and time-independent solutions are described in the literature for determining the onset of local necking [19], but here a new method is introduced based on the DIC technique, using the calculated values of the average strain (AVE), standard deviation (STD) and relative standard deviation (RSD=STD/AVE) of the strains measured in the necking zone. From the effective strain distribution it can be seen that the curves of A2-H22 sheet reach their maximum at L=35 mm (figure 2), so the necking zone can be selected approximately ±5 mm in length from that value, between the ranges of 30 mm and 40 mm. The transition between diffuse and local necking can be marked from the slopes of the relative standard deviation of thickness strain as it is illustrated in figure 3. It can be clearly seen that the relative standard deviation (RSD) indicates very characteristically the onset of local necking. The RSD at -6.5 s (point 6 from left in figure 3.a.) can still be classified as diffuse necking, while the RSD at -5.5 s shows a sharp increase, so this point is already belonging to the local necking stage. According to this, figure 3.b shows the lines fitted to the RSD points both in the diffuse and local necking stage. The intersection point of lines is 18,3% engineering strain (figure 3.b), which corresponds to ε=0.168 logarithmic strain. This value is the upper limit of diffuse necking, while the plane stress condition can be considered, so the relationships for extrapolation of the flow stress curve can be applied.

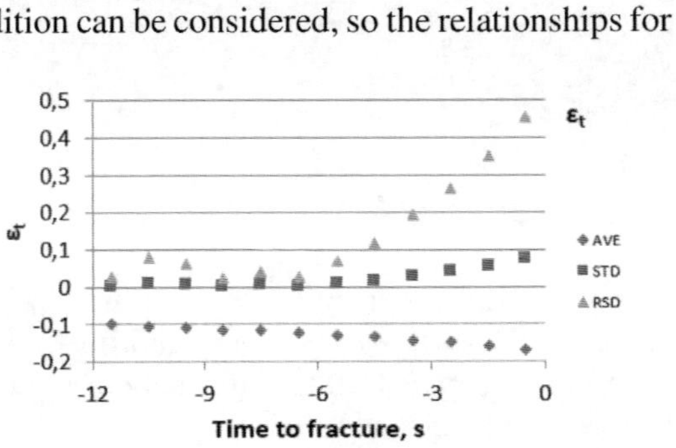

a) characteristics of thickness strain as the function of time

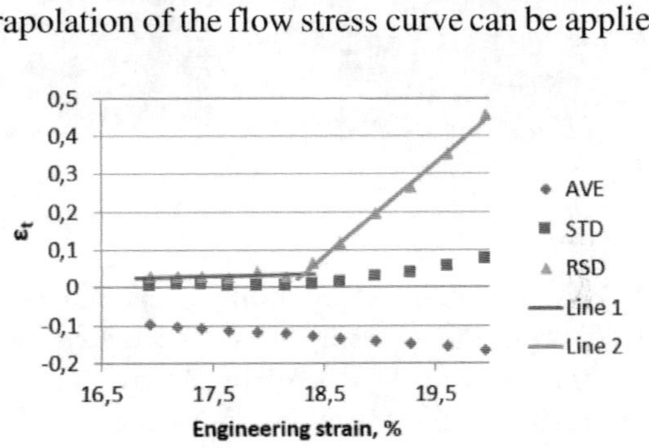

b) characteristics of thickness strain as the function of engineering strain

Figure 3. Marking the limit of diffuse necking

Further characteristics of the necking phase are displayed in figure 4. Local strain increase is defined as the difference between the strain in necking zone and the average strain ($\Delta\varepsilon_{neck}=\varepsilon_{diff}-\varepsilon_{AVE}$). This quantity characterises the local concentration of strain. As figure 4.a shows, the slope of diffuse and local necking range is also distinguishable and about 0,04 strain increase can be detected during diffuse necking related to average strain. At the necking stage, the true cross-section area can be determined from the effective strain ε_e and from the initial cross-section area (50 mm^2) with the relationship of $A=A_0*\exp(-\varepsilon_e)$. Figure 4.b illustrates the calculated values.

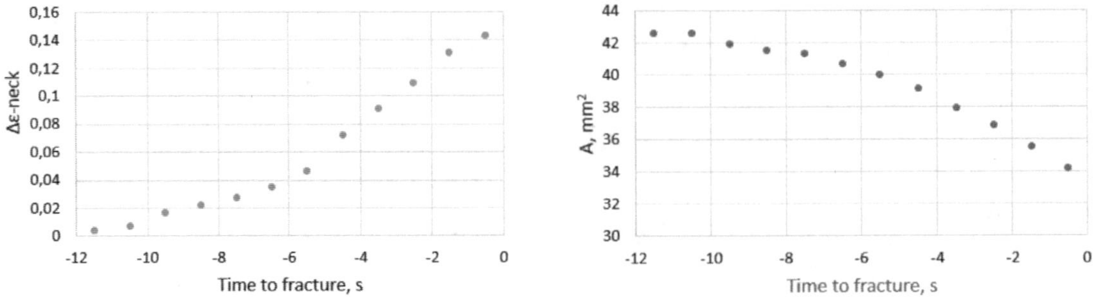

a) strain evolution in necking zone b) reduction of the cross-section area

Figure 4. Change of strain and cross-sectional area during necking

3D measurements – which are completely different from the DIC technique – can also be used to illustrate the necked shape of specimen. An example of this is illustrated in figure 5, which shows the 3D CT model taken after the break from each side of the specimen. Figure 5.a. shows the joined two half-parts of the specimen after break, while images 5.b and 5.c show plane sections perpendicular to the axis. From the 3D model, the area of the plane sections could be determined, the comparison of these values with the results of the area evaluated from DIC strains taken at 0.5 s before the break is shown in figure 5.d. The horizontal axis of the figure shows the distance between the sections recorded in +/- directions from the centre of the break. It can be concluded that the cross-section areas defined by the two procedures are quite well matched in the vicinity of fracture.

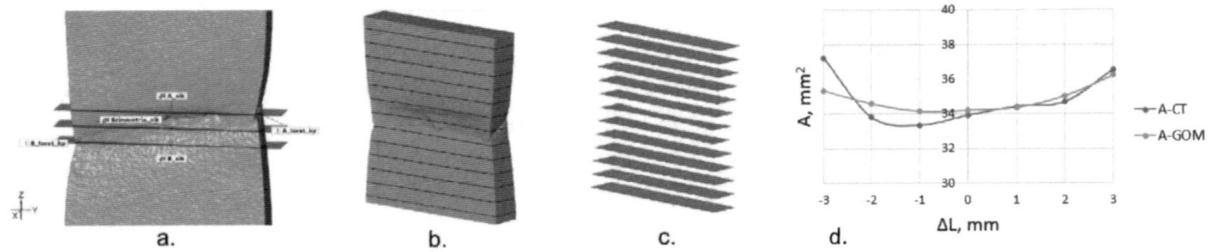

Figure 5. CT model and area sections

GOM-ATOS® photogrammetric technique is also suitable to take 3D recordings during the tensile test. During the necking phase the tensile testing machine was stopped at every millimetre, a 3D recording was made, and then the machine was restarted to create further deformation (figure 6). Repeating this, the entire deformation process could be recorded and evaluated.

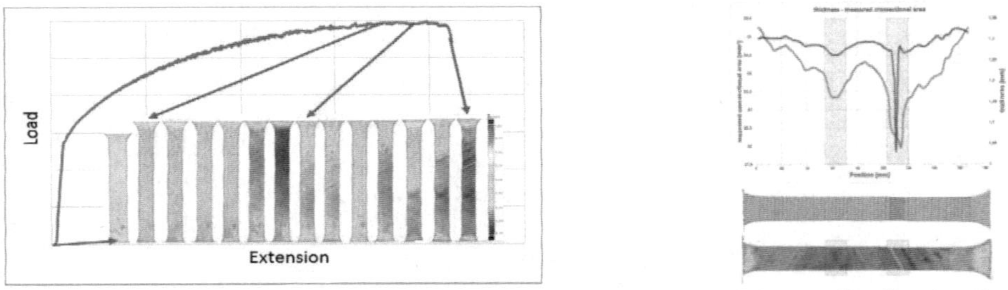

a) change in thickness b) thickness distribution before break

Figure 6. Change in thickness of specimen

Figure 6.a. shows the necking process, colours illustrate the change in thickness of the specimen. The sequence of photos shows the development of diffuse and local necking. The deformation map is almost homogeneous until the fifth image, where local necking is formed, which leads to break. Figure

6.b. gives insight to the evaluation process, where the last image was analysed in detail. The upper chart shows the change in cross-section area and thickness. The original width and thickness of this specimen were 20x1,5 mm, the final area reduced to 22 mm^2 while the thickness decreased to 1,03 mm. These recordings demonstrate that equivalent results can be obtained from the tensile test process using modern 3D technique or DIC method.

4.2. Calculation of stresses and strains in the diffuse necking zone

The most important strength and ductility properties of the tested AA5754 aluminium sheets are shown in table 2. Using the average plastic anisotropy values of the sheets (r=0,7), figure 7 illustrates the comparison of η strain and κ stress ratios (equations in table 1). The curves of the Mises yield criterion for isotropic materials (r=1) differ significantly from the others for both parameters, but the curves of κ calculated on the basis of Hill'48 and Hill'79a yield criteria are quite close to each other, since they are quadratic formulae. According to Hill'79b, curves are slightly different from the previous two, due to the effect of the exponent a=8, they change less as the function of the β strain ratio than the previous ones, in this case the ratios only reach approximately the half compared to the values given by the quadratic formulae.

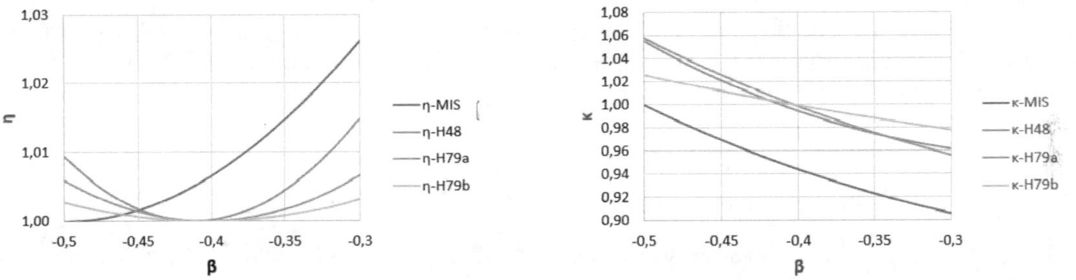

a) change of η strain ratio b) change of κ stress ratio

Figure 7. Comparison of stress and strain ratios calculated by different yield criteria

The starting parameter for the calculations is the ratio of $\beta=d\varepsilon_y/d\varepsilon_x$, which can be replaced by the ratio of $d\varepsilon_w/d\varepsilon_L$ from the measurement results. These two strains shall be determined for the range of diffuse necking. The following calculations are referring to A2-H22 sheet, but in other cases the method of evaluation is the same. It is already known from the former data that the range for calculations lies between L=30-40 mm (figure 2), as the measuring points located here provide a stable average. The extended flow stress curve points were calculated according to equations H79b, the parameters are a=8; $r_0=0.675$ and $r_{90}=0.725$. Preliminary calculations have shown that the scatter of $\beta=d\varepsilon_w/d\varepsilon_L$ ratio is significant, therefore, referring to the findings of [11] literature, the calculations were made by the ratio of $\beta=\varepsilon_w/\varepsilon_L$. During the tensile test, the GOM-ARAMIS® system records the values of force and DIC strains together, therefore the results can be combined over time. At the diffuse necking stage, using η and ε_L the effective strain (ε_e) can be calculated. The true cross-section area can be determined from the effective strain as it can be seen in figure 4.b. Since the force for a given moment is known, this can be used to calculate the axial stress (σ_x=F/A), and the effective stress (σ_e) is derived by multiplying σ_x with the parameter κ.

Figure 8.a shows the measured true stresses (σ1-m) up to the limit of uniform deformation, where the stress serrations caused by the PLC effect are also visible. The fitted Swift and 3-parameter Voce functions are also illustrated as the upper and lower bound of flow stresses. Flow stress curve equations are: $\sigma(\text{Swift})=475.5*(0.0042+\varepsilon_e)^{0.2872}$ and $\sigma(\text{Voce})=114.4+176.2*(1-\exp(-14.95\varepsilon_e))$. The calculated flow stress points (σ-c-H79b) are displayed by yellow dots. The equation for the approximate function σ-H79b is $\sigma=w*\sigma(\text{Swift})+(1-w)*\sigma(\text{Voce})$, where w=0.293 is the weight factor determined on the basis of minimizing the smallest squares. For all other sheets the weighted curves can be seen in the charts of figure 8.b with the calculated weight factors. For the tested AlMg3 sheets the weighting factors are lower than 0,407 – the flow stress curves are nearer to the Voce approach. This statement is in accordance with the findings of the literature, as aluminium alloys fit better to the functions which have an upper bound.

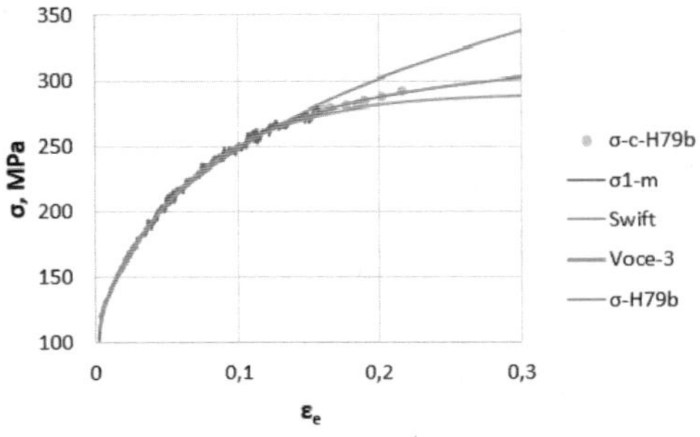

a) Flow stress curves for A2-H22

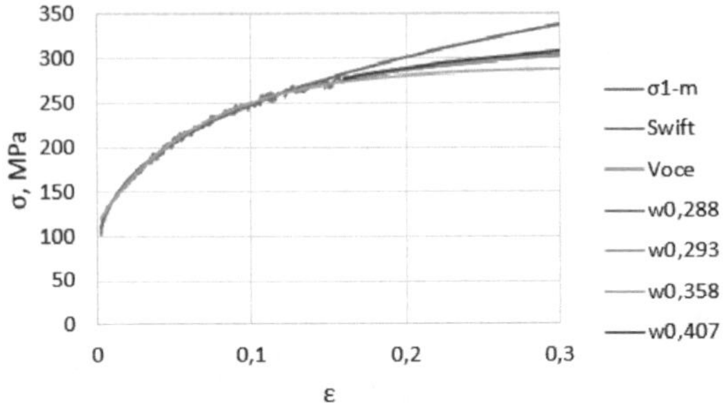

b) Extrapolated flow stress data

Figure 8. Calculated points and extrapolated values of flow stress curves

5. Conclusions

Digital image correlation and up-to-date 3D imaging techniques provide a lot of new information about tensile tests. They can be used to understand the previously discovered phenomena better. For instance, the PLC effect or the necking phase have not been studied so profoundly yet within the given technical conditions. The analyse of these phenomena is indispensable for aluminium-magnesium alloys, since the formability of the sheets depends on these processes largely. The main conclusions of the current research are the followings:

1. The average and local deformations of tensile specimens in length, width and thickness directions can be detected and calculated using DIC technique.
2. A new measurement method has been introduced to determine the transition between diffuse and local necking, based on the calculation of the average elongation, the standard deviation and the relative standard deviation of the necking zone. The appearance of local necking is indicated by a sharp change in the slope of relative standard deviation curve.
3. The minimum cross-sectional areas determined by 3D CT and photogrammetric measurements are well matched to the values calculated from effective strains obtained by DIC technique.
4. For plastic anisotropy coefficients of around 0,7 and in the range of -0,5 to -0,3 transverse/longitudinal strain ratios, the effective stress and strain ratios (κ and η) calculated according to the quadratic yield criteria Hill'48 and Hill'79 are approximately equivalent and the non-quadratic Hill'79 yield criterion results values about half of the former two.
5. The flow stress curve points of the diffuse necking phase determined by Hill'79 yield criterion are laid between the Swift and the Voce functions fitted to the points of the uniform elongation and they are closer to Voce approach, the weight factors determined from the mixing rule is between w=0,3-0,4 for the tested sheets.

Acknowledgments

The research was supported by the program 20523-3/2018/FEKUTSTRAT, Higher Education Institutional Excellence Program – Research on Digital Industrial Technologies at Széchenyi István University.

Literature

[1] Sutton M A, Orteu J J, Schreier H W and Reu P 2012 Introduction to digital image correlation: Best practices and applications *Experimental Techniques* **36(1)** 3–4

[2] Kang J, Wilkinson D S, Jain M, Wu P D, Embury J D, Mishra R K and Sachdev A K 2006 On the sequence of inhomogeneous deformation processes occurring during tensile deformation of strip cast AA5754 *Acta Mater.* **54** 209-218

[3] Kang J, Wilkinson D S, Bruhis M, Jain M, Wu P D, Embury J D, Mishra R K and Sachdev A K 2008 Shear localization and damage in aa5754 aluminum alloy sheets *J. Mater. Eng. Perform.* **17** 395-401

[4] Yang S Y and Tong W 2009 A finite element analysis of a tapered flat sheet tensile specimen 2009 *Exp. Mech.* **49** 317-330

[5] Coppieters S, Cooreman S, Solc H, Van Houtte P and Debruynea D 2011 Identification of the post-necking hardening behaviour of sheet metal by comparison of the internal and external work in the necking zone *J. Mater. Process. Technol.* **211** 545-552

[6] Wang L and Tong W 2015 Identification of post-necking strain hardening behavior of thin sheet metals from image-based surface strain data in uniaxial tension tests *Int. J. Solids Struct.* **75-76** 12-31

[7] Knysh P and Korkolis Y P 2017 Identification of the post-necking hardening response of rate- and temperature-dependent metals *Int. J. Solids Struct.* **115-116** 149-160

[8] Nguyen V T, Kwon S J, Kwon O H and Kim Y S 2017 Mechanical properties identification of sheet metals by 2d-digital image correlation method *Procedia Eng.* **184** 381-389

[9] Bacha A, Daniel D and Klocker H 2007 On the determination of true stress triaxiality in sheet metal *J. Mater. Process. Technol.* **184** 272-287

[10] Zhang Z L, Hauge M, Odegard J and Thaulow C 1999 Determining material true stress-strain curve from tensile specimens with rectangular cross-section *Int. J Solids Struct.* **36** 3497-3516

[11] Wang Y-D, Xu S H, Ren S B and Wang H 2016 An experimental-numerical combined method to determine the true constitutive relation of tensile specimens after necking *Adv. Mater. Sci. Eng.* **2016** 1-12

[12] Murata M, Yoshida Y and Nishiwaki T 2018 Stress correction method for flow stress identification by tensile test using notched round bar *J. Mater. Process. Technol* **251** 65-72

[13] Hyun H C, Kim M, Bang S and Lee H 2014 On acquiring true stress–strain curves for sheet specimens using tensile test and FE analysis based on a local necking criterion *J. Mater. Res.* **29**

[14] Wang L and Tong W 2015 Identification of post-necking strain hardening behavior of thin sheet metals from image-based surface strain data in uniaxial tension tests *Int. J Solids Struct.* **75-76** 12-31

[15] Wang K, Carsley J E, Zhang L, Stoughton T B, Li J and Carlson B E 2014 Forming limits of an age hardenable aluminum sheet after pre-straining and annealing *Int. J. Mech. Sci.* **82** 13-24

[16] Coppieters S and Kuwabara T 2021 Identification of post-necking hardening phenomena in ductile sheet metal *Exp. Mech.* **54**

[17] Harangozó D, Szalai Sz, Kozma I, Czinege I and Dogossy G 2021 *Anyagvizsgálók Lapja* **2021/I** 65-77

[18] Harangozó D, Kozma I, Czinege I and Szalai Sz 2021 *CogInfoCom 2021 Proceedings IEEE* 147-152

[19] Merklein M, Kuppert A, Geiger M 2010 Time dependent determination of forming limit diagrams. *CIRP Annals Manufacturing Technology* **59** 295–298

IOP Conf. Series: Materials Science and Engineering 1246 (2022) 012019 doi:10.1088/1757-899X/1246/1/012019

Finite element simulation of micro-milling of hardened tool steel

B Z BALÁZS[1,*] and M Takács[1]

[1] Department of Manufacturing Science and Engineering, Faculty of Mechanical Engineering, Budapest University of Technology and Economics

[*] Corresponding author: balazs.barnabas.zoltan@gpk.bme.hu

Abstract. The demand for miniature components has been growing steadily in recent years. One of the most important technologies to produce such small components is micro-milling. Although the technology is highly researched, there are still many difficulties due to the size reduction, which makes the finite element simulation, as well as the exploration of cutting process challenging. The aim of the research was to investigate the micro-milling process, especially those features, that can be measured very complicated due to the small sizes at the experiments, such as the cutting temperature, and stresses, by means of finite element simulation. The results show that the cutting temperatures are smaller than in the case of conventional size machining, moreover, the stresses do not show significant changes during the cutting edge impact. Finally, the results were validated by the experimental results through comparing cutting force components, which shows a good agreement.

1. Introduction

Recently the industry requires even more small-sized components. The manufacturing of these parts poses serious challenges to modern manufacturing technology. There are many micro-machining technologies, such as laser micro-manufacturing [1], micro electro-discharge machining [2], and mechanical cutting processes (e.g. micro-turning, micro-milling, micro-drilling). However, one of the most promising technology is micro-milling, because it is able to produce complex 3D geometries with high geometrical accuracy and high material removal rate in a wide range of engineering materials. Although the process is investigated very intensively by researchers, the technology still has many difficulties due to size reduction [3–5]. The main problems of the process originate e.g. from the relatively large cutting edge radius, the minimum separable material thickness, the relatively large vibrations, and the deformation of the micro-milling tool with small stiffness. The paper focuses on the finite element simulation of the micro-milling process, especially on those features, which can be measured very complicated under real conditions due to the small sizes.

Chi-Hsiang et al. [6] investigated the optimal cutting tool design at micro-milling of SKD61 steel using 2D finite element simulations. It was stated that the smaller forces quasi enhanced the stability of the cutting edge due to the smaller loads, and the smaller temperatures are proposed to avoid the thermal softening of the cutting edge. Based on the research the rake angle has the most significant effect on the investigated circumstances. In this scenario, the optimal tool design is the 7° clearance angle, and the 20° rake angle. Furthermore, the authors proposed to use the 188.5 mm/s cutting speed and the 0.01 mm depth of cut values for micromachining of SKD61 material.

IOP Conf. Series: Materials Science and Engineering 1246 (2022) 012019 doi:10.1088/1757-899X/1246/1/012019

Jin and Altintas [7] predicted the micro-milling forces using FEM model on Brass 260. The authors took into consideration the effect of tool edge geometry, the helix angle, the tool run-out, and the trajectory of the cutting edge. The process was investigated at different uncut chip thicknesses, and cutting speed values. The model was validated with experimental results. Based on the results, the prediction of the normal directional cutting force is accurate. In the case of the feed force, the proposed, less accurate estimation can be improved by using a more accurate friction model in the finite element simulations.

Wang et al. [8] investigated the micro-milling using finite element method on hard-to-cut material (Inconel 718). The authors stated that the forces are increased with the increase of the feed per tooth value or the radial depth of cut. In this scenario, the increasing cutting parameters provide even more continuous, i.e. definite, chip formation. The critical value of feed per tooth was determined as 1.5 μm, which is related to the minimum chip thickness.

In one of the previous research made by the authors [9], the 2D finite element modeling of the thin chip removal process was investigated on AISI 1045 steel, where the cutting forces and the temperatures were analysed at different cutting parameter combinations. Moreover, the different material models were compared, such as Oxley's, HRC, and Johnson-Cook material models. Based on the research, the data provided Oxley's model and the Johnson-Cook model was very close to each other.

The main objective of the research was to create a finite element model for the micro-milling of hardened AISI H13 steel and to analyse those characteristics of the process, which cannot be measured in a comfortable and reliable way due to the small sizes, such as the stresses and the temperature. Moreover, another aim was to validate the model with experiments, which were based on the comparison of the measured and simulated active forces.

2. Preparation of the finite element model and the experiments

2.1. Finite element model

In this section, the preparation of the finite element model (FEM) is presented. The FEM simulation was carried out by DEFORM software. The applied tool was a 500 μm diameter, two fluted, AlTiN coated micro end milling cutter, which has a 50 μm corner radius ensuring the enhanced edge stability. The finite element model was prepared based on the features and specifications of the real tool. In this scenario, the clearance angle was α=13,7°, and the rake angle was γ=-10°. Moreover, based on the measurements, the cutting edge radii was set to 4 μm. Figure 1a shows the result of the EDX composition analysis of the coating. Figure 1b represents the accurate CAD model of the tool.

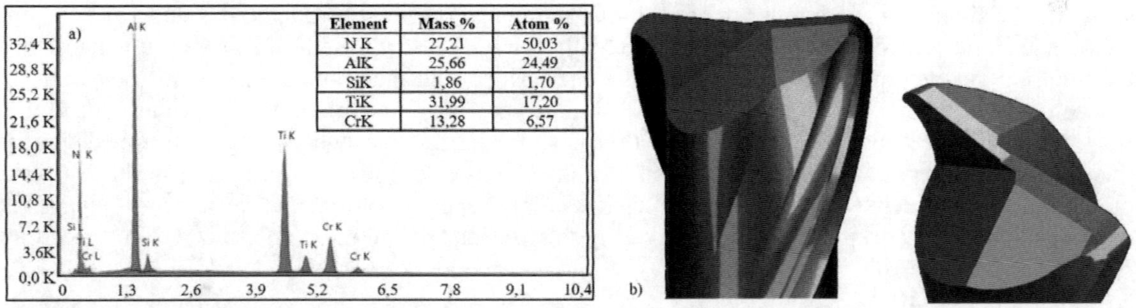

Element	Mass %	Atom %
NK	27,21	50,03
AlK	25,66	24,49
SiK	1,86	1,70
TiK	31,99	17,20
CrK	13,28	6,57

Figure 1. a) Result of the EDX analysis of the tool coating, and b) the accurate CAD model of the applied micro-milling tool.

The workpiece material was AISI H13. The predefined material model and parameters were applied in the software, which calculates the yield stress based on the Umbrello model (Eq. 1.) [10].

$$\sigma(\varepsilon, \dot{\varepsilon}, T, \mathrm{HRC}) = B(T)(C\varepsilon^n + F + G\varepsilon)[1 + (\ln(\dot{\varepsilon}))^m - A)] \quad (1)$$

In the FEM software, the predefined material models were applied also for the tool material and the coating. The values can be seen in table 1, where the data is originating from the Deform software.

Table 1. Specifics of the applied material models.

	Workpiece material	Tool material (WC)	AlTiN coating
Young's modulus (MPa)	210290,09 *	650000	440000
Poisson's ratio	0,3	0,25	0,25
Thermal axpansion coefficient (m/°C)	$1{,}17 \cdot 10^{-5}$ *	$5 \cdot 10^{-6}$	$9{,}2 \cdot 10^{-6}$
Thermal conductivity (W/(m·K))	24,574927 *	59	12 *

* the values here change in the function of the temperature, the presented ones are related to the room temperature

The research covers both 2D and 3D simulations. Figure 2 shows a representative model of both kinds. Since in some cases, the 2D model can give adequate results, calculation time is significantly lower than that of the 3D simulations. However, in many cases, it is required to use complex 3D simulations in order to acquire more accurate results.

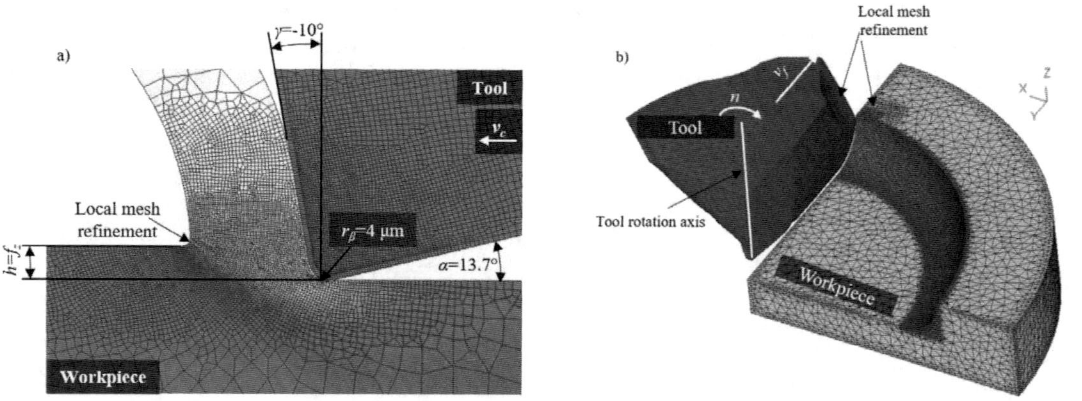

Figure 2. The prepared a) 2D and b) 3D finite element models.

The model of the workpiece already includes the geometry, which was created by the edge impact in the previous rotation of the tool, and the corner radius of the tool was also taken into account. In this scenario, the simulated conditions are very similar to the real ones. According to the study of Póka and Németh [11], the replacement of the cyclois arc with a circular arc represents an error of less than 1.6%. Considering the circumstances presented in the cited paper, and the applied ones in this research, the inaccuracy is less than 1%. Based on this conclusion, the geometry considering the previous edge impact was modeled as a circular arc. In the case of the 3D simulations, the down-milling was applied since in this case the chip separation process about starts immediately after the cutting edge reaches the material due to the adequate size ($>h_{min}$) of the chip thickness. In the region of the chip separation zone and the corner radius of the workpiece, locally smaller mesh size was applied in order to receive more accurate results (see in figure 2b).

2.2. Experimental setup and design

The experiments were conducted on a VHTC 130 micromachining centre, which has 60.000 rpm maximal spindle rotation speed. For the validation, the cutting forces were measured by a three-component dynamometer (Kistler 9257A), which was combined with a Kistler 5080A charge amplifier, and the data acquisition was performed by a NI USB 4431 sound and vibration device and a LabVIEW measuring software. The workpiece material was AISI H13, which was in hardened state (50±1 HRC), and the tool was a 500 µm diameter micro-milling tool. The hardness of the material was ensured by the professional heat treatment process of Böhler Hungary. The sampling frequency during the test was set to 100 kHz.

The full factorial experimental design was applied for both the simulations and the experiments. The applied factors and their levels can be seen in table 2, which are based on the proposed cutting parameters made by the tool manufacturer, and the previous research works of the authors [12,13].

Table 2. Full factorial experimental design.

Parameters	v_c (m/min)	f_z (µm)	a_p (µm)	Milling strategy
Values	90	2, 4, 6, 8, 10, 12	25, 50, 100	Down-milling

3. Results of the finite element simulations

In this section, first, the data (cutting forces, cutting temperature, tool temperature, Von Mieses stress) acquired from the 2D simulations, then, the data originating from the 3D simulations were presented, evaluated, and discussed.

3.1. 2D results

This section is concentrated on the results acquired from the 2D finite element simulations showing. In the following, the temperature distribution of the tool, as well as the workpiece is presented, moreover, the effective stress distribution is also discussed.

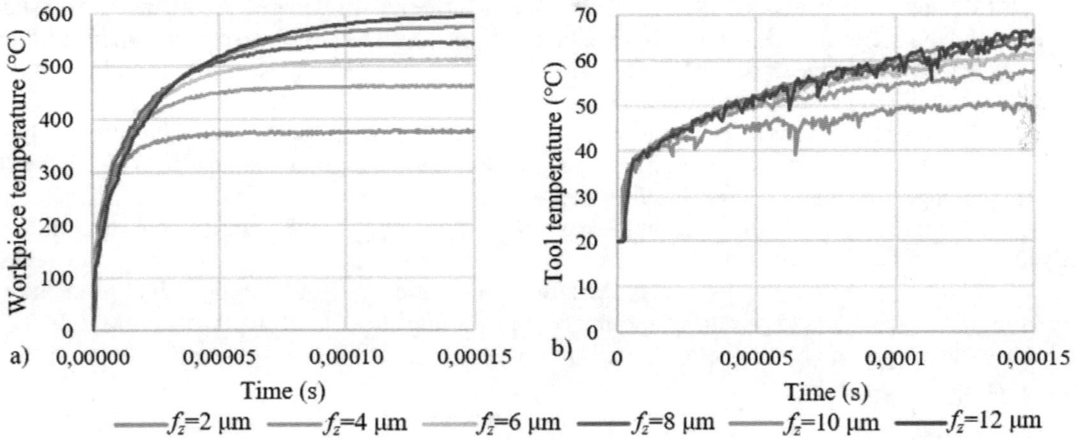

Figure 3. The maximal temperature a) of the workpiece in the cutting zone, and b) of the tool at the different feed per tooth values in the function of time.

The evolution of the cutting temperature and the tool temperature over time can be seen in the diagrams in figure 3. It can be seen that there is a rapid temperature growing at the beginning of the cutting process. Based on the simulations, several 100 °C changes can occur in a few hundred-thousandths of a second. In the case of the workpiece the temperature becomes constant; however, at the tool, the temperature is continuously growing with the cutting length or cutting time. The data was analysed up to about 0.00015 sec, which means 0.25 mm cutting length. The maximal tool temperature was ~70 °C at the greatest feed per tooth value (f_z=12 µm). However, this value is still low enough, therefore it is not expected to cause a significant change in wear behaviour. Basically, the larger the feed per tooth, the larger the temperature. Presumed a constant specific cutting force (k_c), a larger feed per tooth causes larger cutting forces. As it is well known, the frictional force components also increase with a larger cutting force, which is in close relation with the generated temperature. At f_z=2 µm the maximal workpiece temperature is ~378 °C; however, at f_z=12 µm is ~602 °C. As it was expected, the tool temperature is significantly smaller than that of the workpiece (cutting zone), its maximum value changes between ~50 °C and ~67 °C. As it is obvious from the scientific literature, during the cutting process most of the input power (~ 99%) is converted into heat, of which 60-80% is removed by chips, 10-20% is dissipated by the workpiece, and only about 10-18% is used for increasing the tool temperature [14,15].

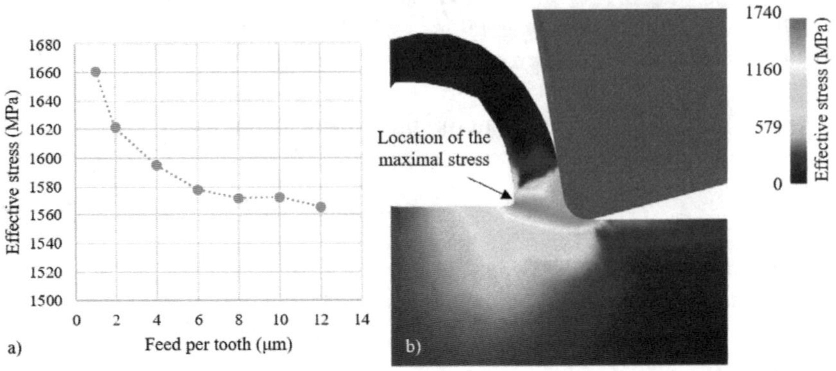

Figure 4. a) The effective stress in the case of different feed per tooth values, b) the simulated results and the stress distribution in the cutting zone.

The diagram of figure 4a shows the Von Mieses stress (effective stress) in the workpiece at different feed per tooth values. The effective stress values here were determined uniformly at 0.25 mm cutting length. At this point, presumably, already steady-state chip removal occurs. The stress values slightly decrease with the increasing chip thickness, which can be explained by the size effect and the decreasing ploughing effect. The range of the simulated values is 1565 – 1660 MPa, therefore, it can be stated, that there are no significant changes in the Von Mieses stress.

Figure 4b presents the stress distribution in the workpiece. It can be clearly seen that the highest stress values occurred in the primary cutting zone, and the single maximum value also arises in the shear plane (at the opposite side of the tool's cutting edge).

The cutting forces were investigated in the case of the 2D model, too. Because of the simplifications, the milling is simulated as an orthogonal cutting process (figure 2a), and in this scenario, the modeling was not provided adequate results. The calculated values are 2.2 – 4.8 times higher than the ones originating from measurements during the micro end milling process at depth of cut of a_p=100 μm. Therefore, the results can be used only for the investigation of trends or for comparing the results with each other.

3.2. Results of 3D simulations and model validation

Since it is clearly seen from the 2D simulations, that the simplified model did not provide adequate results in the case of the forces; therefore, it is required to use more complex 3D models (figure 3b), which contains significantly less neglect. Figure 5a shows the characteristic run of the simulated force data (v_c=90 m/min, a_p=50 μm, and f_z=8 μm). In this paper, the F_a force component was analysed, which is the resulting force of the planar force components, which are F_x and F_y.

Figure 5b shows the simulated F_a values and the measured ones obtained during micro-milling experiments. It can be stated, that the simulated values are smaller than the measured values. The explanation could be based on the fact that the simulation environment does not contain vibrations; however, in the reality, one of the main difficulties of micro-milling is the strong dynamic process characteristic. Therefore, one of the future aims is the implementation of the vibrations and the vibrational behaviour of the machining system in the FEM simulations. Moreover, the microstructure of the workpiece material can also influence the forces. Although an exact CAD model of the micro-milling tool was applied, there can be slight differences (e.g. cutting edge radius) from the instantaneous condition of the real tool. It can be seen that the model provided better results in the case of larger feed per tooth values, which can be explained by the fact, that in this case, the geometrical inaccuracies play a smaller role. Based on the acquired result, there is about an average of 21% difference between the simulated and the measured results, which can be evaluated as a good result in the case of such a complex process, as micro-milling.

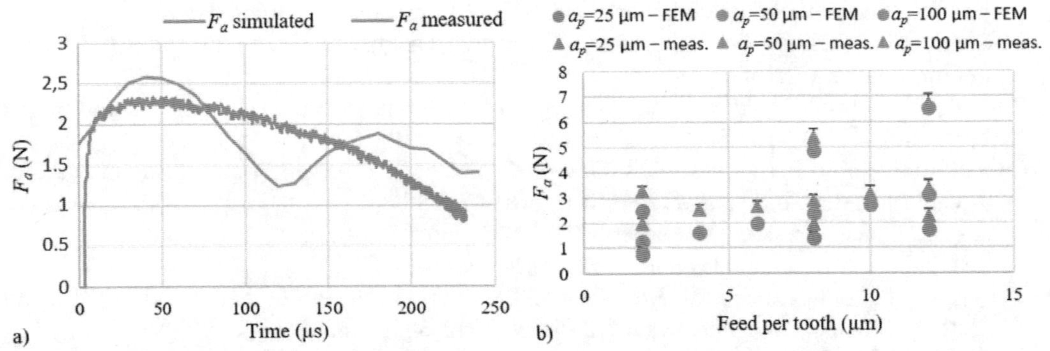

Figure 5. The comparison of the a) characteristic run, and b) the maximal values of the active forces in the case of the simulated and the measured results.

Figure 5 shows a representative run of the active forces at cutting parameter combination of v_c=90 m/min, a_p=50 μm, and f_z=8 μm at micro-milling. It can be clearly seen that the simulation gives a good approximation in the case of the maximal force values; however, there are differences in the characteristic run. It can be explained by the complex behaviour of the micro-milling process, e.g. the vibrational behaviour of the real machining environment, and the continuous presence of the elastic deformation of the small-sized micro-milling tool with small stiffness. In the simulations, only the chip separation process is investigated. Therefore, the results refer to an environment free of external vibrations and influences.

4. Summary

Finite element models were developed to investigate the micro-milling process at machining of hardened hot-work tool steel. Both the 2D and the 3D simulations were investigated and their results were analysed and compared with the measured data. The micro-milling experiments were conducted on a VHTC 130 five-axis micromachining centre with a 500 μm diameter micro-milling tool on hardened AISI H13 hot-work tool steel.

Based on the research work, the significantly faster 2D simulation can be useful to estimate the effective stresses or the temperature in many cases, but it is not suitable for the estimation of the cutting forces. It can be stated that the cutting temperature of the workpiece was between ~378 °C and ~602 °C, which can be considered relatively small compared to the case of conventional milling of hardened materials. The tool temperatures were also relatively low; however, it is continuously growing with the increasing cutting length. Its characteristic range is between ~50 °C and ~67 °C. These values are low enough not to cause thermal softening on the cutting edge, therefore the cutting ability of the micro-milling tool is not influenced by the thermal effect in this scenario.

Moreover, the effective stress values and stress distribution were analysed. The values of the effective stress were calculated to be in a relatively small range (1565 – 1660 MPa).

However, it is obvious that more accurate analysis can be carried out by complex 3D simulations. In this case, the estimation of the active forces is good. There is an average 21% difference between the measured and the calculated force values. The simulated values are always smaller than the measured ones, which can be explained by the complex behaviour of the micro-milling process.

In the future, it is recommended to take into account also the vibrations of the machining environment for achieving more accurate results.

5. References

[1] Weng F, Liu Y, Chew Y, Yao X, Sui S, Tan C, et al. IN100 Ni-based superalloy fabricated by micro-laser aided additive manufacturing: Correlation of the microstructure and fracture mechanism. Mater Sci Eng A. 2020 0 24;788:139467.

[2] Kuzin VV, Fedorov SYu, Szalay T, Farkas B. Micromachining of a High-Density Current-Conducting Ceramic With the Use of Electrical-Discharge Machining. Part 1. Refract Ind Ceram. 2016;57(2):164–9.
[3] Boswell B, Islam MN, Davies IJ. A review of micro-mechanical cutting. Int J Adv Manuf Technol. 2018;94(1):789–806.
[4] Câmara MA, Rubio JCC, Abrão AM, Davim JP. State of the Art on Micromilling of Materials, a Review. J Mater Sci Technol. 2012 0 1;28(8):673–85.
[5] Balázs BZ, Geier N, Takács M, Davim JP. A review on micro-milling: recent advances and future trends. Int J Adv Manuf Technol. 2021;112(3):655–84.
[6] Chi-Hsiang C, Yung-Cheng W, Bean-Yin L. The optimal design of micro end mill for milling SKD61 tool steel. Int J Adv Manuf Technol. 2013;68(1):165–73.
[7] Jin X, Altintas Y. Prediction of micro-milling forces with finite element method. J Mater Process Technol. 2012 0 1;212(3):542–52.
[8] Wang F, Cheng X, Liu Y, Yang X, Meng F. Micromilling Simulation for the Hard-to-cut Material. Procedia Eng. 2017 0 1;174:693–9.
[9] Balázs BZ, Takács M. Finite element modelling of thin chip removal process. IOP Conf Ser Mater Sci Eng. 2018;426:012002.
[10] Davim JP. Machining of Hard Materials. Springer Science & Business Media; 2011.
[11] Póka G, Németh I. The effect of radial rake angle on chip thickness in the case of face milling. Proc Inst Mech Eng Part B J Eng Manuf. 2020 0 1;234(1–2):40–51.
[12] Balázs BZ, Takács M. Experimental investigation and optimisation of the micro milling process of hardened hot-work tool steel. Int J Adv Manuf Technol. 2020;106(11):5289–305.
[13] Balázs BZ, Geier N, Pereszlai C, Poór DI, Takács M. Analysis of cutting force and vibration at micro-milling of a hardened steel. Procedia CIRP. 2021 0 1;99:177–82.
[14] Klocke F. Manufacturing processes 1. Springer-Verlag Berlin Heidelberg; 2011.
[15] Balki NB, Nelge BD, Kale DVM. Investigation of Temperature and Heat transfer during Machining: Review. Int J Sci Res Dev. 2015;3(2):2283–6.

Acknowledgements
This research was partly supported by the National Research, Development and Innovation Office through the projects ED 18-2018-0006 (Research on prime exploitation of the potential provided by the industrial digitalisation), and OTKA-K-132430 (Transient deformation, thermal and tribological processes at fine machining of metal surfaces of high hardness).

IOP Conf. Series: Materials Science and Engineering 1246 (2022) 012020 doi:10.1088/1757-899X/1246/1/012020

State-monitoring and product quality measurement of additively manufactured injection mould inserts

Sz. Krizsma [1] and A. Suplicz[1,2]

[1] Department of Polymer Engineering, Faculty of Mechanical Engineering, Budapest University of Technology and Economics, H- 1111 Budapest, Műegyetem rkp. 3, Hungary

[2] MTA-BME Lendület Lightweight Polymer Composites Research Group, H-1111 Budapest, Műegyetem rkp. 3., Hungary

E-mail: suplicz@pt.bme.hu

Abstract. Additive technologies represent a state-of-the-art and innovative way in today's mould making. They allow production of both metallic and polymeric moulds with relative simplicity and flexibility compared to conventional machining. Despite its possible benefits, this area is scarcely researched. In our work, we analysed in-mould applicability of epoxy-acrylate inserts manufactured by PolyJet technology. We created a comprehensive state monitoring method to quantify important process parameters like cavity pressure, strain and temperature of the inserts. We analysed the effect of holding pressure on the resulting cavity pressure and strain of the insert. We showed that strain of the insert gradually increased after the cycles and we identified characteristic segments of the injection moulding cycle on the strain-time curves. We also applied surface temperature measurements using a thermal imaging camera. The purpose of this measurement was to determine necessary delay time between cycles allowing inserts to cool below heat deflection temperature, so early failure could be prevented. By using thermal imaging camera, we measured cavity surface temperature distribution and demonstrated the effect of low thermal conductivity of the insert material. We also measured thickness and weight variation of injection moulded products to show the effect of holding pressure and mould deformation on final product quality. As we applied higher holding pressures, product mass and average product thickness grew.

1. Introduction

Additive manufacturing (AM) technologies have already revolutionised part making. They are capable of producing functional parts in small series and manufacturing low-volume moulds for well-established plastic processing technologies like injection moulding [1-3]. Materials for AM technologies include metals, thermoplastic and non-thermoplastic polymers. Thermoplastic polymers can be printed from filament that is melted by a heated nozzle at the printer head (Fused Deposition Modeling, FDM) or thermoplastic polymer powder particles are melted and bound together by a high energy laser (Selective Laser Sintering, SLS) [4]. Non-thermoplastic polymer parts can be printed by curing resin using an UV lamp (PolyJet) or laser (Stereolitography, SLA). Metal parts are printed from powder similarly to SLS, that technology is called Direct Metal Laser Sintering (DMLS). [5] Application of additively manufactured mould inserts offers much desired flexibility in mould making. Complex geometries can be manufactured, like conformal cooling channels [6-8]. Research has already been made to analyse the effect of additively manufactured mould inserts on final injection moulded product properties like dimensional accuracy or crystallinity and resulting mechanical properties [9-11]. Yet, only few research papers are available on comprehensive state monitoring of additively manufactured inserts during their in-mould use. [12-13]

In our research we manufactured mould inserts by PolyJet technology from an epoxy-acrylate photopolymer resin. After that we carried out an injection moulding series to measure operational strains, cavity pressures and surface temperature distribution of the inserts. We measured the effect of holding pressure on resulting strains and cavity pressure. We also analysed the thermal state of the insert and showed the effects of low thermal conductivity.

2. Materials and methods

2.1. Materials

Injection moulded material was *Tipplen H145F* (polypropylene homopolymer) manufactured by *MOL Group Public Limited Company.* We chose this material grade because of its low required processing temperature, and good MFI value. Processing and mechanical properties of the material are listed in table 1.

Table 1. Material properties of *Tipplen H145F.* [14]

Physical property	Typical value
Melt flow rate (MFR) (at 230 °C and 2.16 kg) $\left(\frac{g}{10\ min}\right)$	29
Recommended processing temperature (°C)	190 - 235
Flexural modulus (GPa)	1.8
Module of elasticity (in tension) (GPa)	1.99
Tensile stress at yield (MPa)	38
Tensile strain at yield (%)	8

Mould inserts were manufactured by PolyJet using an *Objet Alaris 30* printer (*Objet Geometries Ltd.*). The material was an UV curable epoxy-acrylate photopolymer resin (*VeroWhitePlus RGD835, Stratasys Ltd.*). Important physical properties of the material are listed in table 2.

Table 2. Material properties of *VeroWhitePlus (RGD835).* [15]

Physical property	Typical value
Tensile Strength (MPa)	50 - 65
Elongation at break (%)	10 - 25
Modulus of elasticity (GPa)	2 - 3
Flexural modulus (GPa)	2.2 - 3.2
Heat Deflection Temperature (at 0.45 MPa) (°C)	45 - 50
Shore Hardness (D Scale)	83 - 86

2.2. Mould inserts and mould

For injection mouldings we used a steel mould housing fitted by additively manufactured mould inserts. Figure 1 shows the moving half of the mould and the ejection system a) the stationary half of the mould b) and the product with its main dimensions c). Cavity inserts in both mould halves and the washers underneath them were made of *VeroWhitePlus (RGD835).*

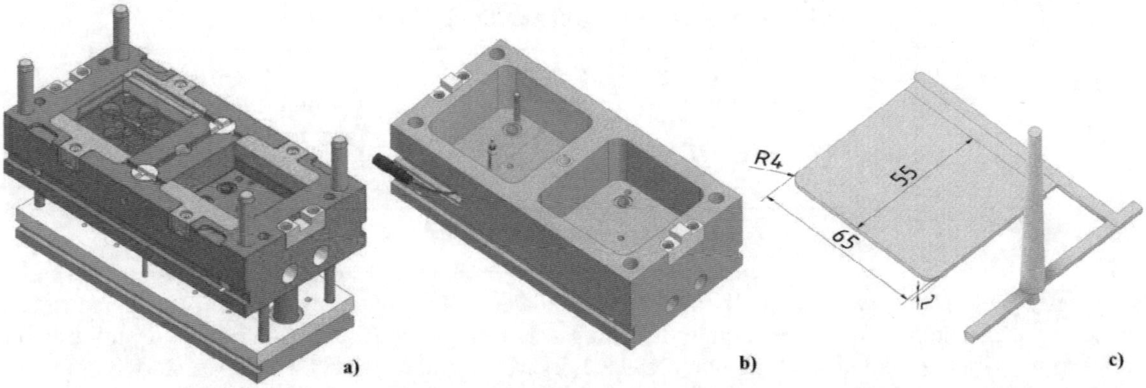

Figure 1. The applied mould: moving half a), stationary half b) and the injection moulded product c).

We injection moulded using an *Arburg Allrounder 270 S 400-170* injection moulding machine (*Arburg GmbH*). Cavity pressure was measured by an *RJG Piezo 6159* piezoelectric direct pressure sensor and its data was collected by a *Como Injection 2869B* unit (*Kistler AG.*). Strains were measured using *KMT-LIAS-06-3-350-5EL* strain gauges (*Kaliber Műszer- és Méréstechnika Kft*) and their data was collected by a *Spider 8* unit (*Hottinger Baldwin Messtechnik GmbH*). Surface temperature of the moving side cavity insert was measured in the delay time between cycles, using a *FLIR A325sc* thermal imaging camera (*Teledyne FLIR LLC.*). Positions of the two strain gauges and the thermal imaging camera are illustrated in figure 2.

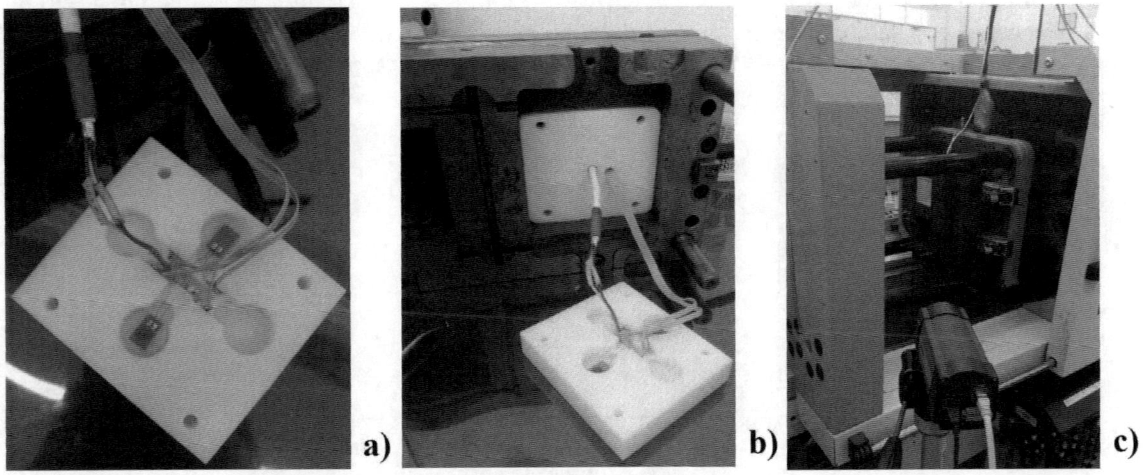

Figure 2. The moving side mould inserts a), b) and the thermal imaging camera location c).

2.3. Injection moulding parameters

The applied injection moulding parameters (constant throughout the cycles) are shown in table 3. As can be seen, melt temperature was chosen approximately in the middle of the recommended processing temperature range. Injection speed, clamp force and injection pressure limit were lowered to protect the inserts from high loads. Long holding time and residual cooling time are needed because heat extraction from the injection moulded product is slow due to low thermal conductivity (approximately. 0.15 - 0.3 $\frac{W}{m \cdot K}$) of the inserts. Also, long delay time is necessary for the cooling of cavity surface after part ejection.

Table 3. Applied injection moulding parameters.

Melt temperature	Injection speed	Clamping force	Injection pressure limit	Holding time	Residual cooling time	Delay time
(°C)	$\left(\frac{cm^3}{s}\right)$	(kN)	(bar)	(s)	(s)	(s)
210	15	50	500	15	30	~300

Injected melt volume was gradually increased by lowering the switchover point from 35 cm^3 to 27 cm^3 using 2 cm^3 steps. Complete volumetric filling was reached at 26 cm^3. In cycles 1-6, no holding pressure was applied. After reaching complete volumetric filling in the 6th cycle, we started applying holding pressure. In the 7th and 8th cycles, we applied 50 bars holding pressure and its magnitude was increased by 25 bar steps in every second cycle. The aim was to analyse the effect of holding pressure on cavity pressure and resulting strain of the moving half cavity insert. Creep of the epoxy-acrylate material was expected to become more dominant as holding pressure increased.

3. Results and discussions

The presented state monitoring method can quantify main mechanical and thermal parameters of the mould inserts during injection mouldings. In this section, relevant results are presented.

3.1. Strain and cavity pressure results

We measured strain in two locations as can be seen in figure 2. One strain gauge was glued into the near-gate slot, the other into the far-gate slot at the back of the moving side cavity insert, using a cyanoacrylate adhesive (*3M Scotch-Weld Plastic & Rubber Instant Adhesive PR100*). Strain results for near-gate gauge are shown in figure 3.

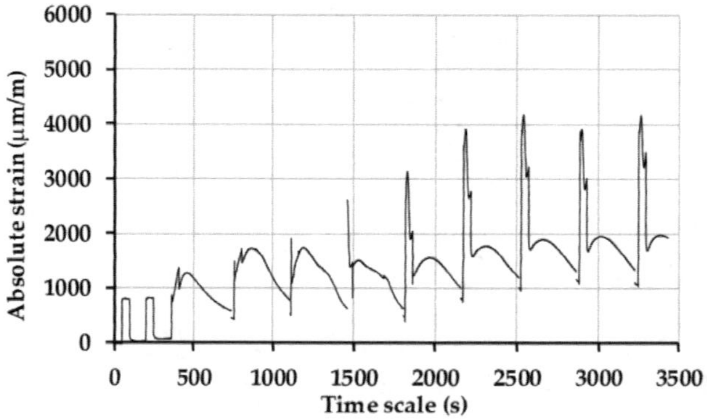

Figure 3. Measured strain results of the gauge near-gate.

In the first two cycles, melt only filled the runner and the gate. Partial filling of the cavity began in the 3rd cycle, therefore the near gate gauge measured a bell-shaped strain-time curve. In the 4th and 5th cycles we increased injected melt volume and it caused an increase on the corresponding strain curves of the near-gate gauge. In the 6th cycle we reached complete volumetric filling but still no holding pressure was applied. The first segment of the 6th cycle's strain curve is missing because strain measurement was started with a delay. From the 7th cycle, the effect of elevated holding pressure can be seen, as maximum points of strain curves show an increasing tendency due to more dominant creep of epoxy-acrylate material in the holding phase. After part ejection, viscoelastic deformation component decreases in the delay time. Figure 3 also demonstrates the importance of delay time when using polymeric injection mould inserts.

Strain curves corresponding to the 7th - 11th cycles have a similar shape. To show the change in strain comparably in each cycle, relative strain was introduced. Relative strain is the difference of measured strain in the current cycle and residual strain after the previous cycle. Relative strain curves of the near gate gauge in the 8th, 10th and 11th cycles are presented in figure 4 with the main segments of the injection moulding cycle.

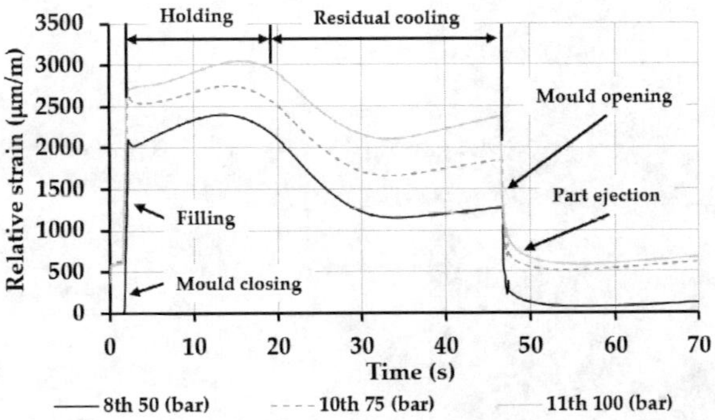

Figure 4. Relative strains near-gate at different holding pressures.

A steep increase is seen during filling but maximum is reached only in the holding phase. This is caused by viscoelastic deformation component of the epoxy-acrylate material. After holding pressure is not applied, strain decreases to a lower, near constant value in residual cooling time. Mould opening and part ejection are indicated by a steep decrease as the cavity has free space to spring back. Figure 4 also shows the effect of increasing holding pressure, as the curve corresponding to the 11th cycle (100 bar holding pressure) shows significantly higher relative strain than that of the 8th cycle (50 bar holding pressure).

Viscoelastic behaviour of mould inserts causes delay between cavity pressure and resulting strain. It can be best observed in figure 5 that shows near-gate relative strains and cavity pressure as function of time. We present near-gate relative strains in figure 5 because we also measured direct cavity pressure there. Cavity pressure maximum is almost immediately reached at the end of the filling phase, but maximal strain is reached only in the holding phase with a significant delay (~11 seconds). In figure 5 the effect of holding pressure on resulting cavity pressure is also illustrated. At 50 bar holding pressure, cavity pressure dropped to zero at 18 seconds, while at 100 bar only a change in the slope of the cavity pressure curve can be observed at 20 seconds. Due to higher holding pressure, cavity pressure can be maintained longer since pressure loss from maximal injection pressure at switchover is lower.

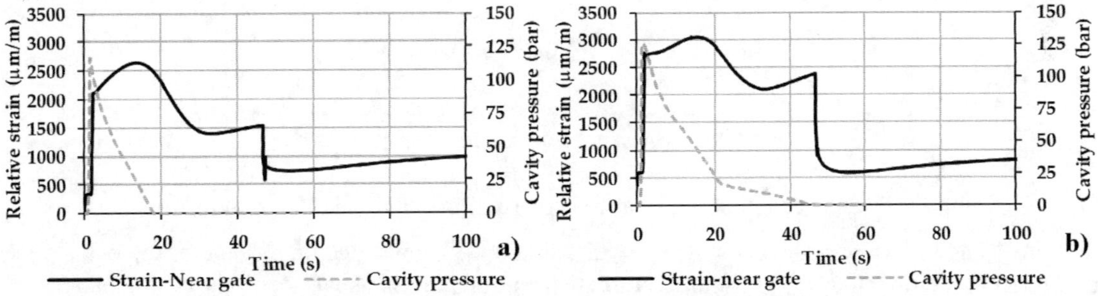

Figure 5. Measured strain results of the gauges and the cavity pressure curves in cycle 7 (50 bar holding pressure) a) and in cycle 11 (100 bar holding pressure) b).

3.2. Thermal imaging camera results

We measured surface temperature distribution of the moving side mould insert to observe cooling of the cavity in delay time. Since HDT of epoxy-acrylate material is low (shown in table 2.) it is vital to allow the cavity surface to cool down between cycles. Figure 6 shows the thermal camera image of a short shot in the 3rd cycle a) and the surface temperature-time curves of the analysed points b). In the thermal camera image, low thermal conductivity of the insert material can be clearly seen as the contour of high temperature zone is almost identical to the actual short shot product. It proves that heat is stuck under the product, and it is not dispersed in the whole cavity. Effect of low thermal conductivity can also be seen in the cooling diagram of the insert. Temperature in the points that are not touched by the melt (P4, P7 and P8) do not change significantly even if they are very close to the higher temperature zone.

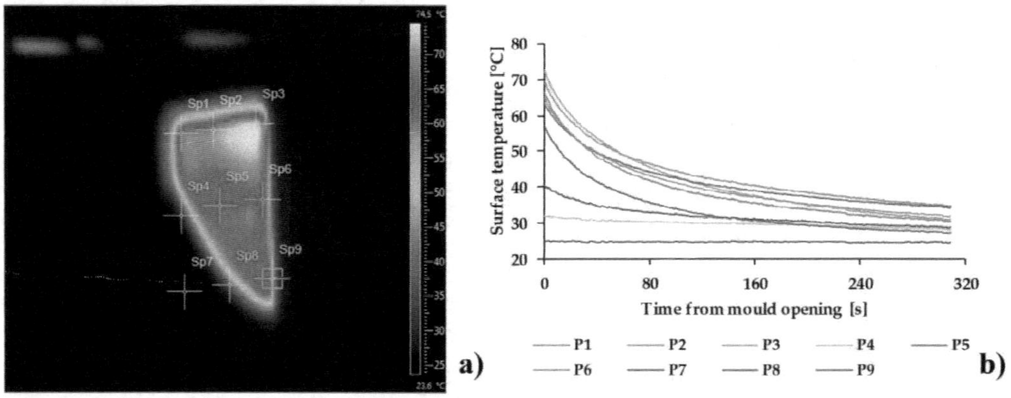

Figure 6. Surface temperature distribution of the mould insert in cycle 3 (short shot) at mould opening a) and the surface temperature-time curves of the characteristic points of the cavity.

Temperature of the cavity midpoint (P5 point in figure 6 a)) is shown as a function of time in figure 7. It can be seen that both minimal and maximal surface temperature values increase over the cycles. In the 1st and 2nd cycles, the melt only filled the runner and in the 3rd cycle it still did not reach the centre of the cavity (P5 point). From the 4th cycle on, the melt reached the centre point and gradual heating of the insert is clearly shown. In the 4th cycle maximal surface temperature (at mould opening) was 63.6 °C and minimal surface temperature (at the end of delay time) was 34.6 °C. Maximal value increased to 83.2 °C and minimal value to 46.8 °C in the 10th cycle. Exponential cooling curves can be observed between the maximal and minimal values. After the 11th cycle we kept longer delay time that resulted in lower cavity surface temperature, reaching 39.9 °C at 4021 seconds.

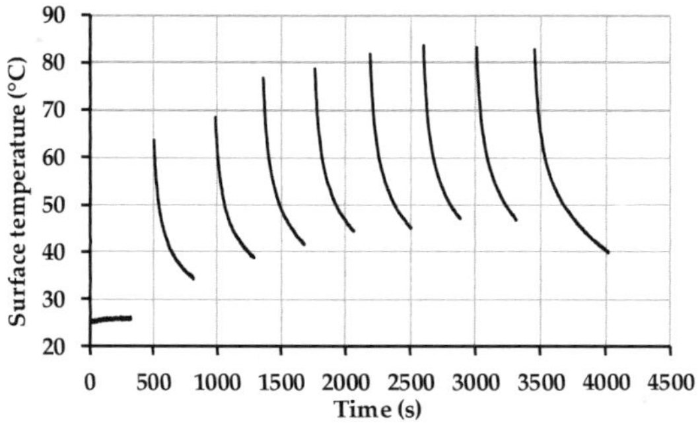

Figure 7. Surface temperature at the centre of the cavity as function of time.

Figure 8 shows cavity surface temperature distribution at mould opening a) and at the end of delay time b). At mould opening, surface temperature distribution correlates well with the filling pattern. It is clear that high temperature zone is identical to the entering location of the melt and cavity surface temperature decreases gradually along the flow length. At the start of the flow length (P3 point) surface temperature is 84 °C (at mould opening) while at the end of the flow length (P7 point) it is only 73.7 °C. Both temperature and pressure loads are higher at the start of the flow length (P3 point) therefore it is a critical part of the mould. However, at the end of delay time high temperature zone is located at the centre of the cavity (P5 point, 45 °C) as the outer surfaces of the insert tend to cool faster because they contact the steel mould housing.

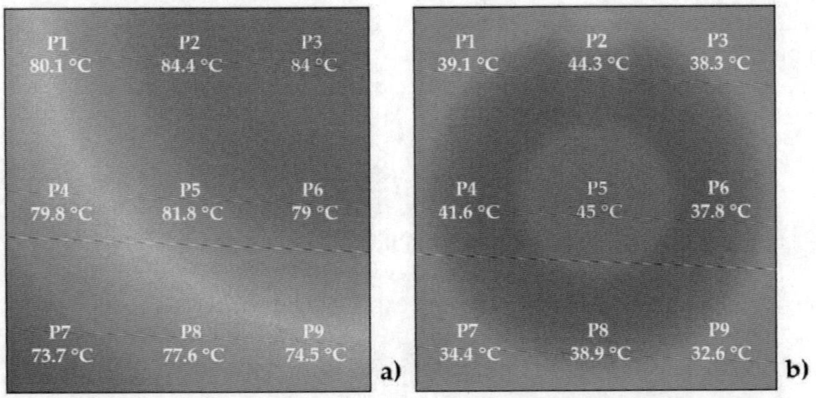

Figure 8. Surface temperature of the cavity at mould opening a) and at the end of delay time b) in the 7th cycle.

3.3. Product weight and dimension

We also measured product mass and thickness and the results are shown in figure 9. Product thickness was measured in 5 points using a *Mitutoyo Digimatic* micrometre and the average of the values with their standard deviation are shown in figure 9. The 1st and the 2nd cycles are not indicated in figure 9 as only the runner and the gate were filled. As the cavity was gradually filled in the 3rd-6th cycles both product mass and average product thickness increased from 1.21 g to 5.25 g and 1.60 mm to 1.74 mm, respectively. In the 6th cycle, complete volumetric filling was reached but no holding pressure was applied. From the 7th cycle, the effect of increasing holding pressure is shown as both product mass and thickness grew. In the 12th cycle, average thickness was closest to the nominal value of 2 mm as it reached 1.90 mm with a standard deviation of 0.15 mm. In our experiment we managed to improve product thickness and mass by applying higher holding pressure.

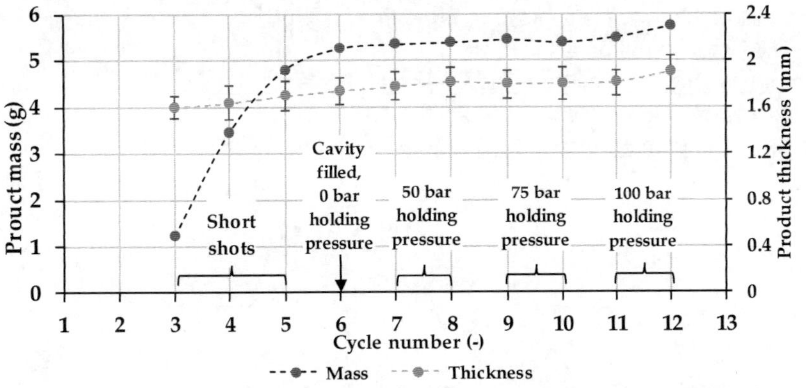

Figure 9. Product mass and average thickness.

4. Summary

In this research paper we analysed the applicability of PolyJet printed epoxy-acrylate injection mould inserts. We created a comprehensive measurement system where we measured strains, cavity pressure and surface temperature of the mould inserts. We presented strain-time curves measured by the two gauges at the back of the moving side insert. We identified the effect of increasing injected melt volume on the strain of the inserts. After reaching complete volumetric filling we also measured the effect of increasing holding pressure on strain of the insert and cavity pressure. On the strain-time curves we identified characteristic segments (mould closing, filling, holding, residual cooling time, mould opening and part ejection) of the injection moulding cycle. During our injection moulding series, we left significant delay time between cycles to allow cavity surface to cool below the epoxy-acrylate material's HDT temperature. In these delay times we measured cavity surface temperature of the inserts by thermal imaging camera and experimentally showed the effect of low thermal conductivity of the insert material. Our injection moulding series was terminated when the fixed side cavity insert failed. After injection mouldings we measured product mass and thickness and found a positive correlation with applied holding pressure.

Funding

This work was supported by the National Research, Development and Innovation Office, Hungary (2019-1.1.1-PIACI-KFI-2019-00205, 2018-1.3.1-VKE-2018-00001, OTKA FK134336, OTKA FK138501). The research reported in this paper and carried out at BME has been supported by the NRDI Fund (TKP2020 NC, Grant No. BME-NCS) based on the charter of bolster issued by the NRDI Office under the auspices of the Ministry for Innovation and Technology.

References

[1] Bagalkot A et al 2019 *Rapid Prototyping Journal* **25/9** 1493-1505
[2] León Cabezas M A et al 2017 *Procedia Manufacturing* **13** 732-737
[3] Boros R et al 2019 *Express Polymer Letters* **13/10** 889-897
[4] Lupone F et al 2021 *Express Polymer Letters* **15/2** 177-192
[5] Keresztes Z et al 2019 *Periodica Polytechnica Mechanical Engineering* **63** 195-200
[6] Kuo et al 2020 *The International Journal of Advanced Manufacturing Technology* **107** 1223-1280
[7] Park H.-S. et al 2017 *Procedia Manufacturing* **10** 48-59
[8] Zink B et al 2019 *International Communications in Heat and Mass Transfer* **108** 104297
[9] Tábi T et al 2016 *Journal of Thermal Analysis and Calorimetry* **123** 349–361
[10] Kovács N K et al 2011 *Műanyag és Gumi* **48** 269-272
[11] Mendible G A et al 2017 *Rapid Prototyping Journal* **23/2** 344-352
[12] Krizsma Sz et al 2021 *Polimerek* **7/5** 155-160
[13] Krizsma Sz et al 2021 *Additive Manufacturing* **42** 102001
[14] *Tipplen H145F* technical data sheet:
(https://www.molgroupchemicals.com/userfiles/products/68/68_tds_hu.pdf) (2022.01.10.)
[15] *VeroWhitePlus (RGD835*) technical data sheet:
(https://support.stratasys.com/en/materials/polyjet/vero-family) (2022.01.10.)

The role of friction in the sheet metal forming numerical simulation

L. A. Carvalho[1] and Z. Lukács[1]
[1] Institute of Material Science and Technology, Faculty of Mechanical Engineering and Informatics, University of Miskolc, Hungary

Abstract. This study aims to fulfil the current needs of the sheet metal forming industry by designing a practical, fast, efficient and low-cost rig to describe friction. Nowadays, the technological tool design process of sheet metal components is unthinkable without the application of finite element modelling. Despite the importance of friction for the metal forming processes, it is frequently overlooked in metal forming simulations. Many manufactures are demanding for a more accurate description of tribology. Therefore, there are some relatively simple devices, based on traditional tests that have been presenting high accuracy on friction prediction. Until now, no universal method for determining the coefficient of friction has been established. This is due to the variety of geometry on the contact between the tool and the deformed material, as well as the presence of various stress and strain states in different areas of the draw piece during different moments of the sheet metal forming. This paper presents the first friction results obtained with the new equipment built at the University of Miskolc. For the analyses it was applied 2 different velocities and 5 different pressures. The results were analysed using the analyses of variance.

1. Introduction

Nowadays, the technological tool design process of sheet metal components is unthinkable without the application of finite element modelling oriented to the results. Due to continuous improvements, the modelling software not only support the process of technological design but also enable the process analysis on the virtual production. One of the essential elements of the current study of manufacturing conditions in sheet metal forming is the numerical investigation of lubrication and friction conditions. Tribology has a critical function to play: to investigates how numerous variables interact with the sheet metal surface.

Despite the importance of friction for the metal forming processes, it is frequently overlooked in metal forming simulations. The current industry standard employs a constant coefficient of friction (Coulomb). This strategy may reduce simulation precision in order to better represent real-world processes. In order to get more realistic simulations with higher accuracy, tribology effects in metal forming simulations must be taken into account. Friction is dependent on local factors including pressure, velocity, strain, and interface temperature, in addition to not being constant in different locations and phases of the process. Some examples of approaches for simulating friction and lubrication in metal forming processes include enhanced Coulomb models, pressure and velocity dependent, and virtual tribology. [1], [2]

Many manufactures are demanding for a more accurate description of tribology. The virtual tribology was successfully applied to the inner trunk component of the Renaut Talisman. To better characterize the tribology aspects, TriboForm, AutoForm, Tata Steel and Groupe Renault, collaborated to solve this problem. They evaluated the virtual tribology model using AutoForm simulation to achieve a more accurate simulation in terms of friction and lubrication, taking into account the variation of friction conditions locally and over time. Friction has been measured as low as 0.03 in the draw bead region, where local contact pressures are quite high. In the blank holder region, on the other hand,

friction coefficients have shown values closer to $\mu = 0.13$. [3] Therefore the virtual tribology may not be the only solution, and requires a detailed description of the involved materials and lubrication.
Since the material flows over the tool's surface at the contact areas on metal forming, it is not possible to analyse a forming process without attention to friction. For this application, friction models are typically based on the Amontons-Coulomb model. [4]

Estimating friction coefficient in a sheet metal forming process is a complicated procedure. A considerable number of factors affect the friction system. Many physical experiments can be mentioned for characterizing friction aiming its application on numerical simulations. Some common examples are the hardness test (indentation test) and the modified Watts-Ford test [4], [2]. As a result, certain very simple devices have been displaying remarkable accuracy on friction prediction, based on traditional testing and dadicated for its particular forming processes.

Figure 1 [5] illustrates the most regularly used sheet metal forming simulation tests.

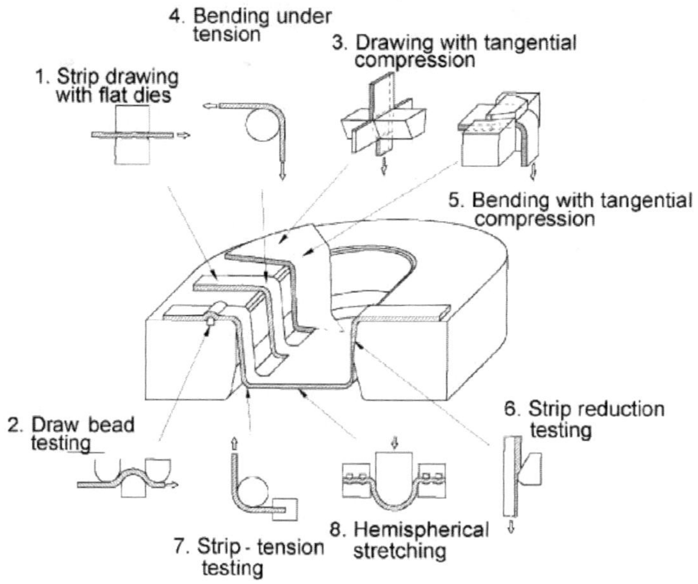

Fig. 1: Sheet metal forming: schematic illustration of simulative tests [5].

No universal method for measuring the coefficient of friction can yet be mentioned in terms of metal forming processes. The diversity of geometries on the contact between the tool and the blank along with the range of stress and strain states varying in different regions and moments on the sheet metal forming can be mentioned as some causes.

The figure 2 shows the tribometer proposed by Kirkhorn et al. based on the strip drawing test [6].

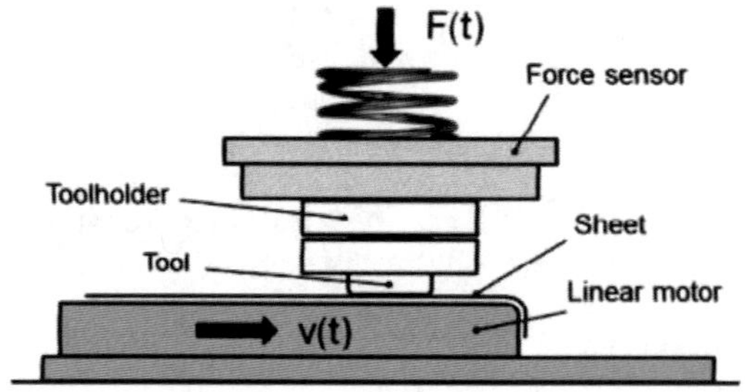

Fig. 2: Tribometer proposed by Kirkhorn et al [13].

The strip drawing test involves pressing a metal strip between the tool and a flat die while the flat die fuse moves. The generated normal load as well as the tangential force opposing the movement are constantly monitored. A load cell is be used to make the measurement, allowing friction coefficient values to be calculated. A 1000 Hz acquisition rate was employed in the research. The apparatus produced by Tavares et al [7] is shown in figure 3.

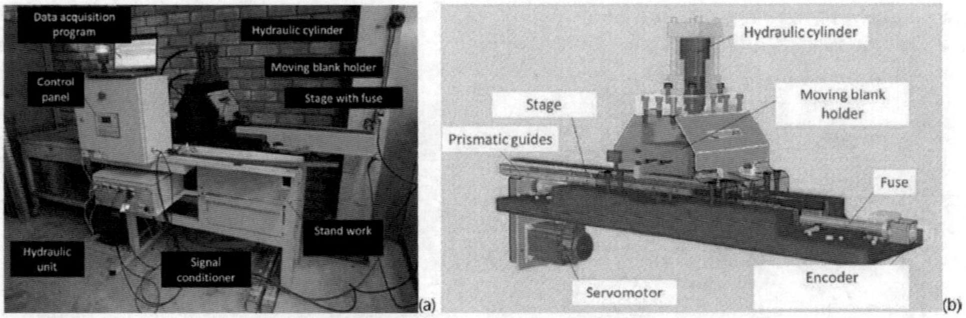

Fig. 3. Strip drawing test rig: (a) general view; (b) scheme.

On the industrial side, the new Tribometer 5100 test equipment (figure 4) is a recent Raziol® Zibulla & Sohn GmbH invention for measuring friction coefficients and analysing its parameters. This model, which is based on the flat line theorem, is dedicated to the analyses of friction parameters during sheet metal forming and also the determination of the coefficient of friction between tool and material. Some well-known clients of The Raziol laboratory lubricating system include Daimler, BMW, and Volkswagen [8].

Fig. 4: Raziol Tribometer 5100 Testing equipment for the ascertainment of friction values.

The tribometer proposed by Raziol allows the lubrication analyses by testing substrates with dimensions of up to 297 mm x 210 mm (DIN A4), along with sheet metal strips with dimensions of 50 mm x 650 mm. The application of various lubricants is made by a relatively fast and simple change of medium. To analyse the results, an applied computer program provides oriented data acquisition, recordings including video and audio, a database system (with the measured data and data management) and also an online diagram representation [8].

This paper presents the construction of a physical friction measurement device based on the previous literature and the application of the results, combined with the enhanced coulomb models, on the metal forming simulations on AutoForm. The aim of the new equipment is to contribute with practical, fast, and low-cost ways to describe friction. By this way, improving the result of the metal forming simulation with better description of the interaction between tool, lubrication, and metal sheet for the numerical simulation.

2. Materials and Methods

2.1. Physical measurement device

Based on the Strip Drawing test and the mentioned literature examples the friction measurement device built at the Laboratory of the Institute of Material Science and Technology at the University of Miskolc is shown on figure 5.

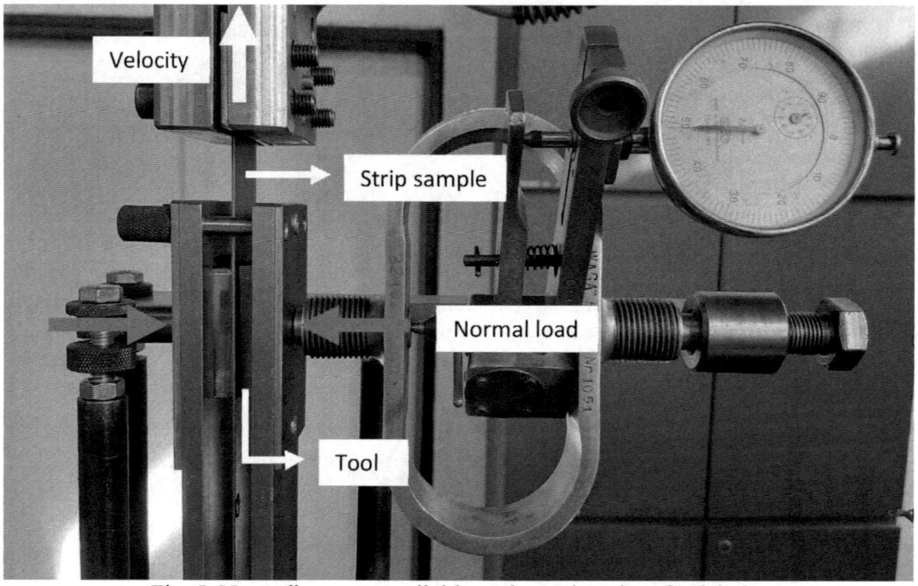

Fig. 5: New tribometer available at the University of Miskolc.

The calculation of the friction coefficient is based on the well-known Amontos-Coulomb friction model, equation 1. With the new rig assembled on the Universal Testing Machine two measurements are made. First, the measurement of controlled pressure between the tools. Secondly, the applied force controlled by the Universal Testing Machine. After the measurements the normal force is calculated using the contact area between tool and metal strip and then applied to equation 1.

$$\mu = \frac{F_R}{F_N} \tag{1}$$

Available friction models for the simulation on AutoForm includes: Constant Coulomb friction; Directional Dependent friction; Pressure Dependent friction; Velocity Dependent friction; and the TriboForm friction model. For the current study both enhanced coulomb models, Pressure and Velocity Dependent, will be discussed.

2.2. Pressure dependent Friction

The equation 2 expresses the pressure dependent enhanced Coulomb model [6], where the p_{ref} is the reference pressure and e is pressure exponent.

$$\mu_{eff}=\mu\left(\frac{p}{p_{ref}}\right)^{(e-1)} \tag{2}$$

2.3. Velocity Dependent Friction

Equation 3 is an enhanced Coulomb model to calculate the effective coefficient of friction, μ_{eff}. [9]

$$\mu_{eff}=\mu-a\ ln\ (\frac{max\ v_{rel},v_{ref}}{v_{ref}}) \quad (3)$$

If the relative velocity is lower than the reference velocity, the effective friction coefficient is equal to the base friction coefficient, as expressed on the equation 4. This equation is a combination of both enhanced models previously mentioned.

$$\mu_{eff}=\mu(\frac{p}{p_{ref}})^{(e-1)}-a\ ln\ (\frac{max\ v_{rel},v_{ref}}{v_{ref}}) \quad (4)$$

2.4. Design of Experiments

The table 1 shows the analysed variables followed by the respective levels and values. The software Minitab has been used for the analyses of variance (ANOVA).

Table 1. Factor description.

Factor	Levels	Values
Velocity (mm/s)	2	10; 20
Pressure (MPa)	5	1; 2; 3; 4; 5

The tool was made of a molybdenum, chromium, vanadium alloy also known as a Molybdenum High Speed Tool Steel. A common austenitic chromium-nickel stainless steel was utilized for the strip sample. It has outstanding drawing characteristics and formability, as well as being highly corrosion resistant.

3. Results

The figure 6 shows the experimental results for friction coefficient. It represents the friction results with normal pressure equal to 4 MPa and the velocity equal to 20 mm/s. Both static and kinetic coefficient of friction can be observed. Therefore, for the current study the analyses will be made using the kinetic friction coefficient. To calculate the kinetic friction coefficient a simple mean was taken from the middle area of the chart disposing 10% on the extremities.

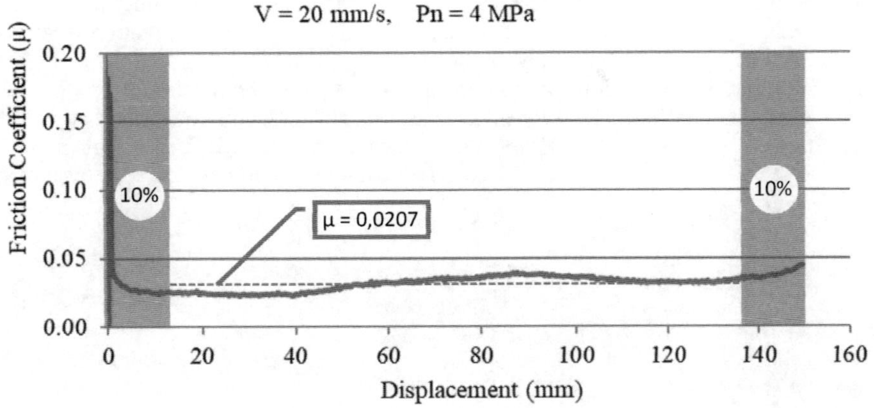

Fig. 6: Friction coefficient.

The main effect chart on the figure 7 represents the mean calculation of all experiments with separated results for both input variables, velocity and pressure.

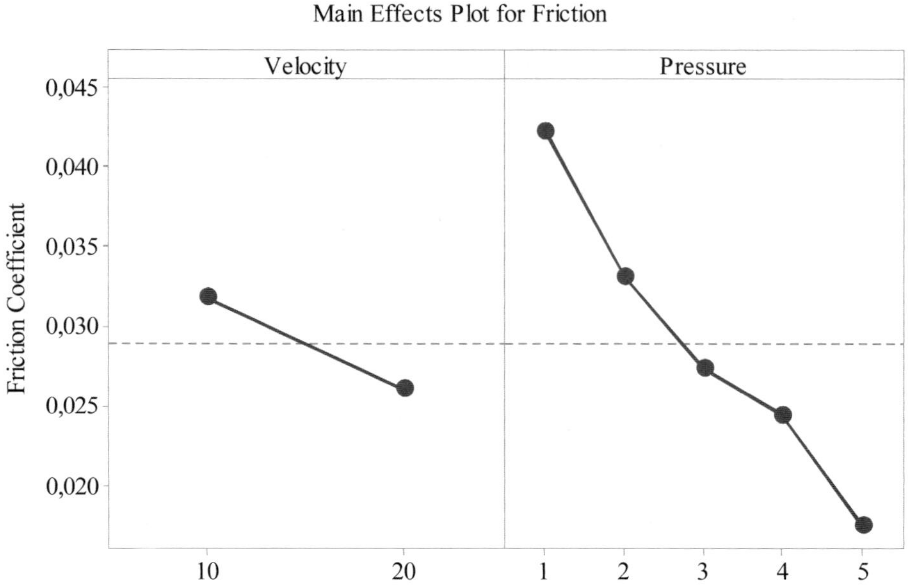

Fig. 7: Main effect chart for friction.

The results shown on figure 6 validate the efficiency of the new tribometer. The friction values obtained with the rig follow the tendencies predicted by the enhanced Coulomb models (sections 2.2, 2.3 and 2.4). The velocity dependent friction model accounts for the decrease in friction coefficient as the relative velocity between the tool and the sheet increases as described on equation 3.

One of the most significant components that determines the coefficient of friction is contact pressure, which may be evaluated at both macro and micro scales. The topography of the in-contact surface flattens out as the contact pressure rises. The contact geometry changes as a consequence of these factors, resulting in a change in the coefficient of friction. This phenomenon is described on the equation 2. The contact pressure put to the sheet rises as the flange area reduces and the material flows to the cavity of the die. In sheet metal forming, this is a crucial factor to consider. An heterogeneous contact pressure distribution can be detected because the thickness of the sheet may vary differently in different places of the sheet during the flow of the material. [9]–[11]

The analyses of the interaction plot on figure 8 allows to observe that the increase in pressure has influenced the friction coefficient by reducing its values. This assumption is likewise valid for both tested velocities. Regarded to the velocity, the expectations predicted by the enhanced Coulomb equations were also fulfilled. It can be noticed a reduction of friction coefficients by increasing the velocity.

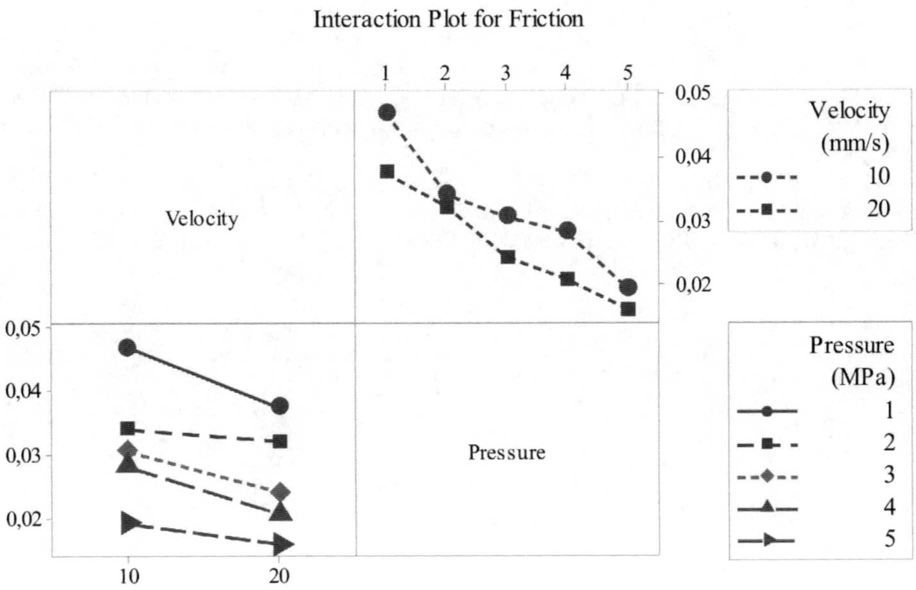

Fig. 8: Interaction plot of the analysed parameters.

4. Conclusions

The results obtained with the new tribometer have sustained the efficiency of the physical measurement of friction using the new tribometer. Velocity and pressure have shown significant difference according to the analyses of variance. The higher velocity has resulted in a decreasing of friction coefficient. An increase on pressure had an effect on the friction coefficient by decreasing friction. This paper is the beginning of a series of studies that aim to investigate not only pressure and velocity but also different materials and lubricants. The final step will also include the definition of a series of friction parameters for the numerical simulation.

5. References

[1] M Sigvant *et al.*, 'Friction in Sheet Metal Forming Simulations: Modelling of New Sheet Metal Coatings and Lubricants', in *International Deep Drawing Research Group 37th Annual Conference*, 2018, doi: 10.1088/1757-899X/418/1/012093.

[2] M. Tisza, 'A general overview of tribology of sheet metal forming', *J. Technol. Plast.*, vol. 6, pp. 11–25, 2001.

[3] J. Lacues *et al.*, 'Friction and lubrication in sheet metal forming simulations: Application to the Renault Talisman trunk lid inner part', in *International Deep Drawing Research Group 38th Annual Conference*, 2019, doi: 10.1088/1757-899X/651/1/012001.

[4] L. Figueiredo, A. Ramalho, M. C. Oliveira, and L. F. Menezes, 'Experimental study of friction in sheet metal forming', *Wear*, vol. 271, no. 9–10, pp. 1651–1657, 2011, doi: 10.1016/j.wear.2011.02.020.

[5] N. Bay, D. D. Olsson, and J. L. Andreasen, 'Lubricant test methods for sheet metal forming', *Tribol. Int.*, vol. 41, no. 9–10, pp. 844–853, 2008, doi: 10.1016/j.triboint.2007.11.017.

[6] L. Kirkhorn, K. Frogner, M. Andersson, and J. E. Ståhl, 'Improved tribotesting for sheet metal forming', *Procedia CIRP*, vol. 3, no. 1, pp. 507–512, 2012, doi: 10.1016/j.procir.2012.07.087.

[7] A. F. Tavares, A. P. O. Lopes, E. A. Mesquita, D. T. Almeida, J. H. C. Souza, and H. L. Costa, 'Effect of transfer layers on friction and wear mechanisms in strip drawing tests of commercially coated forming tools', *Wear*, no. February, p. 203733, 2021, doi: 10.1016/j.wear.2021.203733.

[8] R. Z. & S. GmbH, 'Raziol® Zibulla & Sohn GmbH', 2021. [Online]. Available: https://www.raziol.com/en/company/tribology. [Accessed: 02-Jun-2021].

[9] AutoForm Engineering GmbH, 'Friction - Software Manual', 2020. [Online]. Available: file:///C:/Program Files/AutoForm/AFplus/R8F/help/EN/Friction.html 3/4. [Accessed: 19-Nov-2020].

[10] I. Gil, J. Mendiguren, L. Galdos, E. Mugarra, and E. Saenz de Argandoña, 'Influence of the pressure dependent coefficient of friction on deep drawing springback predictions', *Tribol. Int.*, vol. 103, pp. 266–273, 2016, doi: 10.1016/j.triboint.2016.07.004.

[11] Y. S. Kim, M. K. Jain, and D. R. Metzger, 'Determination of pressure-dependent friction coefficient from draw-bend test and its application to cup drawing', *Int. J. Mach. Tools Manuf.*, vol. 56, pp. 69–78, 2012, doi: 10.1016/j.ijmachtools.2011.12.011.

IOP Conf. Series: Materials Science and Engineering 1246 (2022) 012022 doi:10.1088/1757-899X/1246/1/012022

Analysis of wear surfaces on parts of revisionally removed implants

I Nemes-Károly[1], Dr. Gy Kocsis[2], Dr. G Szebényi[1,3]

[1] Department of Polymer Engineering, Faculty of Mechanical Engineering, Budapest University of Technology and Economics, H-1111 Budapest, Műegyetem rkp. 3., Hungary

[2] Department of Orthopaedics, Faculty of Medicine, Semmelweis University, H-1085 Budapest, Üllői út 26., Hungary

[3] MTA-BME Lendület Lightweight Polymer Composites Research Group, H-1111 Budapest, Műegyetem rkp. 3., Hungary

E-mail: szebenyi@pt.bme.hu

Abstract. Nowadays, with increasing average age - and thus an increase in active years - and the heightened pace of life in our cruel, fast-paced world, our bodies are subjected to extreme stress for ever-longer periods. Unfortunately, this lifestyle does not go away without a trace - it can cause damage to everything from our skeletal system to our nervous system, and our joints are no exception - which is why arthroplasty procedures are becoming increasingly popular. Arthroplasty is one of the most successful surgical procedures, which seeks to re-condition, replace - in some cases remove - the joint surfaces to improve the function of the system or at least preserve its current status. The most common form of arthroplasty is the implantation of joint endoprostheses. These devices have a finite lifespan, like all engineering structures, so their replacement at certain intervals, by means of revision surgery, is essential. The prostheses removed in this way carry a wealth of essential information about their design, their manufacturer and even their user and implantor, which can be incorporated into the design of the next generation of systems using reverse engineering, thus moving closer to the development of the "gold standard" implant. The topic of our article is the analysis and exploration of the failure processes of hip, knee, and shoulder implants removed by revision. We will build a model that adequately represents the wear and tear processes of different types of prostheses, with the help of which the wear and tear processes can be determined and predicted, and the strengths and weaknesses of the constructs can be determined. We consider it essential to distinguish between the different types of damage since, without a comprehensive knowledge of their extent, it is impossible to prevent them effectively. Systems grouped according to implant type and implant modality were examined using 3-dimensional imaging. We determined the degradation of the polymer components caused by use, observing to distinguish between damage caused by revision surgery and damage caused by transportation and storage. During our measurements, particular care was taken to identify the exact extent of creep and wear in order to make the best possible recommendations for modifications to existing systems.

1. Introduction

The subject of our research is to investigate and evaluate the wear process of human hip, knee and shoulder joint prostheses and, in the end, to construct a model that properly represents the wear and failure cycle of the chosen component. Subsequently, the theses are refined and generalised to the different types of prosthesis groups studied, thus highlighting the strengths and possible shortcomings, weaknesses and failures of the implants currently used and analysed. Particular care has been taken to determine the extent of wear and deformation, as these have not been studied separately and in detail before, and their effects have been considered together [1]. Inserts of prostheses were examined using a 3-dimensional imaging technique, and the resulting point clouds were compared with CAD models of the parts using a program. This software plots the differences between the original prostheses and the prostheses that have been removed by revision, similar to the contour lines on a map. This makes it possible to visualize the areas most exposed to wear and tear in a very visual way. These zones were then examined in more detail for characteristic lesions.

2. Materials and methods

In today's most commonly used implant systems, the artificial head is made of a metal alloy (cobalt-chromium alloy, or $TiAl_6V_4$), while the counterpart, the insert (cup), is made of ultra-high molecular weight polyethylene (UHMWPE).

We investigated the inserts, so we will speak first of all about ultra-high molecular weight polyethylene. UHMWPE is a thermoplastic semi-crystalline polymer, which owes its remarkable physical and mechanical properties to its extremely long molecular chains. From our point of view, its most important properties are its high resistance to abrasion, its outstanding impact and compressive strength and toughness, and its chemical resistance to various environments. It is classified as a biocompatible material, so it is likely to be accepted by the body when implanted, where it does not trigger a strong immune response. In addition to its numerous advantages, the relatively long molecular chain also has its disadvantages, as UHMWPE is rather difficult to process [1].

In terms of fixation, a distinction is made between cemented and cementless implants. In the case of cemented implants, the component is fixed with two-component poly(methyl methacrylate) (PMMA) [1]. In this case, stability occurs immediately after the procedure. There is no need for a long rehabilitation period, and drugs can be mixed into the bone cement to achieve a high local concentration [2-4]. In the case of cementless, the recovery is much longer, as it takes longer to achieve secondary stability - osteointegration - as the patient's bone tissue grows on the implant and fixes it by bone ingrowth [1-4].

The measurements were carried out using the ATOS Core 5M measuring system from GOM company. The GOM 3D scanner allows contactless 3-dimensional measurements. The device records the measurement data as a so-called point cloud in a special computer system and generates a model that is displayed on the computer screen and visible to us. table 1 shows the technical parameters of the GOM 3D scanner [5]. As shown in the table below, table 1 shown the technical parameters of the GOM 3D scanner.

Table 1. The technical parameters of the GOM 3D scanner [5].

System configurations	ATOS Core 5M
Measuring range	45 x 45 mm
Measurement distance	170 mm
Measurement point distance	0,02 – 0,03 mm
Sensor dimensions	206 x 205 x 64 mm
Sensor weight	2,1 kg
Supply	90 – 230 V (AC)
Operation temperature	5 – 40 °C

In the evaluation of our studies, we used the so-called Hood's theory, which allows comparison between implants and ordering lesions into classes according to severity and region. This method divides lesions into 7 groups:

- Pitting: Craters and holes 2-3 mm diameter and 1-2 mm deep on the working surface;
- Embedded debris: May come from metal parts, bone chips, or any other debris;
- Scratching: One-dimensional surface error, can be caused by microscopic damage to metal parts, scratches, and debris caught on working areas;
- Delamination: One of the most severe forms of damage, the product falls apart into layers, usually requiring immediate revision surgery, coupled with large amounts of material loss, in severe cases leading to catastrophic outcomes;
- Surface deformation: Cause no loss of material, therefore not considered severe damage, the contact surfaces suffer deformities;
- Burnishing: Burnishing causes the surface to burnish and polish;
- Abrasion: Causes bruising of working surfaces [1].

3. Measurements

In our studies, we had 14 anatomically designed shoulder implants with cementless fixation, 3 hip prostheses with cementless and 3 hip implants with cemented fixation, and 3 total knee systems with cementless fixation. We investigated UHMWPE-based inserts of implants removed during revision surgery using a 3-dimensional scanner.

In the figures, the regions which are covered with warm colors indicate differences in the positive direction from the reference surface, meaning that the part under investigation is beyond the reference. The surfaces covered with cool colours have the opposite meaning in these measurements, i.e. the difference is in the negative direction, so the surface element under test is within the reference CAD model. The green regions represent a perfect fit between the reference model and the working surface. The more pronounced a colour tone, the more significant difference in distance between the two surfaces.

Point corrosion, scratching, and abrasion was observed on all sample surfaces. Also very common was the presence of embedded debris, various cracks and surface deformation, and in some samples, layer-by-layer delamination and abrasion.

3.1. *Results of shoulder implants measurements*

In general, the bottom of the running surfaces of the samples and the corner of the inserts were damaged, as discoloration was detectable in these regions. It is interesting to note that in one area, between the two regions mentioned earlier, all the shoulder implants are intact, as a green band is present in this area. It is important to notice that the lower regions of the ventral sections are abraded - blue regions - while the upper sections are convoluted - yellow and red zones - as these are warm colours. The curling at the edges is due to deformation caused by residual deformation. Thus, the surface along the green regions - which I had previously described as intact - is not intact either, only there the processes responsible for the negative and positive deformations are in equilibrium. It can be observed that the insert - blue - areas are shifted depending on the implantation position of the whole system. Thus, I divided the shoulder implants and the process of failure of shoulder prosthesis into two different groups.

The first is the ring-shaped deformation: the lower part of the contact surfaces is damaged, as a cold discolouration can be detected in these regions, these surfaces are abraded. There is a green band along with one annular band, indicating a perfect fit, where the negative and positive deformations are aligned. This ring is present depending on the degree of damage. The edges are curled up - warm coloured areas - due to deformation caused by residual deformation. As shown in the figure below, figure 1 shown shoulder implants with ring-shaped deformation.

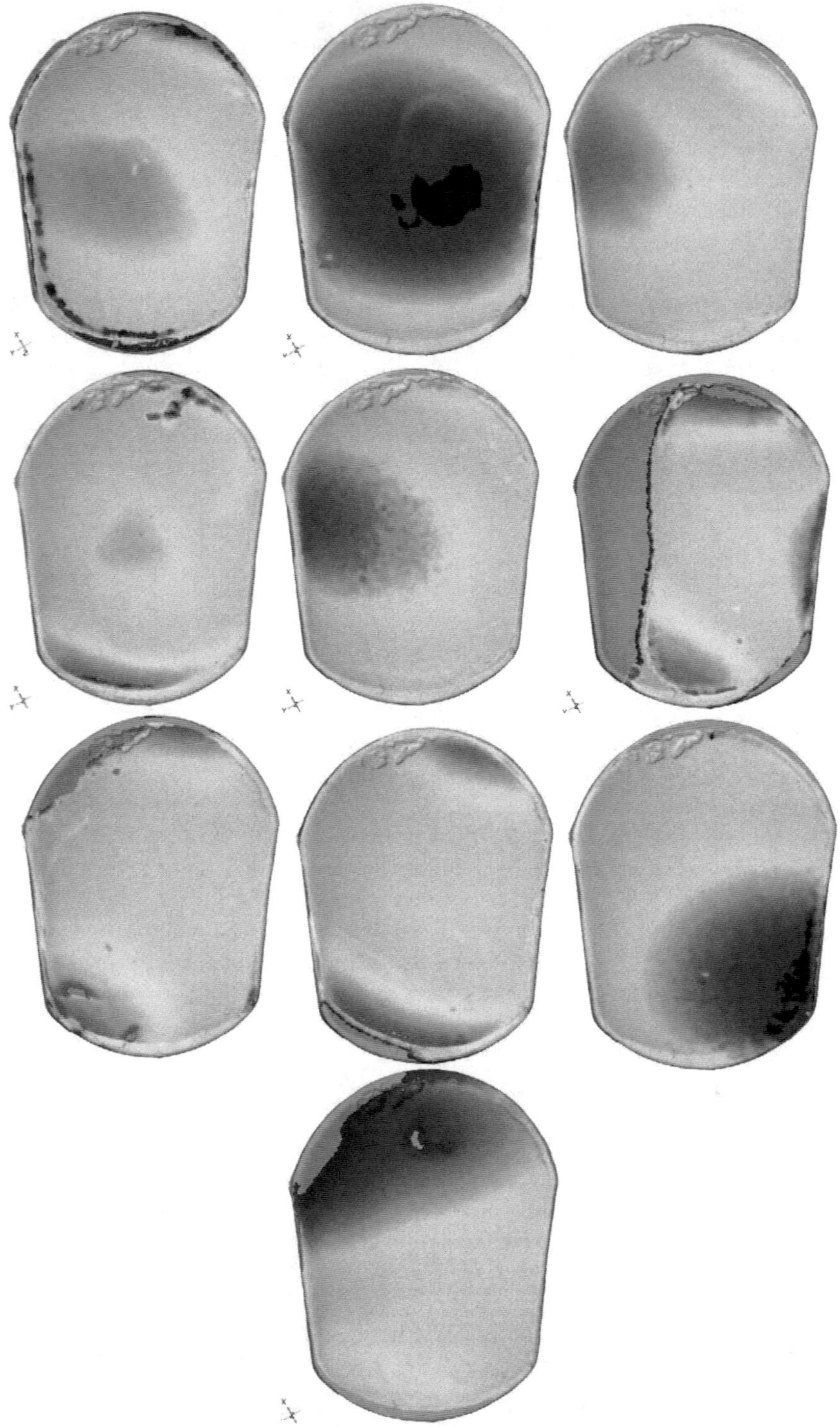

Figure 1. Shoulder implant inserts with ring-shaped deformation. The regions which are covered with warm colors indicate differences in the positive direction from the reference surface. Cool colours have the opposite meaning the difference is in the negative direction. The green regions represent a perfect fit.

The second type of failure is the band-shaped deformation: this type of failure is very similar to the previous one, but the shape and location of the damaged zones are different, the edges are also found to be curled up, and the contact surface has undergone a negative deformation, but in this case, the phenomenon is banding rather than circular. Furthermore, it is important to note that the negative deformations - blue areas - mainly affected the two corners of the system. As shown in the figure below, figure 2 shown shoulder implants with band-shaped deformation.

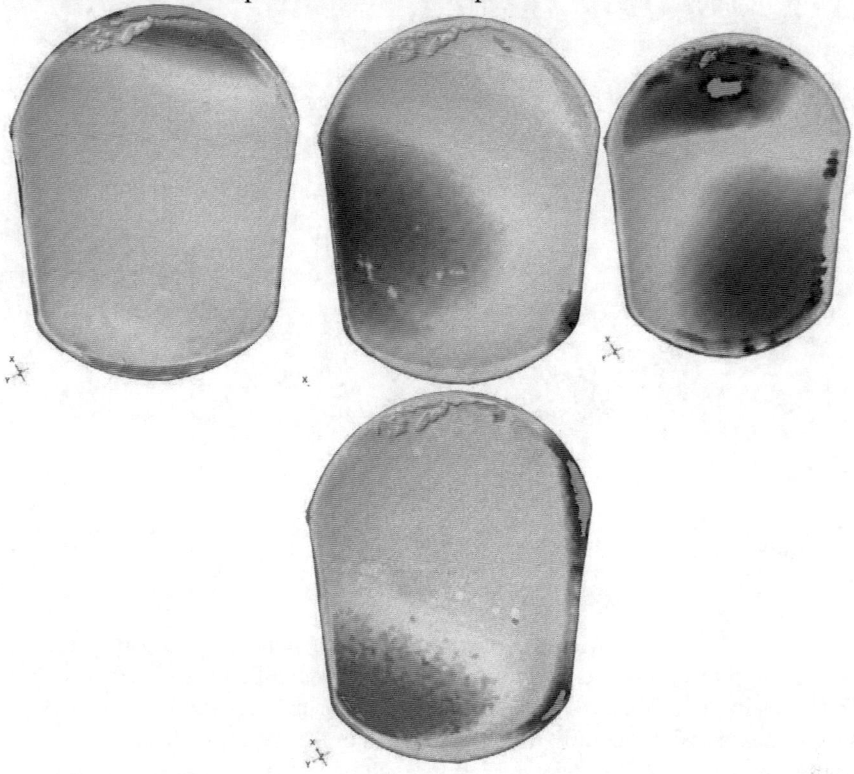

Figure 2. Shoulder implant inserts with band-shaped deformation. The regions which are covered with warm colors indicate differences in the positive direction from the reference surface. The surfaces covered with cool colours have the opposite meaning the difference is in the negative direction. The green regions represent a perfect fit between the reference model and the working surface.

For the inserts we examined, ring-shaped deformation was the most common. There were also some inserts that required immediate revision due to puncture - grey areas - in these cases, the point cloud distance from the CAD model was outside the measurement range. Furthermore, in some samples, severe delamination, as well as cracks and resulting debris, required immediate revision surgery.

3.2. *Results of hip prosthesis measurements*

Hip implants were tested in two groups, those without cement fixation and those with cement fixation. As shown in the figure below, figure 3 shown measurement results for hip implants, with cementless fixation systems in the top row and cemented fixation inserts in the bottom row.

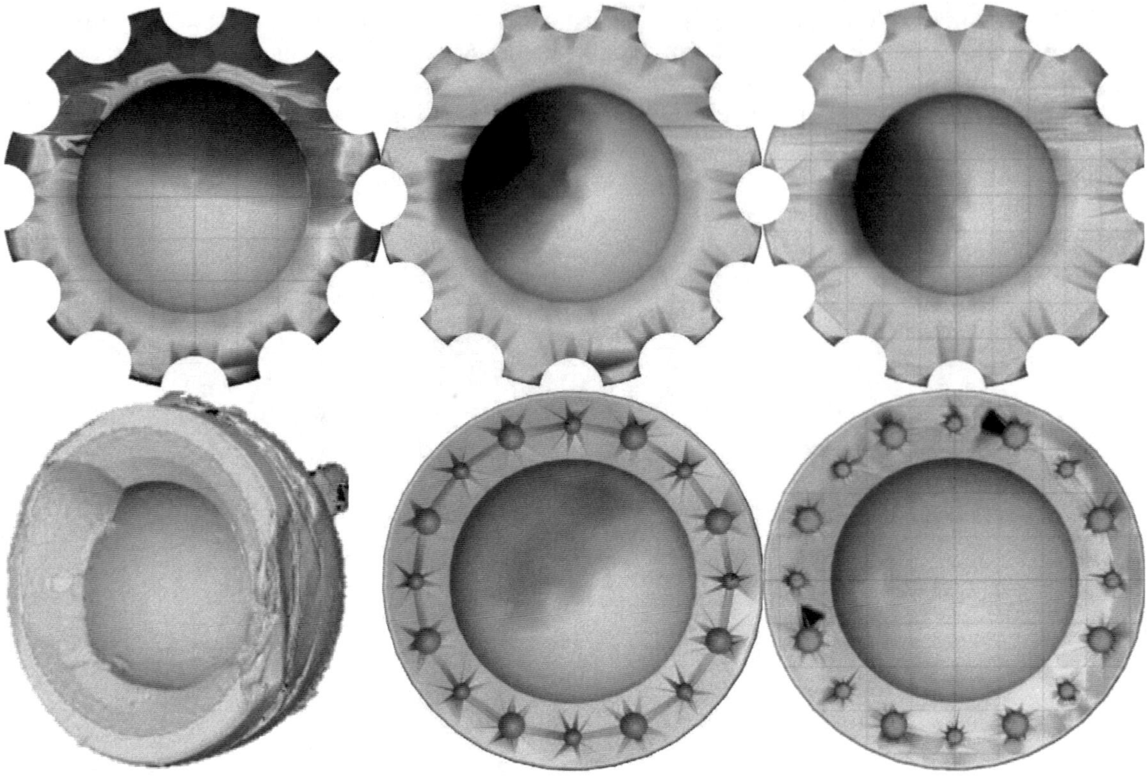

Figure 3. Measurement results for hip implant inserts, with cementless fixation systems in the top row and cemented fixation inserts in the bottom row. The regions which are covered with warm colors indicate differences in the positive direction from the reference surface. The surfaces covered with cool colours have the opposite meaning the difference is in the negative direction. The green regions represent a perfect fit between the reference model and the working surface.

It can be seen that the damage was concentrated around a single point, the location of which was much more favourable in the case of cementless fixation. In the case of cemented fixation, the surgeon's dexterity is a significant factor, and this is clearly shown in the first cemented fixation sample, where the artificial neck is embedded in the insert.

In cemented cases, revision surgery causes much greater damage, with more significant bone loss. In many cases, only quite invasive methods are available to remove the implant from its site.

3.3. Results of knee implants measurements

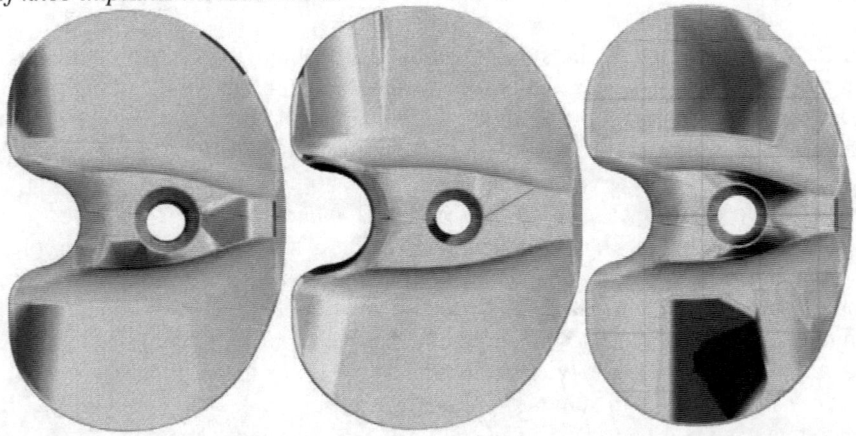

Figure 4. Measurement results for knee implant inserts. The regions which are covered with warm colors indicate differences in the positive direction from the reference surface. The surfaces covered with cool colours have the opposite meaning the difference is in the negative direction. The green regions represent a perfect fit between the reference model and the working surface.

As shown in the figure previous, figure 4 shown measurement results for knee implants. Overall, the knee implants we examined were extremely difficult to position, and in practice, none of the samples were perfect. Furthermore, it can be observed that the contact surface in the vicinity of the indentations are wrinkled due to deformation caused by residual deformations. This may be one explanation for the frequent transverse striations on the working surfaces, which cause a longitudinal undulation that is likely to increase during operation. It is also important to note that in virtually all cases, the central spine has become so worn and deformed that removal of the screw securing the chuck to the tray from the bore has become extremely complicated. In all samples, severe scratching, grinding, and pitting corrosion of the running surfaces was observed, in many cases accompanied by surface deformation. Crushing of the working surfaces was common, and in several cases, delamination occurred, in one case so severe as to require an immediate overhaul of the structure. In two samples, the presence of cracks running through the entire cross-section was observed, which could also result in havoc-like failure.

4. Results and discussion

Shoulder implants may show ring-shaped deformation and band-shaped deformation destruction modes and may be difficult to position. In our opinion, ring-shape deformation is more common than band-shape deformation. Mass measurements have shown that there is no clear separation between wear and deformation, becaue too many other forms of damage with loss of material are present.

For hip replacements, cementless fixation is much more positionable. In these systems, wear is more prevalent based on the lessons learned from mass measurements.

Total hip implants are difficult to position, with deformation dominating over wear in these artificial joints.

5. Conclusion

Mass measurements have shown that there is no clear separation between wear and deformation, because too many other forms of damage with loss of material are present.

For hip replacements, cementless fixation is much more positionable. In these systems, wear is more prevalent based on the lessons learned from mass measurements.

Total hip implants are difficult to position, with deformation dominating over wear in these artificial joints.

Overall, the positioning of all the systems studied is problematic, with practically all the inserts studied showing non-symmetrical wear, particularly shoulder implants and knee replacements. In the case of shoulder implants and hip replacements, there was a tendency for the edges of the systems to show increased damage, with severe deformation and wear. Furthermore, in the case of the working surfaces, some areas were very severely damaged, while others remained completely intact. This tendency was most striking in the case of hip replacements, where the damage was concentrated around a single point, leaving the rest of the surface intact. It was also observed that hip replacements with cemented fixation showed a much more uneven wear pattern and greater bone loss with this fixation than their counterparts without cemented fixation. However, a clear positive aspect of hip implants is that their design means that they do not suffer as much edge damage as the other two groups studied. They are also orders of magnitude easier to position than their counterparts. Overall, it is not by chance that they can be considered the gold standard among prostheses.

The lesson that can be drawn from these studies is that the method can be used to describe the failure process of damaged implants. So, in conclusion, it can be stated that further measurements are important and the data obtained can provide valuable information for reverse engineering.

6. References

[1] Kurtz S M 2016 *Biomaterials handbook* (Oxford: Elsevier)

[2] Szentágothai J 1975 *Az ember anatómiája, fejlődéstana, szövettana és tájanatómiája I. kötet* (Budapest: Medicina kiadó)

[3] Szendrői M 2006 *Ortopédia* (Budapest: Semmelweis kiadó)

[4] Grammont P M, Phadnis J, Huang T, Watts A, Krishan J, Bain G I 2016 *Cemented or cementless humeral fixation in reverse total shoulder arthroplasty? A systematic review.* (The Bone & Joint Journal, **98-B**, 65-74.)

[5] https://www.gom.com/ 2022.01.27.

Acknowledgments

The research was supported by the National Research Fund for Institutional Excellence in Higher Education (BME FIKP-BIO) and the National Bionics Programme. This work was supported by the CELSA Research Fund. The research was supported by the NRDI Office (OTKA K 138472).

IOP Conf. Series: Materials Science and Engineering 1246 (2022) 012023 doi:10.1088/1757-899X/1246/1/012023

Development of drawn polypropylene-based blends and composites

L J Varga[1] and T Bárány[1*]

[1]Budapest University of Technology and Economics, Faculty of Mechanical Engineering, Department of Polymer Engineering, H-1111 Budapest, Műegyetem rkp. 3., Hungary
[*]Author to whom correspondence should be addressed.

Abstract. We developed highly oriented polypropylene tapes with solid-state drawing, and studied the effect of additives—two grades of talc, low-density polyethylene (LDPE), and an amorphous poly-alpha-olefin (APAO). The PP compounds were prepared in a co-rotating twin-screw extruder, and sheet films were extruded. The specimens cut out from the sheet films were drawn in a tensile testing machine, and the drawing temperature of 130 °C was maintained by the heating chamber of the tensile testing machine. A draw ratio of λ=10 was set. We investigated the work of drawing and prepared DSC and tensile tests. Both types of talc acted as strong nucleating agents and increased the tensile strength and modulus of the tapes. APAO reduced the crystalline melting enthalpy only to a small extent. Both APAO and LDPE reduced the tensile strength and modulus and increased the strain at break of the tapes.

1. Introduction

Polypropylene (PP) is one of the most widely used polymers in the world, mainly due to its low price. However, given that it is a commodity plastic, PP has modest mechanical properties. These properties can be greatly enhanced by drawing, as during the drawing process, the PP molecules align to the drawing direction; thus, tensile strength and modulus can be increased by an order of magnitude in the drawing direction, at the cost of reducing strain at break [1].

Polymers can be drawn from the melt (or solution) or in the solid state. During the latter, polymers are first extruded to the desired shape (filament or sheet film), and after solidification, they are drawn between rolls, usually at an elevated temperature but always below the melting temperature of the given polymer [2]. Polypropylene is a semi-crystalline polymer, which contains amorphous and crystalline phases. The molecules in the crystalline phase, originally folded in lamellae, start to move upon drawing and form fibril-like structures while going through micro-necking, as Peterlin described [3]. The molecules in the amorphous phase start to align upon drawing in a more or less parallel manner. Perfect alignment, however, is prevented by the tie molecules, which go through both the amorphous and the crystalline phases. These molecules tighten upon drawing, and eventually, they become taut tie molecules, considerably reducing the capability of the relative movement of the phases, thus they reduce the achievable draw ratio [3].

Additives can be used for two main reasons in the production of drawn PP tapes: (i) to reduce the price of the product with the addition of a cheaper additive, or (ii) to specifically improve at least one property of the product. These properties can be mechanical properties, dyeability, maximal draw ratio, melt viscosity, etc. [4].

IOP Conf. Series: Materials Science and Engineering 1246 (2022) 012023 doi:10.1088/1757-899X/1246/1/012023

Although both polyethylene (PE) and polypropylene are polyolefins, they are insoluble in each other, and they crystallize in a different lattice, which is orthorhombic for PE, and usually monoclinic (α-PP) or hexagonal (β-PP) for polypropylene [5]. PP can, however, facilitate epitaxial crystallization of the polyethylene phase of drawn PP/PE blends [6]. Marcincin et al. investigated drawn blends of PP and low-density polyethylene (LDPE) [7]. They observed that low (below 20 wt%) LDPE content caused an increment in mechanical properties. They attributed the increment to the morphology of the blend, as the LDPE phase melted during the drawing process, and long, fibre-like polyethylene phases were formed upon cooling. The shape of these phases caused a geometrical constraint to crystal growth, thus forming an orientation and strengthening the drawn tape [7]. Similar results were reported for PP/HDPE (high-density polyethylene) blends [8].

Amorphous poly-alpha-olefins (APAOs) were developed to substitute atactic polypropylene (aPP), which originally was a by-product of the production of isotactic polypropylene. They mostly consist of propylene repeating units, residing in an atactic manner, similar to aPP, and they are used as melt adhesives and seals [9]. Although in PP/APAO blends APAOs are almost exclusively atactic and they only occur in the amorphous phase of the PP, the relatively small content of APAOs reportedly increased the crystalline ratio of PP. This was attributed to the fact that small APAO molecules can facilitate the molecular movements of the PP, thus making it easier for the PP molecules to be arranged and form crystalline phases [10, 11].

Talc is a naturally occurring magnesium silicate and a widely used additive in the polymer industry. The incorporation of talc particles reportedly improved tensile strength and modulus at the cost of reducing strain at break. Talc is also an efficient nucleating agent in PP [12, 13].

The main objective of this research is to investigate the effect of LDPE, APAO, and talc on the drawability of PP and the properties of drawn polypropylene tapes. These results can later be a basis of the investigation of a combination of additives, with which the tensile strength and modulus of the drawn PP tapes can be increased without a significant decrement in the strain at break.

2. Materials and methods

2.1. materials

We used Tipplen H681F (MOL Petrolkémia Zrt., Tiszaújváros, Hungary) polypropylene homopolymer as matrix material. This polypropylene grade has a Melt Flow Index (MFI) of 1.7 g/10 min (measured at 230 °C with 2.16 kg).

To improve the properties of the drawn polypropylene tapes, we used the additives listed below:

- HTP05 and HTP1 talc grades (IMI Fabi S.p.A., Postalesio, Italy). Both HTP05 and HTP1 are uncoated, talcs in powder form, and their average particle size is 1.4 and 1.8 μm, respectively, and 99% of their particles are smaller than 10 and 11 μm, respectively. The density of both talc grades is 2.8 g/cm^3.
- Tipolen MF 243-51 (hereinafter referred to as LDPE) low-density polyethylene (MOL Petrolkémia Zrt., Tiszaújváros, Hungary). The MFI value of this LDPE grade is 20 g/10 min (measured at 190 °C, with 2.16 kg).
- Vestoplast 708 (hereinafter referred to as APAO) amorphous poly-alpha-olefin (Evonik Industries AG, Essen, Germany) was also used as an additive. This APAO grade has a mass average molecular weight (M_w) of 75000 g/mol, and a viscosity of 8±2 Pa·s.

2.2. Methods

The PP matrix and the additives were mixed in an LTE 26-44 co-rotating twin-screw extruder (Labtech Engineering Co., Samutprakarn, Thailand). The rotation speed of the screw of the extruder was 90 1/min, and the rotation speed of the screw of the volumetric dosing unit was 12 1/min. The zone temperatures of the extruder were 160, 170, 175, 180, 180, 180, 185, and 190 °C from hopper to die, and the temperature of the die was 190 °C. The extruded filaments were pelletized with an LZ-120/VS pelletizer (Labtech Engineering Co., Samutprakarn, Thailand). We prepared compounds containing 5,

10, and 15 wt% of APAO and LDPE, and compounds containing 0.5, 1.0, and 1.5 wt% of talc. The neat PP reference was also extruded to provide the same thermal history.

0.5 mm thick sheet films were prepared from the compounds with a 25-30C single-screw extruder and an LCR300 sheet film line (Labtech Engineering Co., Samutprakarn, Thailand). The rotation speed of the screw was 90 1/min, the zone temperatures of the extruder were 170, 180, 190, 195, and 200 °C from hopper to die, and the temperature of the die was 200 °C. Take-up speed was 0.6 m/min, and the first roll of the sheet film line was tempered to 85 °C. The rotation speed of the winder was set according to the diameter of the roll (between 6 and 11 1/min).

Dumbbell specimens were cut out from the sheet films (figure 1). The dimensions of the specimen were determined by the geometrical constraints of the heating chamber of the tensile testing machine; we were able to achieve a draw ratio of λ=10. We designed the specimens with the measuring length of 30 mm, and with the width of 10 mm.

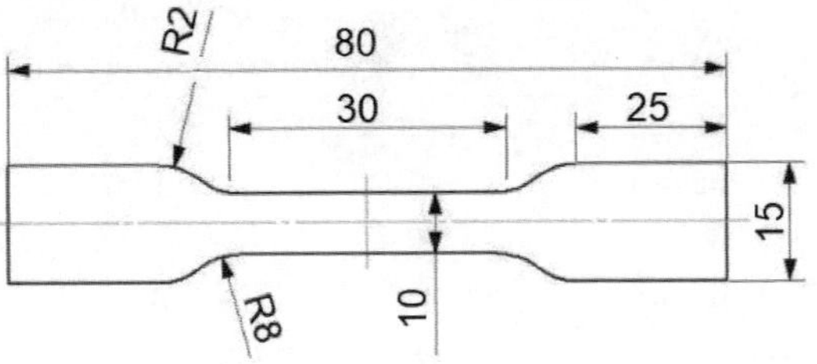

Figure 1. The specimens used in the drawing process.

The specimens were drawn in the heating chamber of a Z250 tensile testing machine (Zwick GmbH, Ulm, Germany). The machine was equipped with 100 kN wedge grips and a 20 kN load cell. The drawing temperature was 130 °C and drawing speed was 600 mm/min. The specimens were drawn to the draw ratio of λ=10. We stored the specimens in the heating chamber for at least 10 minutes prior to drawing to set their temperature, and left at least 2 minutes between closing the chamber door and the start of the drawing to maintain a constant temperature of 130 °C in the chamber. Force–strain curves during the drawing were recorded with the built-in software of the tensile testing machine.

The work of drawing was determined by calculating the area under the force–strain curves recorded during the drawing process. We expressed the work of drawing in terms of the area of the cross-section of the specimens (the work of drawing was divided with the area of the specimens) to make the results comparable.

Differential scanning calorimetry was performed on both drawn and undrawn samples. We conducted the tests using a Q2000 DSC device (TA Instruments, New Castle, United States) with one heating ramp in the temperature range between 0 and 200 °C, with a heating rate of 10 °C/min in a 50 ml/min nitrogen atmosphere.

Tensile tests were conducted with a Z005 tensile testing machine (Zwick GmbH., Ulm, Germany) on both drawn and undrawn samples. In the case of the drawn tapes, we cut out a 140 mm long sample from the middle of the tape, and the tensile tests were performed on these specimens with the measuring length of 60 mm. For undrawn samples, we used the specimens specified in Figure 1. The test was performed with the drawing speed of 20 mm/min at room temperature in both cases. As the drawn width of the drawn samples could not be measured with conventional methods—their width decreased greatly during the drawing process—we measured their width using a VHX-5000 light microscope (Keyence Ltd., Osaka, Japan). At least five specimens were tested in each case.

3. Results and discussion

3.1. Work of drawing

The specific drawing works of the different compounds are listed in table 1. In the case of the talc grades, an unambiguous tendency cannot be drawn due to the relatively large standard deviation. It can be concluded, however, that talc did not have a huge effect on the drawing work of the PP—especially compared to that of the APAO and LDPE.

Both LDPE and APAO acted as a plasticizer, even though they probably caused this effect differently. APAOs—similar to atactic polypropylene—are soluble in the amorphous phase of the PP matrix [10], but LDPE is not, consequently, it forms separate phases when mixed with PP when the melt is cooled [7]. At the temperature of drawing (130 °C), the polyethylene phase was in the melt state, and was consequently easily deformed upon drawing. This greatly reduced the specific drawing work. The APAO used consisted of relatively small molecules—compared to that of the PP matrix—and these small APAO molecules facilitated the movement of the PP molecules during the drawing process, thus reducing the work of drawing.

Table 1. Specific drawing work of the tapes.

Additive	Additive content (wt%)	Specific work of drawing (J/mm^2)
-	-	2.17±0.09
HTP05 talc	0.5	2.13±0.17
HTP05 talc	1.0	2.11±0.12
HTP05 talc	1.5	2.22±0.04
HTP1 talc	0.5	1.99±0.16
HTP1 talc	1.0	2.17±0.15
HTP1 talc	1.5	2.11±0.13
APAO	5	2.04±0.16
APAO	10	1.76±0.13
APAO	15	1.46±0.05
LDPE	5	1.75±0.11
LDPE	10	1.64±0.20
LDPE	15	1.48±0.04

3.2. Differential scanning calorimetry

The drawing process increased both the melting temperature and melting enthalpy of the pure PP, as expected (table 2). APAO, however, reduced both values. The APAO applied is fully atactic, and thus it cannot form crystalline phases. It can reside exclusively in the amorphous phase of the PP matrix. However, the decrement of melting enthalpy is not proportional to the APAO content of the blend. It was reported [10] that small APAO molecules can facilitate the movement of large PP molecules, as discussed above. This effect of the APAO molecules helped the PP molecules to form a larger ratio of the crystalline phase, thus the overall melting enthalpy did not decrease as much as expected due to the addition of the fully amorphous APAO.

The DSC curves of PP/LDPE blends had two distinct melting peaks (figure 2), from which the one at the lower temperature belongs to the polyethylene phase and the other one to the PP phase. The crystalline melting temperature of the polyethylene phase decreased upon drawing. In the literature, if drawing took place above the melting temperature of PE, but below the melting temperature of PP, an increase was reported in the melting temperature of the polyethylene phase of drawn PP/PE blends both for LDPE [7] and HDPE [8]. This difference could have been caused by the high MFI value of the PE applied, which eased the deformation of the PE phase upon drawing. During the solidification of the PE phase, the geometrical constraints of these elongated PE phases reduced the ability of the PE molecules to form crystalline phases, which could have caused imperfect lamellae growth, and thus, lower melting temperature.

Table 2. Results of the DSC tests of drawn and undrawn samples.

Additive	Additive content (wt%)	Draw/undrawn (-)	Melting temperature[a] (°C)	Melting enthalpy[a] (J/g)
-	-	undrawn	164.6	86.4
-	-	drawn	167.6	115.9
HTP05 talc	0.5	undrawn	163.6	86.2
HTP05 talc	1.0	undrawn	163.8	88.0
HTP05 talc	1.5	undrawn	163.1	94.9
HTP05 talc	0.5	drawn	167.9	117.5
HTP05 talc	1.0	drawn	167.1	115.7
HTP05 talc	1.5	drawn	168.5	119.9
HTP1 talc	0.5	undrawn	164.0	91.2
HTP1 talc	1.0	undrawn	164.4	92.4
HTP1 talc	1.5	undrawn	164.3	91.1
HTP1 talc	0.5	drawn	167.0	119.3
HTP1 talc	1.0	drawn	168.3	119.8
HTP1 talc	1.5	drawn	167.8	113.7
APAO	5	undrawn	164.8	86.5
APAO	10	undrawn	164.3	86.7
APAO	15	undrawn	164.4	86.1
APAO	5	drawn	166.9	108.4
APAO	10	drawn	166.4	104.9
APAO	15	drawn	166.4	106.3
LDPE	5	undrawn	111.3; 164.5	1.6; 82.6
LDPE	10	undrawn	111.1; 164.0	4.7; 80.3
LDPE	15	undrawn	110.8; 164.2	9.5; 74.6
LDPE	5	drawn	106.2; 167.4	1.4; 111.8
LDPE	10	drawn	110.2; 167.4	5.4; 103.5
LDPE	15	drawn	109.5; 168.0	8.9; 96.7

[a] In the case of the LDPE additive, the first value belongs to the LDPE phase, and the second to the PP phase

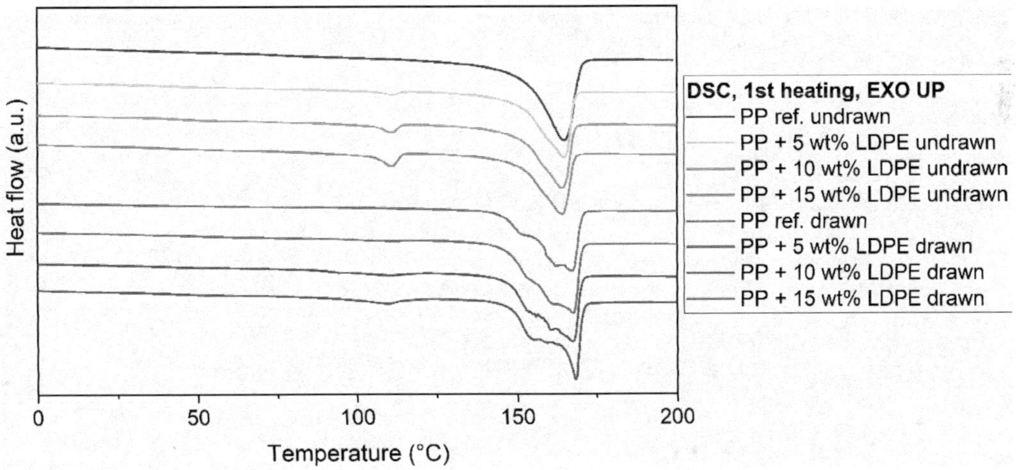

Figure 2. DSC curves of the drawn and undrawn PP/LDPE blends.

3.3. Tensile test

Typical tensile curves are presented in figure 3, and the results of the tensile tests are listed in table 3. HTP1 talc increased both tensile strength and modulus compared to that of the neat PP reference, except when talc content was 1 wt% (note that the melting enthalpy of this compound is also smaller than that of the others, so it is possible that this compound was not mixed homogenously enough). HTP1 talc also increased tensile strength and modulus, but to a smaller extent than HTP05, as it consists of larger particles, which presumably caused a lower level of reinforcement.

APAO reduced tensile strength and modulus, but increased strain at break. The decrement in strength and modulus is less substantial than the increment in strain at break in the case of the drawn tapes. The reason for this can be that APAO did not considerably reduce the overall melting enthalpy—and thus, the crystalline ratio—of the blend.

It is difficult, however, to conclude the overall tendencies of the tensile properties due to the large standard deviations. These could be caused by the micro-cracks originally formed while the specimens were cut out before drawing. This can be overcome by cutting out the specimens more gently and with perhaps the use of a different method.

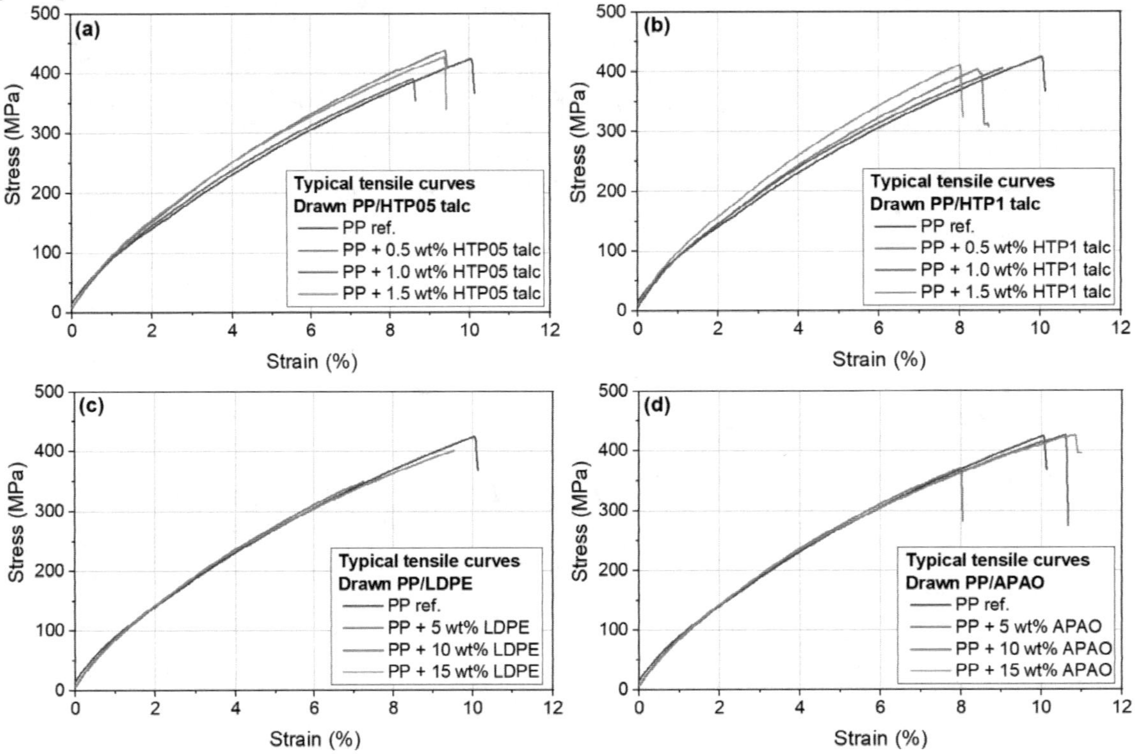

Figure 3. Typical tensile curves of drawn PP/HTP05 talc (a), PP/HTP1 talc (b), PP/LDPE (c), and PP/APAO (d).

IOP Conf. Series: Materials Science and Engineering 1246 (2022) 012023 doi:10.1088/1757-899X/1246/1/012023

Table 3. The results of the tensile tests.

Additive	Additive content (wt%)	Drawn/undrawn	Tensile modulus (GPa)	Tensile strength (MPa)	Strain at break (%)
-	-	undrawn	1.21±0.02	40.1±1.5	333.2±134.1
-	-	drawn	9.44±0.77	398.5±22.2	8.5±0.8
HTP05 talc	0.5	undrawn	1.25±0.11	39.2±0.9	341.1±187.7
HTP05 talc	1.0	undrawn	1.19±0.03	40.8±0.5	179.1±94.5
HTP05 talc	1.5	undrawn	1.22±0.08	39.1±0.9	421.5±94.4
HTP05 talc	0.5	drawn	10.15±0.14	414.2±31.2	9.4±1.8
HTP05 talc	1.0	drawn	9.82±0.18	389.5±60.4	8.1±1.9
HTP05 talc	1.5	drawn	10.21±0.10	421.2±25.1	9.3±1.5
HTP1 talc	0.5	undrawn	1.18±0.06	38.4±0.8	307.4±85.4
HTP1 talc	1.0	undrawn	1.23±0.10	38.2±0.8	406.8±86.3
HTP1 talc	1.5	undrawn	1.31±0.07	38.3±0.6	270.3±142.8
HTP1 talc	0.5	drawn	9.69±0.26	415.8±44.1	10.4±1.4
HTP1 talc	1.0	drawn	9.61±0.25	412.9±28.1	10.2±2.2
HTP1 talc	1.5	drawn	9.73±0.40	362.5±39.3	8.3±1.8
APAO	5	undrawn	1.01±0.08	35.7±0.2	134.3±62.1
APAO	10	undrawn	0.91±0.03	32.5±0.7	430.1±187.1
APAO	15	undrawn	0.87±0.06	31.0±0.4	404.5±129.7
APAO	5	drawn	9.38±0.21	374.2±25.5	10.4±1.7
APAO	10	drawn	8.75±0.20	384.1±41.2	10.3±1.9
APAO	15	drawn	8.60±0.34	386.5±49.1	11.9±1.8
LDPE	5	undrawn	1.16±0.10	40.9±3.0	626.4±102.6
LDPE	10	undrawn	1.00±0.05	35.4±2.1	632.9±234.7
LDPE	15	undrawn	0.94±0.04	33.1±1.2	511.5±127.5
LDPE	5	drawn	9.32±0.59	338.2±28.5	6.6±0.9
LDPE	10	drawn	10.50±0.72	401.2±59.4	7.3±2.2
LDPE	15	drawn	9.72±0.53	420.1±63.3	9.3±3.6

4. Conclusions

Polypropylene-based compounds were prepared with a twin-screw extruder with LDPE, APAO, and two different grades of talc as additives. The specimens cut out from the sheet film made from these compounds were drawn in the heating chamber of a tensile testing machine to the draw ratio of $\lambda=10$ at 130 °C. DSC and tensile tests were performed, and we also investigated the work of drawing. The following conclusions can be drawn from the results of the tests:

- Both LDPE and APAO reduced the specific work of drawing, as LDPE was in the melt state at the drawing temperature, and small APAO molecules facilitated the movement of the PP molecules. The talc grades showed only a slight effect on the work of drawing.
- Both talc grades acted as nucleating agents, but both APAO and LDPE reduced the crystalline melting enthalpy. The fully amorphous APAO, however, reduced the melting enthalpy less than it could be expected from its weight ratio in the blend. This could also be linked to the small APAO molecules facilitating the movement of PP molecules, thus making it easier for the PP molecules to form crystalline phases.
- Talc grades increased tensile strength and modulus and reduced strain at break—this was more significant in the case of HTP05 talc, but both LDPE and APAO had an adverse effect on tensile properties. APAO, however, increased strain at break more substantially than it reduced strength and modulus, as it reduced the crystalline ratio of the PP phase less than LDPE did.
- Due to the above-mentioned properties of the drawn PP/APAO blends, APAOs can improve the properties of highly drawn PP tapes. When combined with other additives, it can improve the strain at break of the drawn tapes without reducing their tensile strength and modulus.

5. Acknowledgments
This work was supported by the National Research, Development and Innovation Office, Hungary (2019-1.1.1-PIACI-KFI-2019-00335). The research reported in this paper is part of project no. BME-NVA-02, implemented with the support provided by the Ministry of Innovation and Technology of Hungary from the National Research. Development and Innovation Fund, financed under the TKP2021 funding scheme.

6. References
[1] Fourné F 1999 *Synthetic fibers: Machines and equipment, manufacture, properties* ed Fourné F Cincinnati, United States, Hanser
[2] Krassig H 1977 *Macromol. Rev.* D **12** 321–410
[3] Peterlin A 1971 *J. Mat. Sci.* **6** 490–508
[4] Marcincin A 2002 *Prog. Polym. Sci.* **27** 853–913
[5] Teh J W, Rudin A and Keung J C 1994 *Adv. Polym. Technol.* **13** 1–23
[6] Na B, Wang K, Zhao P, Zhang Q, Du R N, Fu Q, Yu Z Q and Chen E Q 2005 Polym. **46** 5258–5267
[7] Marcincin A, Ujhelyiova A, Zemanova E and Marcincinova T 1998 *J. Macromol. Sci.* **A35** 1239–1247
[8] Schmidt P, Baldrian J, Scudla J, Dybal J, Raab M and Eichhorn K. J 2001 *Polym.* **42** 5321–5326
[9] Sustic A. 2016 *ASC Hot Melt Short Course*, New Orleans, United States
[10] Chen J H, Tsai F C, Nien Y H. and Yeh P H 2005 *Polym.* **46** 5680–5688
[11] Chen J H and Chang Y L 2007 *J. Appl. Polym. Sci.* **103** 1093–1104
[12] Rane R H, Jayaraman K, Nichols K L, Bieler T R, and Mazor M H 2014 *J. Polym. Sci.* B **52** 1528–1538
[13] Castillo L A, Barbosa S E, and Capiati N J 2013 *J. Polym. Res.* **20** 1–9.

IOP Conf. Series: Materials Science and Engineering 1246 (2022) 012024 doi:10.1088/1757-899X/1246/1/012024

Demonstration of the effect of vibration stress relief in the case of metal castings in acoustically made measurement images.

Zs Bársony, A Rutscher, A Kiri

T.E.T.T. Mérnökiroda Ltd. R&D Laboratory, Baja, Hungary
http://www.tettkft.hu

Abstract. This article demonstrates the effectiveness of vibration stress relief (VSR). VSR has been described in a number of written literature and its efficacy has been investigated by many researchers. The interesting thing about this article is that we do not use a common test method to detect the residual stress of a casting part. The diagnostics are performed acoustically, from which the natural frequency of the component is determined, from which, the residaul stress can be deduced. The aforementioned data play an important role in setting up a VSR. The results are significant because the new method significantly reduced the time of diagnostic. Last but not least, this type of diagnostics is easy to implement in industrial work areas, even factories and foundries, as it does not require a large area, is not energy and investment-intensive.

1. Introduction

Stress relief of raw cast parts is still an almost unavoidable technological step to get to the finished part. As a possible alternative to classical stress relief methods, vibration stress relief (hereinafter VSR) appeared around 1970. Even then, many people realized that the procedures used so far did not meet the needs in many ways. Of all the processes, the most common is the heat treatment process and aging. If the processes are considered from several aspects, such as time, cost, energy demand, pollution effects and effects on the condition of the workpiece (eg corrosion), the heat treatment is not the most favorable. This is discussed in several articles.

In their study, A Jurčius, A V Valiulis, V Kumšlytis study the VRS process in connection with the welding seam, they find that the form of energy communication is not a central issue, so there is no justification for heat to occur in the end result [1]. Several articles examine the effect of VSR on the macrostructure and microstructure of a substance. In several studies, R A Claxton considers VSR to be effective, or even more effective, than heat treatment when we consider the mechanical properties of the material [2] [3]. It also examines the industrial suitability of VSR and the technologies of the manufacturers present in the market, as well as the parameters used during VSR, such as frequency or treatment time. Several manufacturers recommend VSR with 100% reliability.

A S M Y Munsi, A J Waddell, and C A Walker also examined the effects of treatment in connection with welding seams, primarily on the service life of the seam, with particular emphasis on possible fractures. Better results were obtained after vibration treatment than with heat treatment. Heat-treated pieces showed a decrease of about 30% in terms of fatigue life, while those treated with vibration showed an increase of between 17% and 30%. Taking this into account, VSR-treated workpieces have a longer service life than heat-treated workpieces, and even with aging, the service life has increased [4]. J Yang et al studied, among other things, the time, frequency, amplitude of shaking, the role of support, and the effect of workpiece aging, which can be related to the rest we also studied [5]. Y P Yang studied longitudinal and transverse stresses in the material separately for both resonant and non-resonant VSR. Its result shows that there is a minimum voltage state at which the voltages can only increase overall. It makes an important distinction between transverse and longitudinal stresses, as beyond a certain treatment time, their studies show that transverse stresses begin to increase and increase to a greater extent than the longitudinal ones would decrease, so it is advisable to complete stress relief at this point. He also clearly considers the time, frequency and amplitude of the vibration treatment as key parameters [6].

Numerous comparative studies have been conducted to compare VSR and heat treatment (TSR) in addition to those listed so far, of course. In several studies, X Dong et al highlight the advantages of VSR over TSR [7] [8]. In one of their studies, they analyze a cylinder wallboard, which in shape may serve as an analogy to the disc-like scheibe castings we studied. A number of factors are taken into account in the tests, thinking here about the support and the mounting of the VSR equipment on the workpiece. Their results are compared with those of heat-treated workpieces and it is concluded that a well-adjusted VSR treatment may be more effective than TSR.

So far, we have seen that the VSR procedure has been studied in a number of studies, and based on these, it is at least as effective as the TSR, but in spite of this, there is still some accepted procedure in place. One of the main limitations of using VSR is therefore the lack of a thorough understanding of the phenomenon, which means a lack of trust in a wide range of applications. Without them, however, it is difficult to determine when, under what circumstances, and how VSR can be applied. Experience to date suggests that it is particularly well suited to use where fatigue and fracture cause serious problems. A H Hassan subjected the workpieces he examined to various mechanical tests after treatment and his results also confirmed the effectiveness of the VSR procedure, but his results showed that only at certain settings. At some settings, there was no change in the material with respect to internal stresses [9].

In one study, H Gao et al investigate the occasional adverse effects of VSR. For a Ti-6Al-4V alloy, it is shown that VSR can have a negative effect on the fatigue life of the workpiece [10]. This only confirms the importance of setting the appropriate parameters of the VSR, which have been listed several times before, with pre-treatment diagnostics playing a key role.

More and more studies are dealing with VSR equipment such as sensors, support and last but not least the vibrating equipment, the vibromotor. In a relatively new study, I P Lyan, G Y Panovko, and A E Shokhin examine these as a whole as technological machines. The article provides a comparative analysis of the required energy consumption, shaking technology machines for both resonant and non-resonant VSR [11].

In the case of resonant VSR, it is very important to determine the natural frequency of the workpiece, for which several methods have already been developed, but there is no generally accepted method for it. A C C Barbosa, R R F Pinto, J Landre Jr Measurements have clearly shown that the self-frequency can shift to certain effects [12].

S Lou and Q Yang wrote about natural frequency measurement by the sound signal, which is very similar to our method. A significant advance is that this type of diagnostics is completely non-destructive, moreover, there is no need for sensors mounted on the workpiece. The data is recorded using an electronic device (eg a tablet or mobile phone). The data can also be displayed in the form of a spectral image for evaluation, where the eigenfrequency values are displayed as peaks. Based on their results, they report high accuracy and, in addition, ease of use [13].

Relatively only some internal studies can be found in this topic, but some can be relevant. Lots of studies are about welding and the occurred stress, but laser welding and treatment appear only in a few paper in connection with residual stress. J Hlinka et al investigated laser welding and surface treatment, particularly wetting properties, recrystallization Phenomena and interfacial reactions [14].

They used SEM and XRD for validation and studying the structure of the material. G Béres et al investigated predictable deformation in case of automotive thin sheets, which caused by welding [15]. Z Weltsch also wrote about a comparative study of the joining technologies of vehicle bodywork sheets [16]. According to Z Weltsch in the case of base-strength steel, mostly for all bonding types, the bond strength for low-strength steel can be provided.
The purpose of this article is to present an acoustic study that demonstrates the success of VSR treatment.

2. Experimental

2.1. Describing of the process of measurement and VSR parameters
Several experimental measurements and the experience gained from them helped to design the measurement layout. The parameters of the VSR treatment and their position in relation to the part to be treated were also determined empirically. The fugure 1. below shows a placement that was included in the experiment being shown.

Figure 1. The physical arrangement of the measurement.

The subject of the experiment is a spheroidal graphite gray cast iron with GG30 material quality, D520xd95x25 in terms of dimensions. The casting is fixed to the workbench with a closed section clamp, which can be seen in the middle of figure 1. The importance of this is unquestionable, as both diagnosis and shaking are done in a system. Clamping the part to the workbench also means attaching the part to the system. The vibration generator is shown on the right side of figure 1., while on the left side the measuring system is presented. The whole system was placed on a special work station.

The effect of shaking can be clearly examined and demonstrated by comparing the measurement images taken before and after. After suppression, diagnostics are performed at several points, using a device capable of generating an acoustic wave, a device capable of receiving that wave, and a device for controlling the two and processing the signals. The acoustic wave is created by an exciter that is connected to the part by a magnet so that it can more easily transmit the generated waves to it. The reception of the wave was solved with a microphone not far from the exciter. The control of the two devices and the processing of the signals were solved with a tablet and the suitable software on it.

IOP Conf. Series: Materials Science and Engineering 1246 (2022) 012024 doi:10.1088/1757-899X/1246/1/012024

After capturing the measurement image, it is necessary to analyze it. From the obtained data we can deduce the magnitude of the stresses in the casting and the natural frequency of the part. Based on the latter, enter the input frequency of the VSR. In order to target the most dominant natural frequency range of the component, we try to create an input frequency that is an integer divisor of the aforementioned frequency. The smaller order of magnitude is able to harmonize in the material and reach the natural frequency range of the material. This frequency at the time of the experiment was 45.33 Hz. Duration is also an important issue, as there may be a case of overvoltage, which means that voltage is already being applied to the system. This parameter was also determined experimentally at 45 minutes. The vibration was generated by driving an eccentric shaft provided by an electric motor. A cabinet equipped with a frequency converter was used to set the exact frequency. After the treatment, the "after" measurements were taken and compared with the "before" measurements, which are presented in the following chapters.

During the examination of the workpieces, the acoustic measurement images were initially recorded in several positions, from which the clamped position was selected for the next stage of the measurements. Vibration stress relief was also performed in this position, so that the workpiece was not moved between the treatment and the recording of the measurement images.

2.2. Acoustically made measurment images

The pattern of the acoustic measurement images follows a similar analogy in all cases, regardless of the position of the measurement and the workpiece. This means to us that the natural frequency range is always a wider peak or band, so we can talk about the initial value, the maximum intensity value and the final value of the natural frequency. This is one of the reasons why it is not enough to draw conclusions based only on the read frequency values, but the recorded measurement images themselves are of paramount importance.

If we compare the measurement images, we can see that the peaks appearing in each image move, change their width, stabilize or occasionally disappear or new ones appear. It also shows that the analysis of the measurement images is a rather complex task, but at the same time we can observe the tendencies that can be well observed in the individual machining phases or as a result of the treatments.

In connection with vibration stress relief, it is extremely important to be able to continuously diagnose the condition of the workpiece before, after and during the treatments. In the following, we specifically examine the effect of VSR by comparing the pre-treatment and immediate post-treatment conditions.

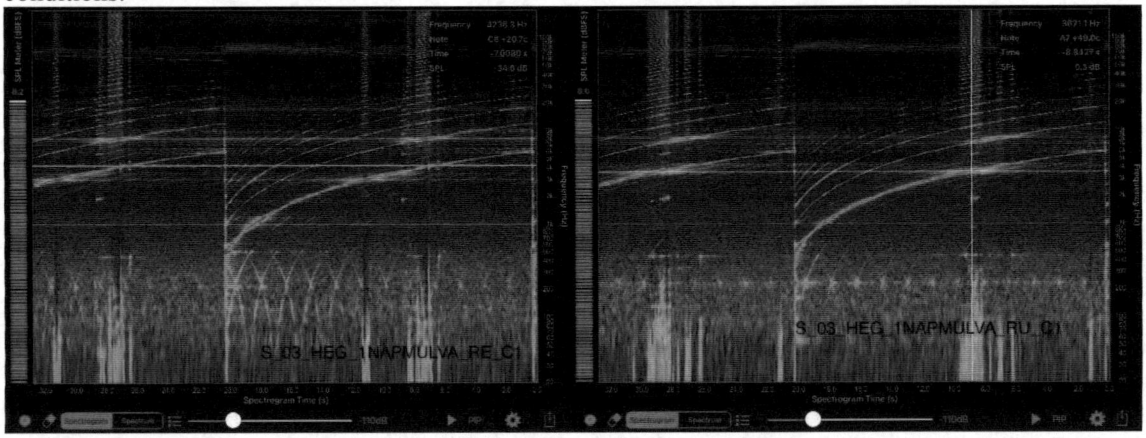

Figure 2. Comparison of pre-shake and post-shake acoustic measurement images, S_03 scheibe.

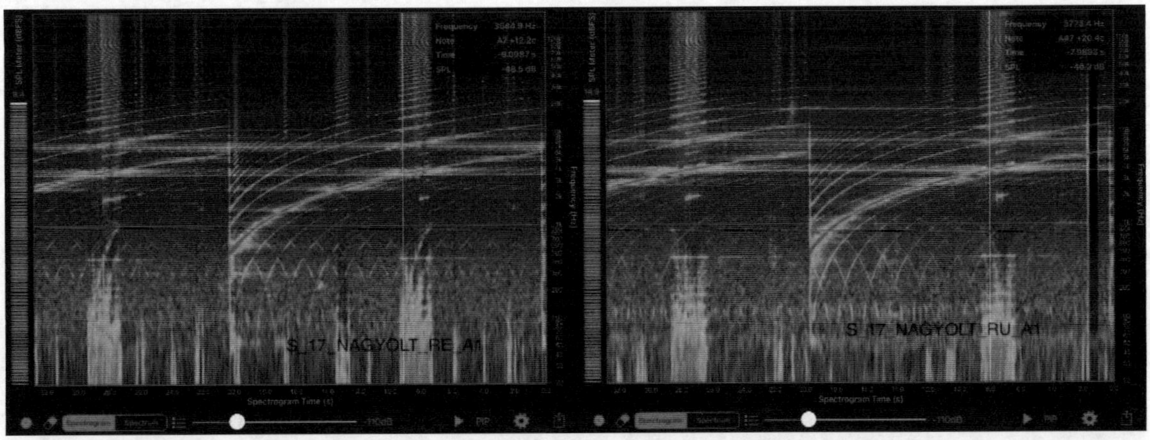

Figure 3. Comparison of pre-shake and post-shake acoustic measurements, S_17 scheibe, treatment 1.

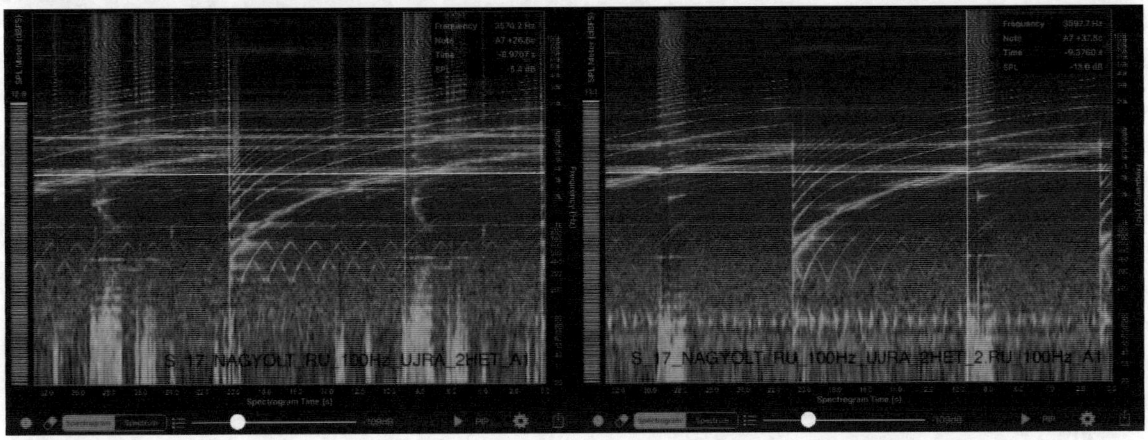

Figure 4. Comparison of pre-shake and post-shake acoustic measurement images, S_17 scheibe, treatment 2.

In figure 2., figure 3. and figure 4., the effect and changes can be seen before and after VSR treatment in case of different workpieces. This kind of spectral images plot the frequency as a fuction of time and the harmonics look like columns in the picture. When we want to make conclusion we have to investigate these frequency bands, for example the value of frequency, the width or the intensity and their changes. As the tendency shows, the main, natural frequency band becomes thinner after the VSR treatment or relaxing, and the number of other bands decreases, if the treatment is successful.

3. Result and Discussion

We have already stated that it is not advisable to examine the values alone without the measurement images, but a number of conclusions can be drawn together with the measurement images. After the analysis of the measurements, the measured natural frequency values were summarized in tables, and further analyzes were subtracted with the help of the measurement images.

In the case of S_03 scheibe, looking at the values, we would see that they have risen. However, if we look at the measurement images in relation to each other point by point, we cannot observe this increase, which is highly dependent on human factors. In many cases, the marker was not retracted in the wrong place, but the hitherto wider peak disintegrated into several narrower ones, and the person taking the measurement selected the widest of them. This can also be seen as a kind of displacement, but so far in the classical sense it has not been a displacement. If, on the other hand, it is not considered a displacement, the stabilization of the peaks as a result of the shaking can be deduced from the measured images. After shaking, similarly to the sche of S_03, the frequency values of S_17 seem to stabilize or show a further decrease. The relatively turbid pre-shake measurement images were also clearly clarified, which occurred primarily after the natural frequency band, as shown in figure 2., figure 3„ and figure 4. above. Table 1. shows that the value of frequency for many points, seems to stabilise, and in some cases the accuracy of the measurement is very high.

Table 1. Frequency change before and after VSR treatment in the first treatment of S_17 scheibe at point A1-A3.

Point	Frequency (Hz) before the treatment	Frequency (Hz) before the treatment
B1	3697,3	3697,3
B2	3621,1	3621,1
B3	3878,9	3644,5

The diagnostics developed by us have been continuously validated with X-ray diffraction equipment, so that the efficiency of the vibration treatment can be determined with acoustic images. Although the relationship between acoustic measurements and X-ray diffraction measurements cannot be described by a simple formula, trends have been observed in many cases. After the vibration treatments, the stress values measured by XRD usually ranged from 0 to 30 MPa, at which time the stabilization and purification already described above were observed in the acoustic images, in some cases a decrease in the frequency values. Our additional objectives include a deeper understanding and exploration of relationships, even through mathematical models and analysis.

4. Conclusion

On the one hand, we proved the effectiveness of vibration stress relief in connection with X-ray diffraction validation, and on the other hand, we developed a new type of diagnostics that allows the use of VSR in industrial environments. With the help of acoustic diagnostics, the input frequency of the treatment can be determined, and the success of the stress relief can be verified on the basis of the measured images, thus the end of the treatment process.

We are currently developing our acoustic diagnostic and vibration system based on market needs, with the aim of developing a complete diagnostic and control system with electronic and software controller. The software is under development, with an online interface for remote monitoring of diagnostics and control from any internet-connected device with a database, because of the needs of the 21st century.

5. Acknowlegdement

We are grateful for the professional support of dr. Zoltán Weltsch is an associate professor, deputy rector of science at the János Neumann University and for the publication of the article to 13th Hungarian Conference on Materials Science.

We would like to thank the management of the T.E.T.T. Mérnökiroda Ltd. for providing the location and equipment for the measurements.

6. References

[1] A Jurčius, A V Valiulis, V Kumšlytis, 2008, *Vibratory stress relieving – It's advantages as an alternative to thermal treatment* (Vilnius: Journal of Vibroengineering) pp 123-127

[2] R A Claxton, 1991, *Vibratory stress Relieving an effective alternative to thermal treatment for component stabilisation?* (Heat Treatment of Metals) pp 53-59

[3] R A Claxton, 1991, *Vibratory stress relieving an effective alternative to thermal treatment for component stabilisation?* part II (22nd International School of Welding) pp 1-6

[4] A S M Y Munsi, A J Waddel, C A Walker, 2001, *The influence of vibratory threatment on the fatigue life of welds: A comprasion with thermal stress relief* (Glasgow: Strain) pp 41-49

[5] J Yang, H Wang, J Shi and X Gao, 2004, *Choice of vibration frequency for vibratory stress relief based on ANSYS* (Hong Kong: Materials Engineering and Environmental Science) pp 746-755

[6] Y P Yang, 2009, *Understanding of Vibration Stress Relief with Computation Modeling* (Journal of Materials Engineering and Performance) pp 1-8

[7] X Dong, X Zhao, S Yu, X Zhou, D Ren, Y Zhai, 2011, *Comparative Study on Vibratory Stress Relief (VSR) and Thermal Stress Relief (TSR) of Cylinder Wallboard* (Advanced Materials Research Vol. 331) pp 460-464

[8] X Dong, J Han, D Ren, K Mu, Z Fu, Y Zhai, 2012, *Study on Vibratory Stress Relief of Cylindrical Welded Component with Energy-saving and Environment* (Applied Mechanics and materials Vols. 174-177) pp 1168-1172

[9] Abdallah Hage Hassan, 2014. *Fundamentals of Vibratory Stress Relief* (Asian Journal of Applied Sciences 7) 317-324.

[10] H Gao, Y Zhang, Q Wu, J Song, 2017, *Experimental Investigation on the Fatigue Life of Ti-6Al-4V Treated by Vibratory Stress Relief* (MDPI) pp 1-13

[11] V Gursky, 2020, *On the issue of energy consumption of vibration technological machines* (IOP Conference Series: Materials Science and Engineering) pp 1-9

[12] A C C Barbosa, R Rocha, 2020, *Influence of plastic deformation over the natural frequency of steel structures* (International Journal of Advanced Engineering Research and Science) pp 1-10

[13] S Luo, Q Yang, 2019, *Natural frequency measurement of steel components by the sound signal* (Journal of Low Frequency Noise, Vibration and Active Control) pp 1-12

[14] J Hlinka, Zs Fogarassy, Á Cziráki, Z Weltsch, 2019, *Wetting Properties, Recrystallization Phenomena and Interfacial Reactions Between Laser Treated Cu Substrate and SAC305 Solder* (Applied Surface Science) pp 1-12

[15] G Béres, Z Weltsch, Zs Lukács, M Tisza, 2018, *Prediction of stress- and strain-based forming limits of automotive thin sheets by numerical, theoretical and experimental methods* (AIP Conference Proceedings) pp 1-7

[16] Z Weltsch, 2018, *Comparative study of the joining technologies of vehicle bodywork sheets* (IOP Conference Series: Materials Science and Engineering) pp 1-7

Recording acoustic spectrum for residual stress determination and assigning natural frequency ranges for vibration stress relief

A Rutscher, A Kiri

T.E.T.T. Mérnökiroda Ltd. R&D Laboratory, Baja, Hungary
http://www.tettkft.hu

Abstract. We have developed a new technique for diagnostics in connection with vibration stress relief. The measurement images developed by us with a unique acoustic spectrum recording are not only suitable for determining the natural frequencies of metal castings, but can also be deduced from them the dislocation in the crystal structure and residual stress, which can cause deformation later. From the changes between the measurement images, the effect of different machining even resting and aging on the casting can be seen. Furthermore, the required input frequency for vibration stress relief can be easily determined from the measured images. Some of the main advantages are the process obtained quickly, easily and immediately. The research will be continued, we are working further development, including the acoustic diagnostic system, the VSR equipment, the recording software with an online platform.

1. Introduction

A Vibration Strain Relief (VSR) has discovered for decades as a stress relief technique, but it has not been widely used, due to a lack of understanding of the phenomenon. The technology is not a completely developed process besides heat treatment (TSR) or ageing, but it spreads all over the world slowly and becomes more accepted. This is mainly due to the fact that ageing is time consuming and heat treatment is space, investment and energy intensive. Several articles have discussed the efficiency and advantages of VSR over heat treatment.

In a study A Jurčius at al examined the advantages of VSR over TSR. According to the study, the emphasis is on the right energy input, not on the way of energy input, because there is no difference at the microscopic level to the crystals and atoms if this energy intake to the system by electromagnetic way or heat treatment or even by vibration. The properly injected energy into the system reduces the residual stress thus the effects of deformation [1].

R A Claxton had already written several publications and articles in connection with VSR and its application by the early 1990s, referring to much earlier articles, even from 1968. In several cases he reported significant residual stress reductions, generally between 30-80%. R A Claxton also discusses the drawbacks of VSR, which are mainly related to the potential fatigue of treated workpieces, which is a concern for users in industrial applications [2] [3].

Besides R A Claxton, several studies have already shown that VSR has no negative effect on the mechanical properties of the material. A S M Y Munsi et al investigated the effect of treatment on weld life and the causes of possible fractures in welds. Their studies showed that better results were obtained with vibration treatment than with heat treatment. The heat-treated pieces showed a reduction in fatigue life of about 30%, while those treated with vibration showed an increase of between 17% and 30%. This suggests that VSR-treated workpieces have a longer lifetime than heat-treated ones, and even than against ageing [4].

Already in the studies mentioned earlier, the possible concepts of VSR procedures are discussed, that two basic cases can be distinguished in conventional VSR procedures, namely non-resonant VSR and resonant VSR. In both cases, there are some important parameters must highlighted, which means the adjusted frequency or amplitude. In the case of non-resonant VSR, the amplitude is controlled, while the frequency is set at a specific value. The effectiveness and comparison of the two methods has been discussed by Y P Yang also. His results show that both resonant and non-resonant VSR are effective, but that resonant VSR is more appropriate for heavy workpieces, while non-resonant VSR may lead to better results for smaller workpieces [6].

For the resonant VSR, which makes us more interested, the key parameter is the frequency, which should be determined by the natural frequency of the workpiece. The natural frequency is the frequency at which a left-alone-system will vibrate. If the amplitude is plotted as a function of frequency, we see a jump in amplitude at the natural frequency. A metal cast or a welded structure may have several natural frequencies or natural frequency bands. Resonance is the phenomenon where the excited frequency approaches or becomes equal to the natural frequency, at which point the amplitude of the system is at a maximum, so to determine the natural frequency of a casting or system, we need to excite as close to the natural frequency as possible to produce resonance. The excitation causes the energy exceed the resistance of the metal at certain frequencies and the amplitude suddenly increase or jump up. This is the point where the resonance condition is established, thus allowing the natural frequencies of the workpiece to be measured. The importance of natural frequency and treatment frequency has also been analysed by J Yang et al as well as treatment duration, applied force or amplitude [5].

G Cai, Y Huang and Y Huang compared using pure lateral resonance and coupled lateral-torsional resonance. There is some advantages of using the second, coupled lateral-torsional resonance, for example the natural frequency (as lateral natural frequency as torsional natural frequency) can be adjusted. Or the two kind of natural frequency can be excited by this method of resonance and the specimen is subjected to a multidirectional excitation [7].

S Kwofie also analysed the parameters of VSR treatment and he presented a mathematical model for simulating the VSR process. His cyclic plasticity model has shown that VSR can be described with a cyclic stress relaxation and redistribution process. In the cycle, when its peak stress decreases, the inelastic strain increases [8].

In a study by A C C Barbosa at al, the relationship between natural frequency, load force and internal stresses was investigated, mainly during tensile strength tests on steel alloys. Their results show that the natural frequency of the system changes when applies force is varied, so the natural frequency of the system may shift in response to some external force [9].

S Li et al wrote about critical natural frequency (CNF) as a method to predict the state after VSR treatment. They found that VSR treatments in some cases decrease the natural frequency, but in some cases because oh the complexity of the phenomenon, can not be described the change this kind of simple way, but they presented a simulation with a surface to illustrate this [10].

S Luo and Q Yang, in a relatively recent paper, in 2019, described a new method for determining natural frequencies by sound signal. Their method does not require sensors installed on the workpieces, and the response signal is recorded on some device (for example a mobile phone). The determination may require several measurements, which are plotted on a spectrogram from which the natural frequency values can be read. This method is very similar to the one we have developed, as it works on the same principle [11].

Relatively only some internal studies were written in this topic, but there are few that can be relevant. It is already known from several studies that during welding, significant stresses can be generated in and along the weld, but there is one more interesting opportunity, the laser welding and treatment. Laser welding and surface treatment has been investigated in studies by J Hlinka et al, with reference to wettability and its characteristics [12] [13]. In their studies, the micro and macro structure of the materials under investigation was also important, which was analysed, for example, by XRD or SEM. G Béres and Z Weltsch investigated the variation of hardness in dual phase steels with different martensite volume fraction to deduce to strength properties [14]. G Béres et al also investigated the stress-induced and even predictable deformation of automotive thin sheets [15].

The aim of this research and article is to provide a practical and easy-to-use natural frequency determination method to set the parameters of the resonant VSR to the most efficient one, both in terms of time and cost. The primary reason for using VSR is to save time and energy, as mentioned earlier, but also to avoid deformation of the workpiece, as this can significantly reduce the number of rejects. For vibration stress relief, it is necessary to analyse the workpiece, because the frequency of the treatment must be adjusted correctly to reach the natural frequency and achieve the most effective stress reduction.

With the diagnostics developed at T.E.T.T. Mérnökiroda Ltd., it is not only made possible to discover the natural frequencies, but also to gain a more complete knowledge of the residual stress of the workpiece and their variation. From changes between measured images, the effects of various machining operations and even resting and ageing on the casting can be seen, all quickly, easily and with immediate results.

2. Describing the process of recording the acoustic spectrum

Acoustic measurement images were taken at the R&D laboratory of T.E.T.T. Engineering Ltd. in Baja, Hungary.

Traditionally, sensors, usually an accelerometer or dynamic strain gauge, are used to measure the natural frequency. The technical development of information technology has made it possible - by means of a method developed and applied by T.E.T.T. Mérnökiroda Ltd. - to replace these sensors with an acoustic diagnostic system and the associated software, which has the advantage over the traditional accelerometer diagnostic system that the workpiece can be excited over a wider range, and the diagnosis of complex geometry with multiple cut-outs is not an insoluble problem.

The acoustic measurement process can provide a quick and efficient way to determine the natural frequencies of the workpiece, and the measurement images can be used to infer the internal residual stress of the workpiece.

For acoustic measurements, the disadvantages include the sensitivity of the diagnostic system to external noise sources and an improperly fitting microphone on the workpiece, which can affect the clarity of the measurement images. In addition, diagnostics require the collection of data on multiple castings of all different weights, geometries and materials to determine which workpiece has a lower or higher residual stress. Furthermore, it is important to note that, in order to ensure comparability of acoustic measurement images, the diagnostics should always be carried out with the same equipment and in the same position.

During the measurement, an electronically controlled exciter and an ultrasonic microphone are placed on the workpiece. The exciter can be controlled with the desired parameters within the defined frequency range, while the diagnostic software connected to the microphone can be used to detect the harmonized frequency values, frequency bands, generated within the frequency range. In this way we can obtain the natural frequency values.

The natural frequency appears as a wide peak in the measurement images, which can be characterized by three frequency values, the starting frequency, the one at the intensity maximum or the final frequency. We have used the starting frequency in most cases, because for a successful VSR treatment it may be enough to shake at the lowest natural frequency or even lower frequencies, since our measurements and experience have shown that the input shaking frequencies are harmonised, so they have an effect in a higher frequency range.

3. Recordings of the acoustic spectrum

The acoustic measurement images were recorded at several points on a workpiece. It can be clearly observed that acoustic measurements allow us to examine the workpiece globally, yet each measurement image also carries a sense of locality. The globality in the measurement images is that the natural frequency peak, which is important for us, is at nearly the same location at each point and that, compared to subsequent measurement images, the individual points tend to behave similarly, for example, decreasing or increasing in frequency. In most cases, local properties are indicated by other peaks outside this main peak. These may vary in number, width and location at each point, but in many cases globalities can be detected. A good example of this is shown in the following figure 1., where the parts of the numbered frames represent the same ranges.

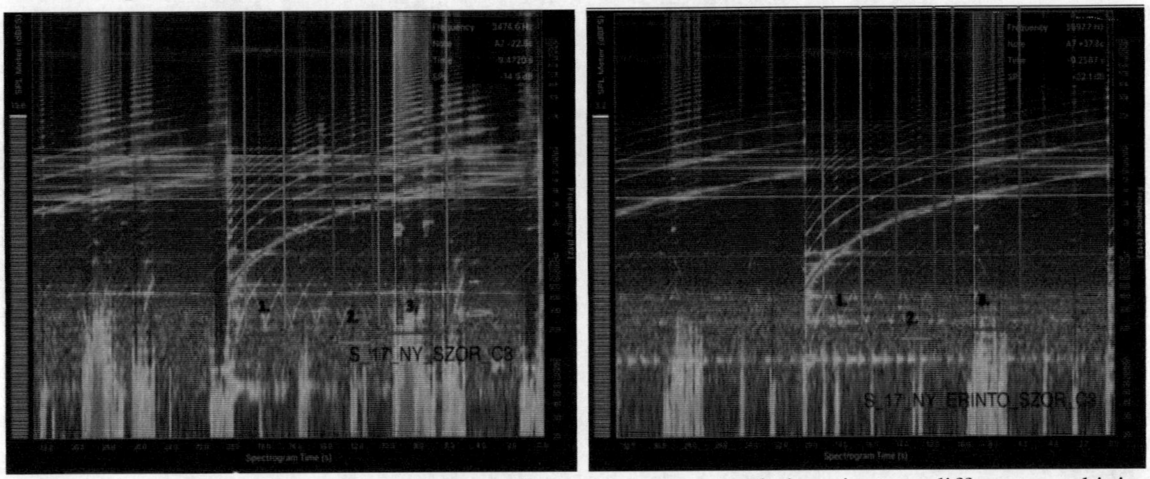

Figure 1. Acoustically made measurement images about a workpiece in two different machining stages.

Spectral images recorded the frequency as a function of time, while spectrum images plot the amplitude as a function of frequency.

In the first phase of the measurements, the workpieces were tested in four conditions, clamped, rubberised, standing and suspended. In each condition, analogous measurement images were recorded with small deviations from each other. The clamped condition was selected for further measurements, firstly for industrial use and secondly for evaluability and cleanness of the images. The workpieces were subjected to various machining operations, the changes in the acoustic measurement images were investigated, which can be divided into 3 main groups. These general observations are either displacements of the natural frequency band, or changes in the band width, or in the third case, changes in the peaks outside the natural frequency band. This can be seen in figure 2., which shows how the measurement images change at each machining stage. The machining phases are the followings, raw and not machined, turned, rested for 2 weeks, then rested for a further week and treated with VSR. The effects of resting and VSR treatment are similar, the natural frequency bands are narrowed, and the frequency value is stable. These have been verified on other workpieces and with much more measurements.

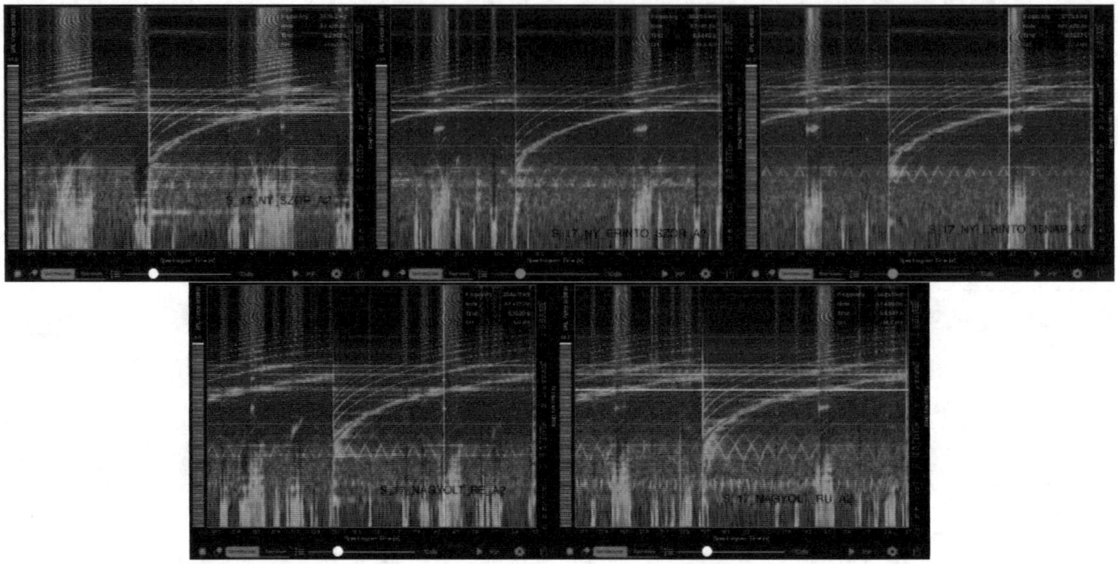

Figure 2. Acoustically made measurement images about a workpiece in five different machining stages.

A spectrum can be assigned to each point of each spectral image. These were used to analyse the harmonics. In figure 3., an acoustic spectrum can be seen and the peaks on the spectrum show the harmonics and from this kind of images the amplitude can be measured easily.

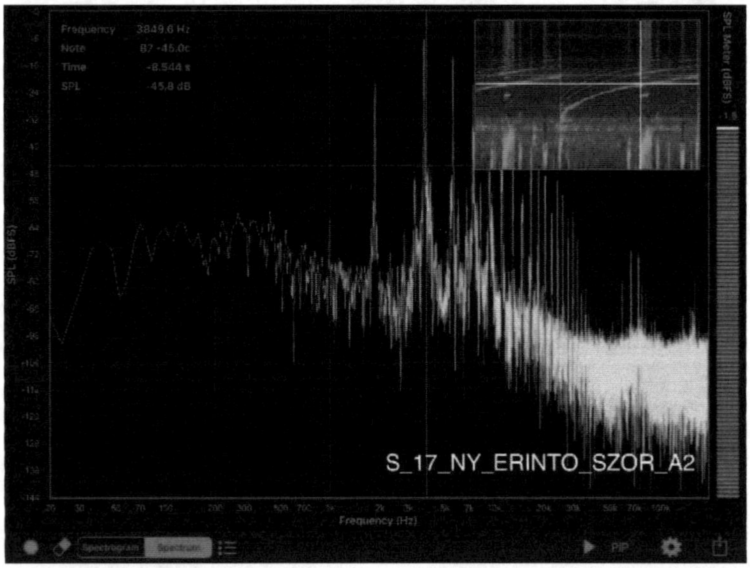

Figure 3. Acoustic spectrum.

4. Result and discussion

Acoustic spectral images can be used not only to determine the natural frequencies, but also to infer the dislocation in the crystal structure, the amount of residual stress that may cause deformation in the later stages. Furthermore, from the images, the required input shaking frequencies for vibration stress relief can be well determined. As mentioned earlier, this frequency does not have to be achieved directly, but can be achieved by using harmonics. In many cases, natural frequency values alone are not suitable for drawing clear conclusions, since the value only gives the starting frequency of the natural frequency range. For the disc-like castings we have investigated, the natural frequency most often falls between 3,000 and 4,500 Hz, with the white cross marker in the measured images showing the beginning of the dominant frequency band.

Other important information is therefore the turbidity of the measurement images and the number and width of the peaks that appear, which could be caused by external effects, but these have been effectively ruled out by the developments and tests carried out so far. In this case, the answers have to be sought in the internal structure (microstructure or macrostructure) of the workpiece under test, which may indicate inclusions, dislocations, cracks or even high stress peaks.

In general, it is observed that both VSR treatment and relaxing have a positive effect on the measurement images, which means that the measurement images become clearer, non-dominant peaks become thinner or even disappear, and when examining the natural frequency, a decrease and then stabilisation is generally observed.

For acoustic spectrums, peaks regularly occur in the same range, typically around 2000 Hz, 3700 Hz, 5500 Hz, 7200 Hz and 8000 Hz, as figure 4. shows.

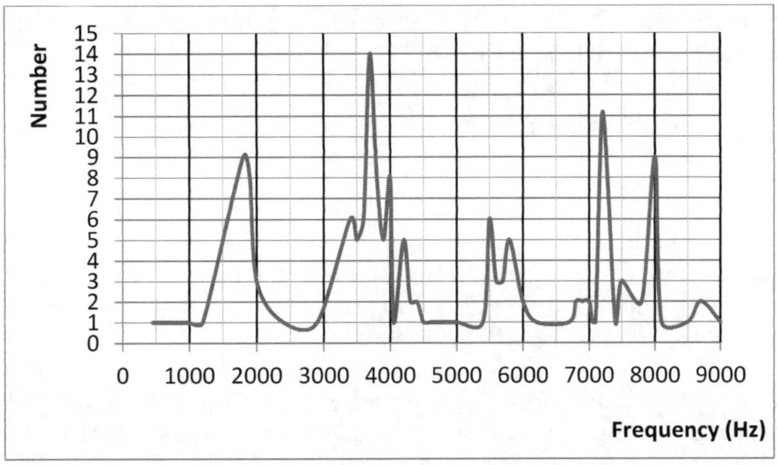

Figure 4. Distribution of frequency ranges of harmonics based on spectrum images.

Two methods were used to validate and verify the measurements, one is X-ray diffraction measurement, and the other is the Brüel & Kjær Photon instrument, which works in a similar way to our acoustic measurements, using an impact hammer to measure the frequency. The results of the Brüel & Kjær Photon measurements are obtained as shown in figure 5.

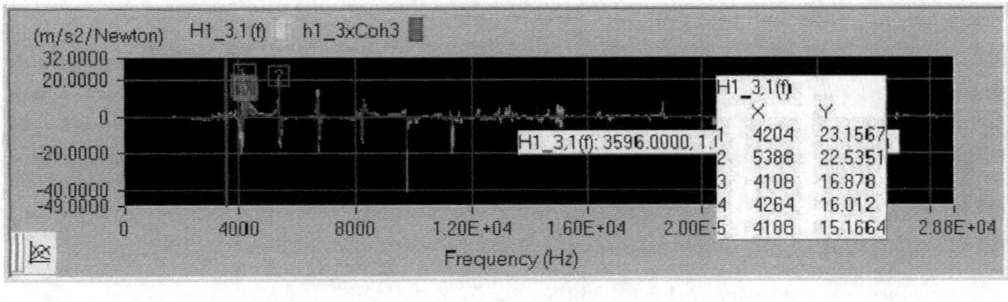

Figure 5. Measurements with Brüel & Kjær Photon.

The discrete values of the Brüel natural frequency determination can be compared with the values of the acoustic measurement, the frequency values fall within the dominant frequency band of the acoustic measurement images. We have compared our result with XRD stress measurements, relaxing and VSR treatment also had similar effect, the residual stress decreased.

5. Conclusion

With the acoustic diagnostic system developed by T.E.T.T. Mérnökiroda Ltd., the natural frequencies of a workpiece can be determined easily, quickly and accurately, and last but not least, the measurements can also be used to deduce the residual stress of the workpiece. The results can be used for the optimal parameterisation of the vibration stress relief, and in addition, a picture of the stress variation introduced during machining can be obtained, even as a feedback to the machining process.

We are currently developing our acoustic diagnostic and vibration system based on market needs, with the aim of developing a complete diagnostic and control system with electronic and software controller. The software is under development, with an online interface for remote monitoring of diagnostics and control from any internet-connected device with a database, because of the needs of the 21st century.

Acknowledgment

We are grateful for the professional support of dr. Zoltán Weltsch is an associate professor, deputy rector of science at the János Neumann University and for the publication of the article to 13th Hungarian Conference on Materials Science.

We would like to thank the management of the T.E.T.T. Mérnökiroda Ltd. for providing the location and equipment for the measurements.

References

[1] A Jurčius, A V Valiulis, V Kumšlytis, 2008, *Vibratory stress relieving – It's advantages as an alternative to thermal treatment* (Vilnius: Journal of Vibroengineering) pp 123-127

[2] R A Claxton, 1991, *Vibratory stress Relieving an effective alternative to thermal treatment for component stabilisation?* (Heat Treatment of Metals) pp 53-59

[3] R A Claxton, 1991, *Vibratory stress relieving an effective alternative to thermal treatment for component stabilisation?* part II (22nd International School of Welding) pp 1-6

[4] A S M Y Munsi, A J Waddel, C A Walker, 2001, *The influence of vibratory threatment on the fatigue life of welds: A comprasion with thermal stress relief* (Glasgow: Strain) pp 41-49

[5] J Yang, H Wang, J Shi and X Gao, 2004, *Choice of vibration frequency for vibratory stress relief based on ANSYS* (Hong Kong: Materials Engineering and Environmental Science) pp 746-755

[6] Y P Yang, 2009, *Understanding of Vibration Stress Relief with Computation Modeling* (Journal of Materials Engineering and Performance) pp 1-8

[7] G Cai, Y Huang, Y Huang, 2017, *Operating principle of vibratory stress relief device using coupled lateral-torsional resonance* (Nanning: Journal of Vibroengineering) pp 4083-4097

[8] S Kwofie, 2009, *Plasticity model for simulation, description and evaluation of vibratory stress relief* (Kumasi: Elsevier) pp 154-161

[9] A C C Barbosa, R Rocha, 2020, *Influence of plastic deformation over the natural frequency of steel structures* (Belo Horizonte: International Journal of Advanced Engineering Research and Science) pp 1-10

[10] S Li, H Fang, W Cui, 2018, *Critical natural frequency: an improved empirical effectiveness criterion in vibration stress relief of rectangle welded plates,* (Changzhou: Vibroengineering Procedia) pp 85-90

[11] S Luo, Q Yang, 2019, *Natural frequency measurement of steel components by the sound signal* (Journal of Low Frequency Noise, Vibration and Active Control) pp 1–12

[12] J Hlinka, Zs Fogarassy, Á Cziráki, Z Weltsch, 2019, *Wetting Properties, Recrystallization Phenomena and Interfacial Reactions Between Laser Treated Cu Substrate and SAC305 Solder* (Applied Surface Science) pp 1-12

[13] J Hlinka, M Berczeli, G Buza, Z Weltsch, 2017, *Wetting properties of Nd:YAG laser treated copper by SAC solders* (Emerald Insight) pp 69-74
[14] G Béres, Z Weltsch, 2017, *Estimation of Strength Properties from Microhardness Results in Dual Phase Steels with Different. Martensite Volume Fraction* (Periodica Polytechnica Transportation Engineering) pp 1-6
[15] G Béres, Z Weltsch, Zs Lukács, M Tisza, 2018, *Prediction of stress- and strain-based forming limits of automotive thin sheets by numerical, theoretical and experimental methods* (AIP Conference Proceedings) pp 1-7

IOP Conf. Series: Materials Science and Engineering 1246 (2022) 012026 doi:10.1088/1757-899X/1246/1/012026

Automated determination of the optimal manufacturing direction of polymer composite shell structures

L. Takacs[1,2], F. Szabó[1]

[1] Department of Polymer Engineering, Faculty of Mechanical Engineering, Budapest University of Technology and Economics, H-1111 Budapest, Műegyetem rkp. 3., Hungary
[2] eCon Engineering Kft., H-1116 Budapest, Kondorosi út 3., Hungary

E-mail: szabof@pt.bme.hu

Abstract. There is a promising potential in polymer composites in the automotive industry; therefore, design methods aimed at cost and weight efficiency have become increasingly important. The tool cost is a considerable amount of the total cost, which is affected by the direction of the manufacturing planes. In this paper, a method is introduced to automatically determine the optimal manufacturing direction of an arbitrary shell structure. The method is implemented in Python environment. A meshed surface model is read as input, and the geometrical complexity factors are calculated from all directions in the space in a discretized way. A new 3D branch diagram is shown with which the calculated values can be visualized and evaluated. After that, an undercut factor is introduced and calculated, and it is demonstrated that the minimum of the product of these measures can give the optimal manufacturing direction in a fully automatized way. The paper presents a study about a train seat to show the industrial applicability of the method by evaluating the effect of the manufacturing directions and geometrical complexities at different partitioning of the composite shell on material cost.

1. Introduction

Polymer composite parts have become increasingly important in the automotive industry due to the regulations of emission reductions and to make vehicles more energy-efficient [1]. The large production volume of this industrial sector drives the conduction of extensive research of cost-effective manufacturing technologies [2]. The manufacturing systems, in general, have undergone significant development in the last decades [3]. Focusing on the automotive industry and the cost-effectiveness, Khan and Mehmood [4] accomplished a comprehensive review of composite manufacturing technology. The authors confirmed that the traditional technology in this sector is resin transfer molding (RTM). Still, they showed two success stories to prove the potential of the vacuum-assisted resin infusion (VARI). Here, it is worth mentioning that the RTM technology is developed nowadays by using composite matrix materials other than epoxy, like polyurethanes [5] or thermoplastic matrix with in-situ polymerization [6].

By the spread of composite materials, not only the manufacturing technologies have been developed, but new design methods have emerged like design-to-cost (DtC) [7], design for manufacturing (DfM) [8], or even recycling by design [9]. DtC and DfM methods are based on using cost estimation techniques. Cost-estimation has three main approaches: analogous, parametric, and bottom-up [10]. The analogous method is based on the data of similar projects from the past and includes expert judgment, and the parametric method defines mathematical relationships between cost and past events and trends. Both methods can be used, e.g., in the aerospace industry where enough data is available from the past. In the automotive industry, the bottom-up method can be helpful to examine the work at the most granular level of detail. Kaufmann et al. [11] enhanced this method with a combined cost and weight optimization and showed that if the total lifecycle cost is taken into account, the weight and its magnitude of penalty strongly affect the results. The authors later developed their method to target the best draping strategy of a composite part [12]. On the principle of the bottom-up approach, Bacharoudis et al. [13] developed a process- and feature-based cost modeling method in Matlab/Simulink that can be used on complex composite assemblies, while Irisarri et al. [14] developed a new method called Quilted Stratum Design (QSD) and implemented it in a software tool which became a commercial software product with the aim of the partitioning of the structure into small numbers of constant stiffness areas which can serve as a trade-off between manufacturing complexity and mechanical performance.

Martensson et al. [15], by also using a bottom-up cost estimation approach, showed that larger and more complex composite parts might become more cost-effective when divided into several sub-parts and then joined together. The authors showed that the assembly cost has a shallow effect on total cost, and the partitioning of the part can lower the complexity and thus reduce scrap volume and tooling cost. The tooling cost they proportionated with a geometrical complexity factor. This method was later enhanced by combining the partitioning with multi-material diversity [16]. Besides, an analysis routine was also introduced [17] to investigate the partitioning of a part from the perspective of costs and mechanical performance. These methods calculate the complexity of the parts by knowing the manufacturing direction in advance because the partitioning has to be made manually. To automatize the partitioning in an optimization framework, the optimal manufacturing direction of an arbitrary shell structure has to be determined automatically.

We aim to develop a method that can automatically determine the optimal manufacturing direction of a composite shell structure. This would allow to automatically calculate the complexity factor of an arbitrary shell structure. That would mean a further step to have a fully automated optimization framework with the aim of cost- and weight-efficient partitioning.

2. Method

To determine different geometrical properties and then manufacturing characteristics of composite shell structures, we developed a software application in Python environment. Python is an open-source programming language that is very powerful and has an application programming interface (API) for every finite element software. This is beneficial because, in the conceptual design phase, not only the manufacturing properties but also the stiffness and strength of a composite structure are to be evaluated, and the most effective tool of that is the finite element analysis.

Based on this, it is appropriate to work with a geometry discretized by a finite element mesh throughout the whole process. In the case of composite structures, this means a finite element mesh of layered shell elements that handles the geometry as a surface model. In addition, if a mesh can be used as input, a 3D-scanned surface model in STL-format can also be used.

As we investigated composite structures produced with resin transfer molding (RTM) manufacturing technology, we used the complexity-factor C as the main measure, quantifying the geometry's deviation from a flat plate [15]. The complexity factor is described with formula (1) and shown in Figure 1.

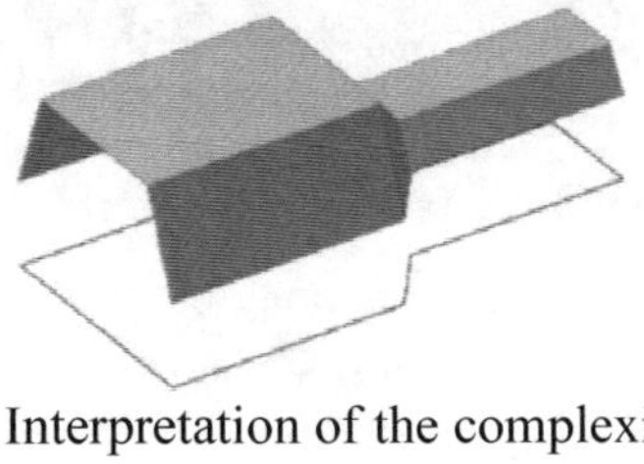

Figure 1. Interpretation of the complexity-factor.

$$C = \frac{A_{total}}{A_{proj}}, \tag{1}$$

where *A_total* is the total area of the component, i.e., the sum of the areas of each element that make up the geometric model; *A_proj* is the area of the perpendicular projection of the geometry in the manufacturing direction. It is clear from the definition that $C \geq 1$ and equality are obtained for *A_total* = *A_proj*, which means producing a completely flat part. The lower the value of C is, the less the shape of the part deviates from a flat plate, the less complex and expensive the tool required for production is, and the less production waste will be produced [15]. Since C depends on the projection (i.e., manufacturing) direction, it is worth plotting its value as a function of the possible three-dimensional directions. Since the value of C is invariant for translation and mirroring in a plane perpendicular to the view, all possible values of C can be obtained if we investigate it in a spherical coordinate system in the range of $\{(r, \theta, \varphi) | r = 1 \wedge \theta \in [0,2\pi) \wedge \varphi \in [0, \pi/2]\}$. This range includes the vectors of a unit radius hemisphere in the +z half-space which is shown in Figure 2.

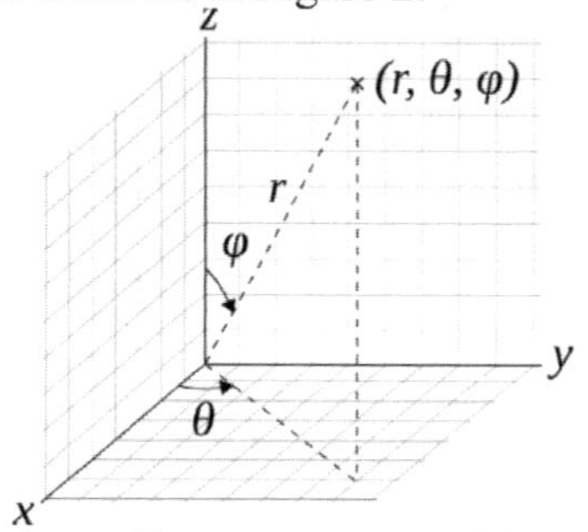

Figure 2. The spherical coordinate system used for selecting directions.

The assignment (θ, φ) → C can thus be represented in plane on a polar graph, and the values of C belonging to different theoretical production directions can be displayed in a spatial diagram. These are shown in Figure 3, where we plotted the complexity factor C of a five-sided rectangular shell from different directions in space. In the figure on the right, each radius indicates different projection directions, and their color indicates the C value calculated from that direction.

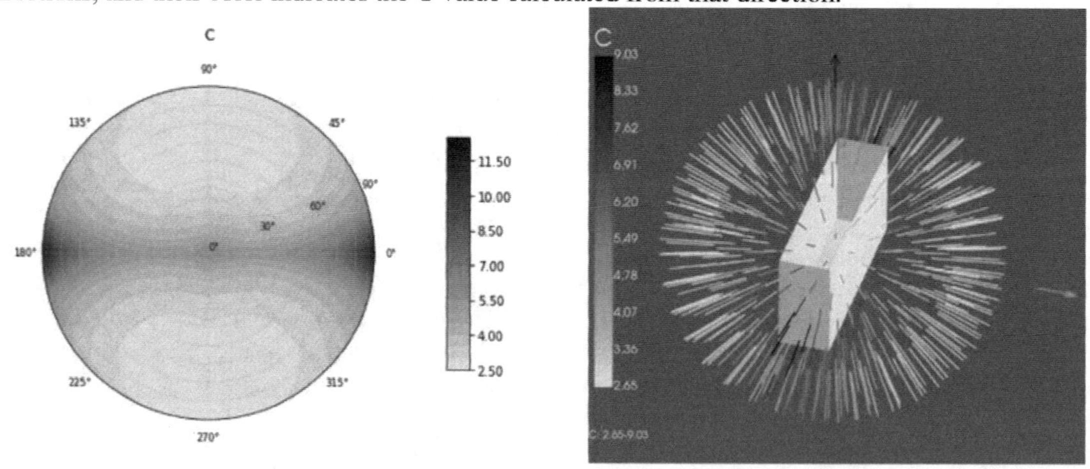

(a) (b)

Figure 3. Polar graph (a) and branch-diagram (b) of complexity factor C.

The general trends are more observable in the polar diagram, while in the branch diagram, the spatial directions are more illustrative. The branch diagram shows that – according to engineering considerations – C is maximal when the projection direction is perpendicular to the side with the smallest area. But it does not take its minimum value in the real manufacturing direction: it would be parallel to the z-axis in this case, which corresponds to point (0,0) in the polar graph. In addition, the minimum and maximum C values can be in a significant range for such a simple part, so it is necessary to consider the only C value that belongs to the manufacturing direction as a measure of the complexity of the part.

Examining the C value alone does not provide a method for selecting the direction of production but considers it to be predetermined. The direction of manufacturing has a significant effect on the projected surface and thus on the value of the complexity factor C: different views give a completely different C value, but the aim would be to assign a single C value to the part that characterizes its complexity and can be used in cost estimation methods.

During the traditional development process, the manufacturing direction is determined by an expert technologist. To define it automatically, we supplemented the complexity factor with another essential quantity: the undercut factor.

In the case of the production of composite shell structures, it is essential to consider whether their geometry includes undercuts: a geometry with undercuts makes the production of the part significantly more difficult or impossible in the given production technology. That is why it is worth considering whether the geometry has undercuts from that direction when examining the production directions. It is advisable to investigate this to the extent that it indicates not only the existence of the undercut but also its degree. With such a measure, the assignment of the view to the undercut factor can be made continuous, which can be helpful, especially for optimum search.

To quantify the undercut according to the above criteria, we introduced a value, denoted by U, which is the quotient of the projected area (A_proj, "shadow") and the projected surface of the part surface ($\tilde{A}\_proj$, "multiple shadows"). This can be illustrated by applying N projection beam perpendicular to the projection plane in the direction of the part, each of which intersects the surface model at a total K point. This is shown in Figure 4.

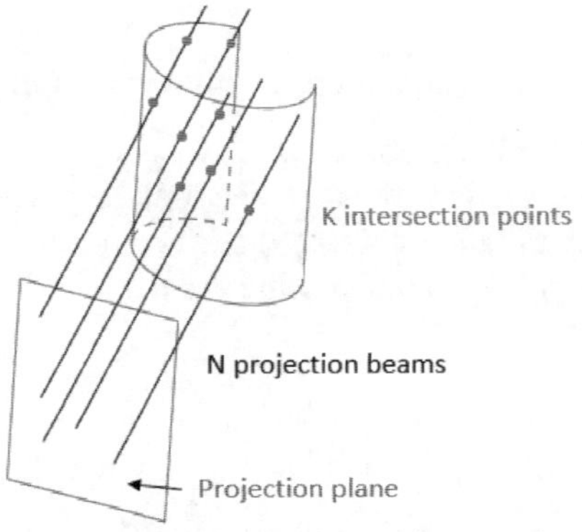

Figure 4. Determining the undercut of a part.

The quotient of the two values for a sufficiently large N gives a good approximation of the value of U (2):

$$U = \frac{\tilde{A}_{proj}}{A_{proj}} \approx \frac{K}{N} = \frac{number\ of\ intersections}{number\ of\ projection\ beams} \quad (2)$$

Similar to C, for any geometry U ≥ 1. The value of U for a given view of some simple geometries is illustrated in Figure 5.

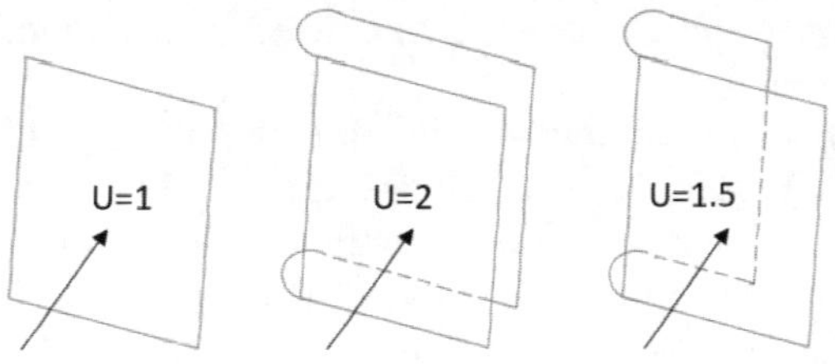

Figure 5. Illustration of the undercut-factor on three simple geometries.

Equality occurs from a given direction in the case of an entirely undercut part, in which case each projection beam intersects the geometry exactly once. U > 1, if there is an undercut in the geometry from the examined view; U = 2, if it is undercut from a given direction with respect to the entire projection of the part, and so on, in the case of multiple total undercuts, U > 2 may be obtained. A further similarity with the complexity factor C is that smaller values are also preferred here.

But the value of U can give surprising results as Figure 6 shows: for this kind of U-shaped part, a low U-value belongs to a direction that has no undercut but is still not suitable for manufacturing since we see the inner and outer surfaces of the part from the same viewpoint; thus, e.g., a press would not be lockable. Such a direction is indicated with a red arrow in Figure 7 (a), and on the right-hand side (b), the part is seen from this view. The inner surface is shown with grey; the outer surface is shown with yellow. Based on this, we quantified whether both sides of the part are visible from the manufacturing view. The definition of the outer and inner surfaces is simple in the case of a properly prepared surface model since the surface normal vectors assigned to the elements obviously give the orientation of the entire surface.

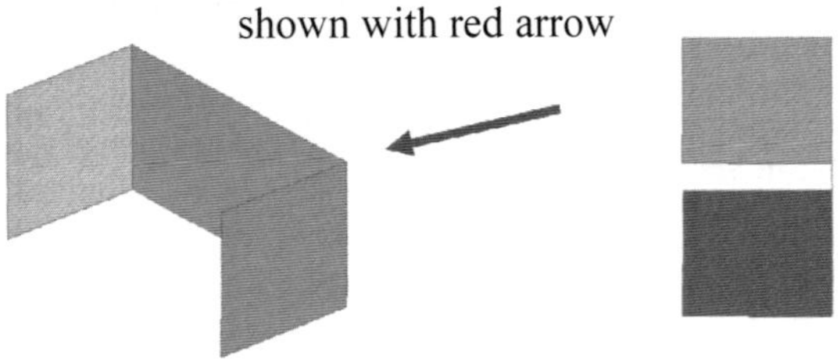

Figure 6. Example for a particular case with a misleading evaluation of the value U.

By using the information of surface normal, we extended the definition of U with a sign: we don't simply take the number of intersections K into account, but separate K+ intersections, where the direction of the projection beam and the intersected surface normal is the same, and K-, where they are opposites. This way, a measure of U_{eqv} can be defined (3).

$$U_{eqv} = \frac{K^+ - K^-}{N} \tag{3}$$

$U_{eqv} = U$ if $K+ \cdot K- = 0$, i.e., only one side of the surface is visible from the examined view.

This can be used to define a value of U_s, which can also be considered as an extension of U (4):

$$U_s = U + (U - |U_{eqv}|) \tag{4}$$

As the definition shows $U_s = U$ if only the outer or only the inner of the surface is seen from a given viewpoint, and $U_s > U$ if the outer and the inner of the surface are both seen from a particular viewpoint. Thus, this measure can be used to quantify the undercut and the orientation of a surface with a single measure. Furthermore, the nature of U_s stays the way that the lower values are preferred for manufacturing reasons. Thus, we have defined two measures above, which make it possible to quantify manufacturing-relevant properties of a component represented by a surface model from certain directions. However, it is also necessary to automatically select one of the possible directions corresponding to the manufacturing direction. By knowing such a direction, the value C will already be considered characteristic of the part.

Our proposed way to choose the manufacturing direction is to select the direction corresponding to the best value of the measures defined so far. But as shown above (e.g., see Figure 3), the minimum C often does not belong to a valid manufacturing direction, so this method is not appropriate. On the other hand, minimum U_s is neither enough alone, because if it is applied to the simple geometry shown in Figure 3, for example, it will not give a good result. This is because the value of C has a local maximum in the directions perpendicular to the sides, since there is a smaller projected area than from the oblique directions. Based on the above, we proposed the method to examine what directions we can find in the case of minimizing the product C·U$_s$: since neither too large C nor too large U$_s$ is suitable for manufacturing. Still, one can see that neither is suitable separately.

3. Case study

We performed a case study on the composite shell of a train seat. This is a glass fiber reinforced, polyester resin composite produced with RTM technology. We investigated the effect of the manufacturing direction on the material costs with two different partitioning variants shown in Figure 7.

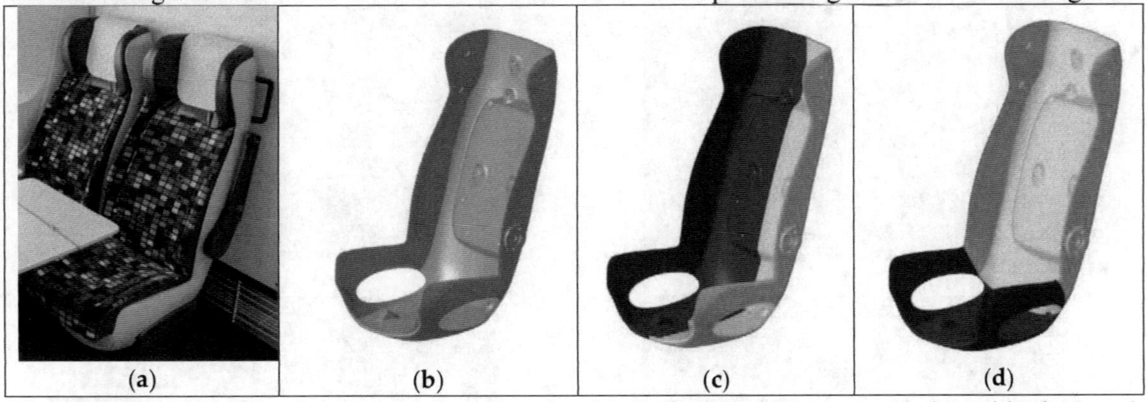

Figure 7. The investigated train-seat a) end-product b) shell geometry as one part c) partitioning ver 1 d) partitioning ver 2.

We assessed the shell as a single part and with two different partitioning variants as well. We calculated the C values for each part from the optimal direction of the original shell and from the optimal direction of the smaller parts themselves as well. With the calculated C values, we estimated the material costs. The input parameters of cost estimation are summarized in Table 1. Specific costs and scrap levels come from industrial experience; weight portions of fiber and resin are calculated with the densities of glass and resin, the volumetric fiber content of 40%, and a total weight of the complete shell of 12 kg.

Table 1. Input parameters of material cost estimation.

Abbreviation	Interpretation	Value	Unit
C_{fiber}	Specific cost of fiber	8	EUR/kg
C_{resin}	Specific cost of resin	3	EUR/kg
w_{fiber}	fiber weight of the complete shell	7.2	kg
w_{resin}	resin weight of the complete shell	4.8	kg
$s_{init\_scrap\_fiber}$	initial scrap level of fiber	0.15	1
$s_{init\_scrap\_resin}$	initial scrap level of resin	0.02	1

With the bottom-up cost estimation we used, the total cost includes investment, tool, material, and running costs [15]. In this case study, the seats were produced in a working composite plant with RTM machines and operational staff, so we investigated only the material cost estimated with formula (5), where C_{mat} is the total material cost of 1 piece of product.

$$C_{mat} = C_{fiber} w_{fiber} \left(1 + s_{scrap_{fiber}}\right) + C_{resin} w_{resin} \left(1 + s_{scrap_{resin}}\right) \quad (5)$$

The scrap levels of fiber and resin are calculated as the product of the initial scrap level and the calculated complexity factors. And note that the assembly cost's effect is negligible [15].

4. Results and Discussion

With the proposed method we calculated the complexity and the undercut factors on the train seat, and with the minimum search, we determined the optimal manufacturing direction of the shell structure. The

3D branch diagrams are shown in Figure 8 with the optimal direction highlighted with a purple arrow. For an illustrative visualization, the parts contain only the beams for which the U value falls in the lowest 20 percent.

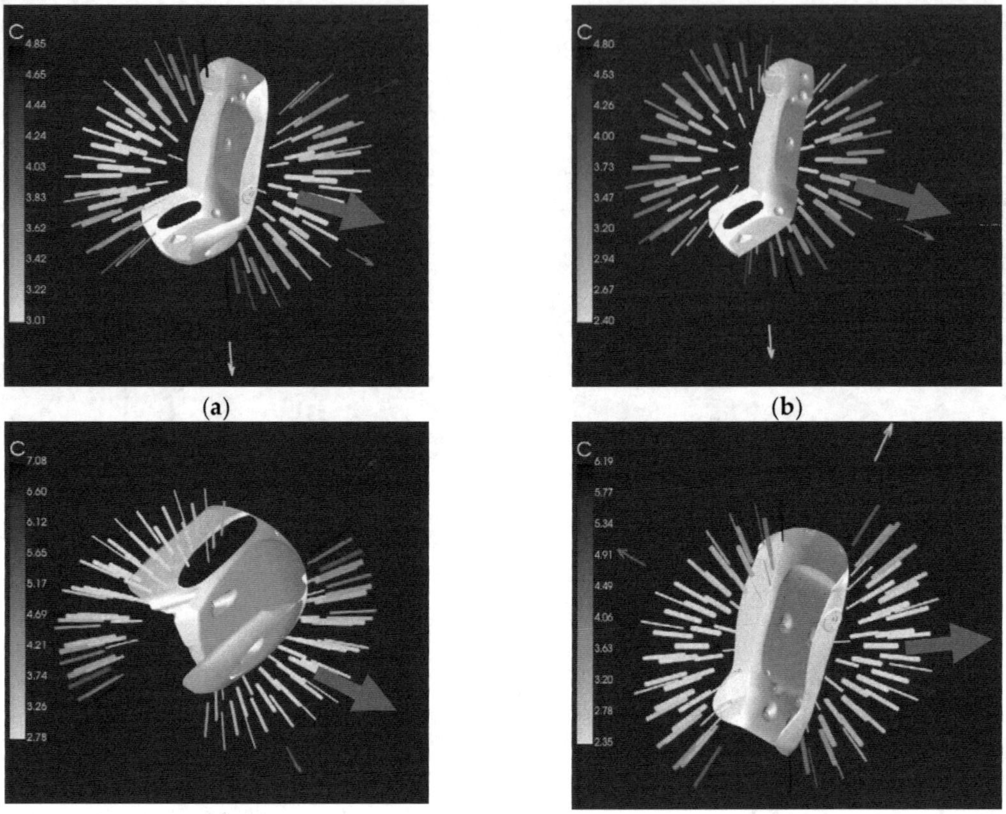

(a) (b) (c) (d)

Figure 8. Complexity factors and optimal manufacturing directions (a) train seat as a single part (b) partitioning ver 1, right side (c) partitioning ver 2, seat bottom (d) partitioning ver 2, backrest.

As for the 1st partitioning variant, only one side is shown in Figure 8 (b) since the structure is symmetrical. The purple arrows indicate the optimal manufacturing direction obtained. According to engineering judgment, the results are reasonable. But to have a quantitative comparison, Table 2 summarizes the complexity factors and material costs, that are calculated in two different ways.

Table 2. Complexity factors and material costs.

		C [-]	**C_{mat} [€]**	**C [-]**	**C_{mat} [€]**
		optimal direction of the seat as single part		**optimal direction of the subpart itself**	
Seat – as single part		3.02	98.9	n/a	n/a
Seat – partitioning ver 1	left-side	3.19	100.5	2.87	97.6
	right side	3.19		2.87	
Seat – partitioning ver 2	backrest	2.56	101.4	2.36	94.6
	seat bottom	4.39		2.78	

On the one hand, from the optimal direction of the seat as a single part, this means that even if the seat is divided into partitions, the manufacturing direction of each part remains the same direction as the seat had as a single part. On the other hand, as the seat is divided into partitions, each part has its own optimal direction. After partitioning the structure, the values show that we must consider the parts' optimal manufacturing directions; the complexity values are naturally always smaller from these directions. The results also show that partitioning a complex part gives subparts with smaller complexity, which is beneficial in manufacturing.

According to these complexity factors, the material costs were also calculated. It can be clearly seen that the partitioning would mean, in this case, 1.3-4.3% saving when comparing material costs 97.6 € and 94.6 € to the original 98.9 €. But on the other side, taking the optimal manufacturing directions of the subparts into account would mean a 2.8-6.7% saving compared to the case when the manufacturing direction of the subparts remains the same as the undivided seat. With an annual production volume of 20 000 parts, these saving would mean a significant difference in manufacturing costs.

5. Conclusions

This paper aimed to introduce the method we developed to determine the optimal manufacturing direction of composite shell structures. We implemented the method in Python software environment that reads in a discretized surface model as input then calculates geometrical complexity factors and undercut factors from any direction in space. The new undercut factors we introduced characterize the surface in terms of undercuts. One can also evaluate the values visually since the software illustrates them with a 3D branch diagram. We showed a case study to demonstrate that by minimizing the product of these two factors, the optimal manufacturing direction can be determined automatically. By applying a bottom-up cost estimation approach, we pointed out how the manufacturing direction affects the material cost. In the case study of a train seat, we proved that partitioning the shell structure and determining the optimal manufacturing directions separately for each part can save more than 6% in the material cost. Based on this, the method can be a very efficient tool in the hands of composite technologists.

Acknowledgement

The project is funded by the National Research, Development and Innovation (NKFIH) Fund, Project title: "Production of polymer products by a short cycle time, automatized production technology for automotive applications, with exceptional focus on the complexity and recyclability of the composite parts"; The application ID number: NVKP_16-1-2016-0046. The developers are grateful for the support.

References

[1] Holmes M. 2017 *Reinforced Plastics* **61** 294-298
[2] Sarfraz et al 2021 *Composite Structures* **266** 113864
[3] Rozs R.; Andó M. 2020 *Periodica Polytechnica Mechanical Engineering* **64** 51-66
[4] Khan L.A.; Mehmood A.H 2016 *Lightweight Composite Structures in Transport* **5** 93-119
[5] Echeverria et al 2022 *Express Polymer Letters* **16** 234-247
[6] Ageyeva et al 2018 *Polymers* **10** 357-404
[7] Amer et al 2020 *Energy and Buildings* **223** 110170
[8] Anderson et al 2014 *Procedia CIRP* **17** 362-367
[9] Toldy A. 2021 *Express Polymer Letters* **15** 1113-1113
[10] Hueber et al 2016 *Advanced Manufacturing: Polymer and Composites Science* **2** 1-13
[11] Kaufmann et al 2009 *Structural and Multidisciplinary Optimization* **41** 325-334
[12] Kaufmann et al 2010 *Composites Part A: Applied Science and Manufacturing* **41** 646-472
[13] Bacharoudis et al 2021 *Procedia CIRP* **104** 1143-1148
[14] Irisarri et al 2019 *Composites Part B: Engineering* **158** 364-372
[15] Martensson et al 2014 *Journal of Composite Materials* **49** 2897-2908
[16] Martensson et al 2015 *Proceedings of the Institution of Mechanical Engineers, Part D* **230** 49-60
[17] Martensson et al 2017 *Polymer Composites* **38** 2174-2181

Optimization of bonding technology for aluminium foams using femtosecond laser beam pulses

F Tajti[1] and B Petrovics[1]

[1] Department of Materials Technology, GAMF Faculty of Engineering and Computer Science, John Von Neumann University, Hungary

E-mail: tajti.ferenc@gamf.uni-neumann.hu

Abstract. Weight reduction is becoming increasingly important in the automotive industry. To achieve this goal, special materials are needed. We are no longer just using solid materials in production, but also metal foams. Metal foams, in particular aluminium foam, have high strength but their mass is a fraction of that of solid metals. Aluminium foams are usually not used on their own, but as a reinforcement for some structural element. The joining of these materials needs to be improved, as low weight is also important here. Therefore, bonding is a technology often used. For metal foams, use a little adhesive to avoid filling the pores. So we are only bonding on a very small surface area, so it is very important to increase adhesion. We used laser beam surface treatment to increase adhesion. The interfacial properties of the laser beam surface treated aluminium foam were first investigated by distilled water wetting and then, after setting the optimum parameters, the wetting of the adhesive was also investigated.

1. Introduction

Manufacturers have to meet increasingly stringent standards in vehicle production. Increasing chassis strength is a strict requirement, but so is reducing emissions. The greatest reduction in emissions is achieved by reducing the weight of the vehicle. To achieve this, manufacturers use a variety of materials such as high-strength steels, aluminium alloys, magnesium alloys, composites and, metal foams. Metal foams are porous materials containing air or some kind of protective gas bubbles. This results in a fraction of the density compared to solid metals. The low density results in low mass, but the strength of the material is still high. Aluminium foams are used in applications where, when filled into structural elements, they provide an increase in strength with a minimal increase in mass. To be able to strengthen a structural element, a stable connection between it and the foam must be created. As the aim is to reduce weight, heavy fasteners cannot be used, so adhesive bonding is a good solution [1-4].

In such structures, the components are subjected to quite high loads, so it is important to create high quality, high strength bonded joints. To achieve this, we need surfaces with good adhesion. There are several methods to increase adhesion, but for metal foams, traditional methods (surface roughening, etching) are difficult to use, as particles may remain in the pores. In the case of metals and metal foams, laser beam surface treatment can be used for this purpose. Lasers are increasingly used in industry. These lasers are becoming more and more advanced, capable of operating not only in continuous mode but also in pulsed mode. The latest lasers are capable of emitting ultra-short pulses, with pulses as short as 300 fs. At such short pulses, new kinds of laser-matter interactions are created. A cleaning process, an activation process, takes place on the surface, creating a surface structure created by the laser beam. These processes greatly influence the quality of adhesion bonding technologies [5-7].

Surface adhesion can be tested by wetting measurement. The process involves dropping liquid onto the surface and observing how it spreads across the surface. It is observed that if a liquid is more spread out on the surface, this means more adhesion. This test is easier to perform with more dilute liquids, such as distilled water, but also works with more viscous liquids such as adhesives [8-10].

Our aim was to use the laser beam to create a surface treatment that increases the adhesion on the aluminium foam surface and increase the strength of the bonded joints. In the course of this research, we have verified that wetting measurements can be performed well on closed cell aluminium foam. Thus, the effect of surface treatment was verified using this method. First, distilled water and then an anaerobic adhesive were used for the tests.

2. Materials and Methods

In this study we used aluminium foams which is based on 6061 aluminium alloy. The density of the aluminium foam used is 676 kg/m^3. This type of foam has a closed cell structure, which was an important consideration as it prevents the adhesive from flowing between the cells, as shown in figure 1.

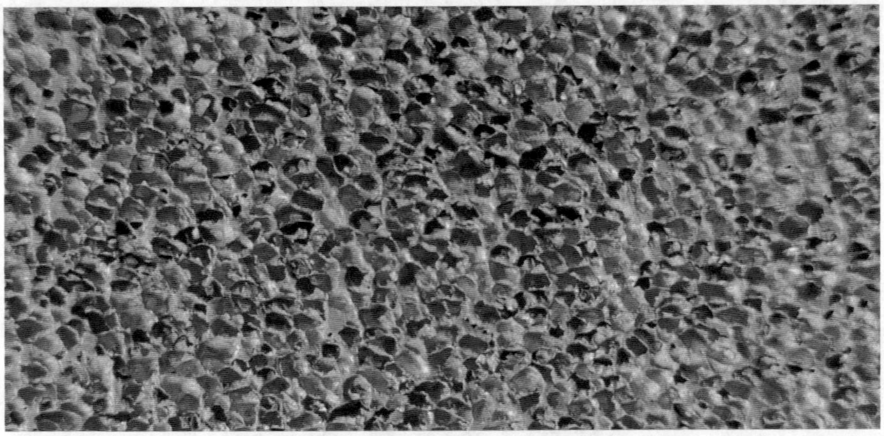

Figure 1. Closed cell aluminium foam.

Smaller samples were machined from the aluminium foam block so that the cells would not break, thus ensuring surface treatment and testability.

The laser beam surface treatment was carried out with a Coherent Monaco Femtosecond laser with wavelength of 1035 nm. The laser beam pulse width was 277 fs and the maximum average power was 40 W. The beam was focused on the foam surface with a diameter of 80 μm. There are some parameters that we cannot change. These include the wavelength and frequency of the laser beam. However, there are some parameters that we could change, such as scanning speed and laser average power. We created 10 x 10 mm^2 squares on the surface of the foams with different parameter combinations, as shown in figure 2.

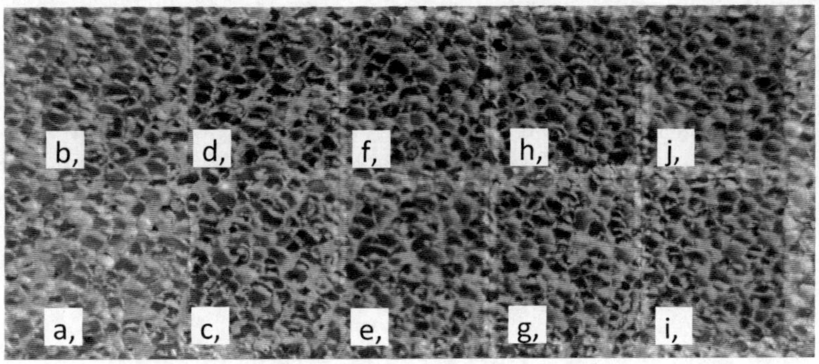

Figure 2. Laser beam surface treated squares on the aluminium foam with various laser power levels (a=10 %; b=20 %; c=30 %; d=40 %; e=50 %; f=60 %; g=70 %; h=80 %; i=90 %; j=100 % of average laser beam power and scanning speed of 5 m/s).

We tested 5 scanning speeds (25 m/s; 10 m/s; 5 m/s; 1 m/s; 0,1 m/s) with combinations of 10 power levels (10 %; 20 %; 30 %; 40 %; 50 %; 60 %; 70 %; 80 %; 90 %; 100 % of average laser beam power). The laser beam frequency was set to 750 kHz. We have tried to find parameter combination that increase the adhesion but do not damage the surface.

The laser beam treated surfaces were subjected to a wetting test on a specially developed device as shown in figure 3. The unit has an object table that can be moved vertically. The samples are placed on it. Above it, a micropipette is placed in the space provided for this purpose, with which the test liquid is dropped onto the surface. It is important that the liquid is not dripped onto the surface from above but is scooped out onto the tip of the pipette and touched to the surface. This is important because if the liquid droplet were dropped from a height, the kinetic energy would deform the droplet on impact and the result would not be accurate. So we touch the drop to the surface and it will flow through. The test used 10 μl of liquid to ensure that the liquid droplet was clearly visible on the porous aluminium foam. During the measurement, a side-view photo of the droplet was taken immediately after the drop was taken.

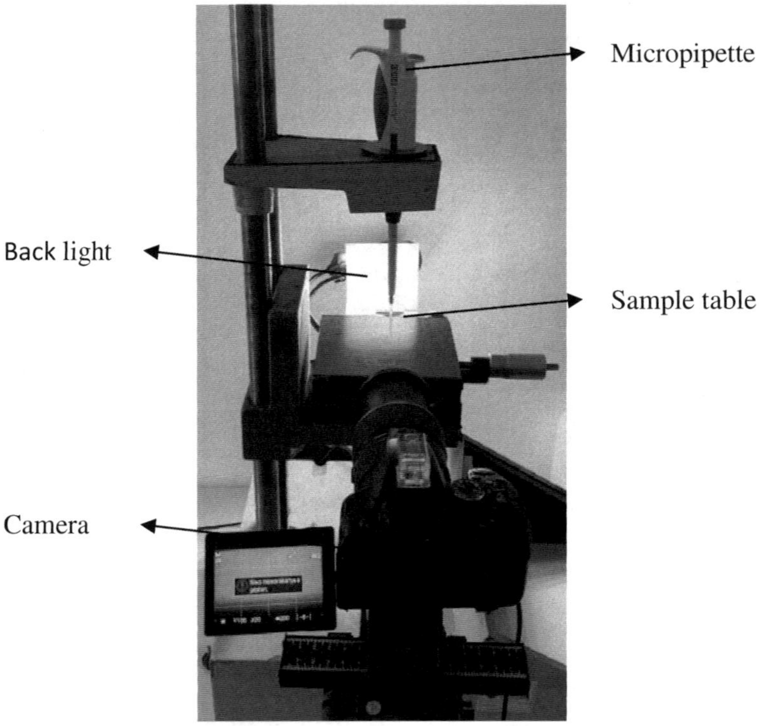

Figure 3. Wetting measuring equipment.

The first test liquid was distilled water because it is easy to dispense and can be accurately measured on the surface. Once the results were visible in distilled water, we performed the test with anaerobic adhesive. Loctite 648 adhesive was used, as it provides sufficient strength for metals and also acts as a sealant between the surfaces. It's much more viscous than distilled water so it was harder to drop with the pipette, and it has a lower surface tension than water, so it was more spread out on the surface.

The wetting measurement was done with a software, in which we open the captured image, select the surface of the work piece and draw a tangent from the two edge contact points of the liquid drop to the drop. The angle closed between the tangent on the drop and the surface is called the wetting angle, as shown in figure 4. The contact angles of the two sides can be asymmetric so the average of the two contact angles was taken when evaluating the measurements.

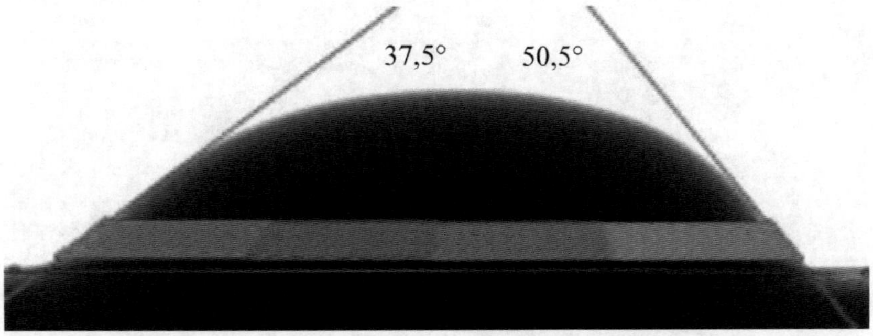

Figure 4. Static contact angle measurement.

3. Results

As expected from the tests, it was found that by increasing the average power and decreasing the scanning speed, the wetting on the aluminium foam surface was improved. However, it is observed that increasing the average power of the laser beam and decreasing the scanning speed darkens the surface, a phenomenon we would like to avoid. Therefore, an optimum parameter should be found where the surface is not yet discoloured but the wetting is already greatly improved.

In figure 5 we can observe how the aluminium foam is affected by the laser beam surface treatment when the adhesion is tested with distilled water.

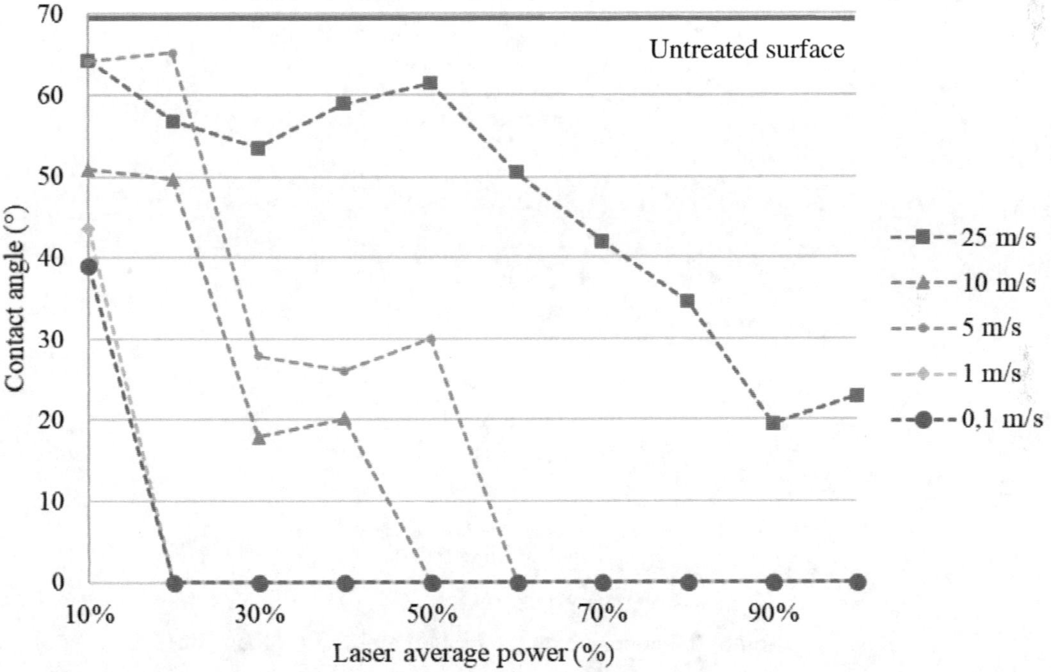

Figure 5. Wetting of the distilled water on the laser beam treated aluminium foam surface.

Figure 5 shows that the wetting angle for untreated aluminium foam is 69°. It is observed that when the scanning speed is reduced and the average power of the laser beam is increased, the surface contact angle values decreased. It can be seen that there are parameters where the distilled water behaves as a hydrophilic film on the aluminium foam surface. Unfortunately, the more intense the treatment, the more the treated surface darkens as shown in figure 6.

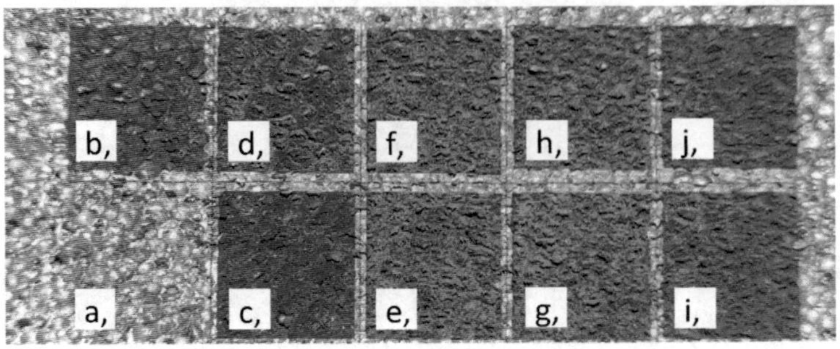

Figure 6. Darkened Al foam surfaces from laser surface treatment with various laser power levels (a=10 %; b=20 %; c=30 %; d=40 %; e=50 %; f=60 %; g=70 %; h=80 %; i=90 %; j=100 % of average laser beam power and scanning speed of 0,1 m/s).

Adhesion is also very good on these darkened surfaces, but we cannot allow such a large optical change in the treatment.

The surface tension of the distilled water is high, so it can be used to investigate quite large interfacial energy changes. However, if we are interested in the effect of surface treatment on adhesive technology, it is advisable to carry out tests with the adhesive itself. The result of this is shown in figure 7.

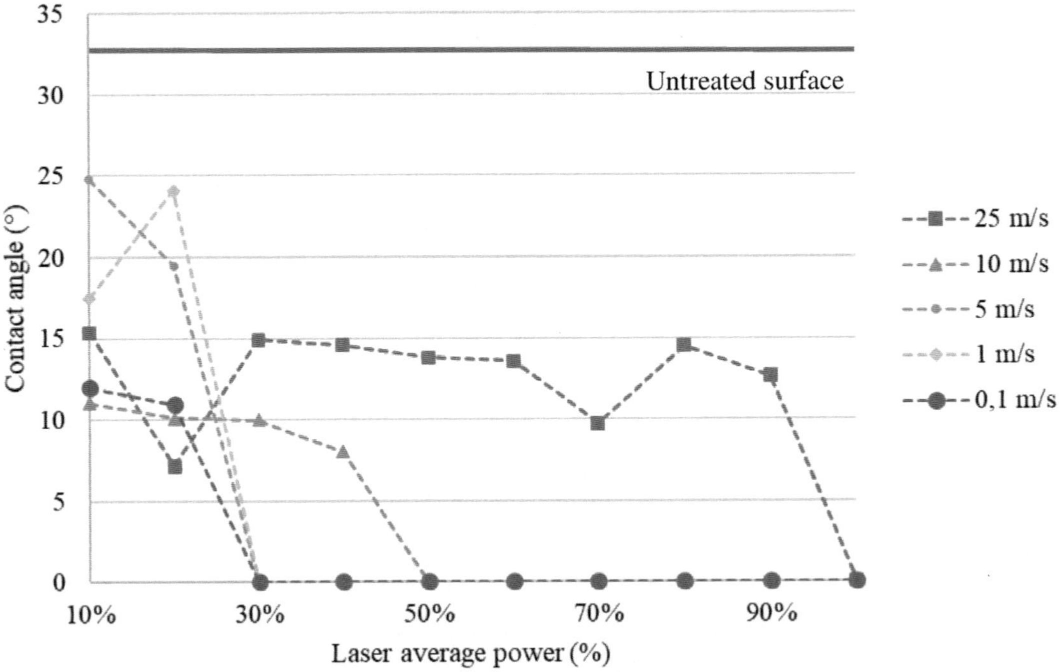

Figure 7. Wetting of the adhesive on the treated aluminium foam surface.

Figure 7 shows that on an untreated aluminium foam surface, the adhesive wetting angle was 33°. It is immediately apparent that the adhesive has a much lower surface tension than distilled water. It can therefore spread better on the surface and exhibits better adhesion properties. However, the laser beam surface treatment also had a good effect on this, as can be seen with increasing laser beam power and decreasing scanning speed, the wetting contact angle of the adhesive material is reduced to a fully hydrophilic level for certain parameters. In laser beam surface treatment, an important aspect is the time required for the treatment, in addition to the effectiveness of the treatment. It is therefore worth choosing an optimal parameter that meets all our criteria. For aluminium foam, this parameter is a scanning speed of 10 m/s and an average power of 50%. With this treatment, a fully hydrophilic state is achieved but no visible surface lesions are formed, and the scanning speed is high enough to quickly treat large surfaces.

4. Conclusion

Reducing emissions is a major challenge for car manufacturers today. The most widely accepted method to reduce emissions is to reduce vehicle weight. Weight reduction is achieved by using aluminium foam as a reinforcing material in the body panels. Bonding is the simplest and lightest weight bonding technique for joining these elements. In this research, we investigated the adhesion of bonded joints to increase the surface adhesion of the aluminium foam by laser beam surface treatment. We have investigated a wide parameter matrix and selected an optimal parameter with which we can increase the adhesion economically and to a large extent. This improvement was verified with distilled water and with the adhesive itself.

In our study we found that laser beam surface treatment can improve surface adhesion. Lower average laser power and high scanning speed laser surface treatment can also improve adhesion slightly compared to the untreated condition without visible surface changes. We were able to achieve very good results with high laser power and slow scanning speed, but these parameters discoloured the surface, which was an unacceptable criterion for us.

Finally, we found optimal parameter, that a scanning speed of 10 m/s and an average laser beam power of 50% were the optimal surface treatment parameters. This allows us to quickly perform the surface treatment on the aluminium foams and create a fully hydrophilic surface.

In the following, the resulting surface will be examined by scanning electron microscopy to see what surface structure the laser beam has created.

Acknowledgments

This research was supported by 2019-1.1.1-PIACI-KFI-2019-00075

References

[1] Claar T D at al 2000 Ultra-lightweight aluminum foam materials for automotive applications *SAE Technical Paper*

[2] Chege A et al 2017 Design and development of impact energy absorbing bumper *International Journal of Scientific & Engineering Research* **8** 326-330

[3] Fuganti A et al 2000 Aluminium foam for automotive applications *Advanced Engineering Materials* **2(4)** 200 - 204

[4] Garai F et al 2020 Development of tubes filled with aluminium foams for lightweight vehicle manufacturing *Materials Science and Engineering A* **790** 139743

[5] Yongchao Y. et al 2018 Ultra-short pulsed laser manufacturing and surface processing of microdevices *Engineering* **4** 779-786

[6] Berczeli M and Weltsch Z 2018 Improvement of adhesive joining of hybrid aluminum – GFRP using surface modifications *IOP CONFERENCE SERIES: MATERIALS SCIENCE AND ENGINEERING* **448** 012050

[7] Mishra S and Yadava V 2015 Laser beam micromachining *Optics and Lasers in Engineering* **73** 89–122

[8] Weltsch Z 2016 The relation between the surface tension and the surface texture formation *Gradus* **3** 38-44

[9] Berczeli M and Weltsch Z 2021 Enhanced Wetting and Adhesive Properties by Atmospheric Pressure Plasma Surface Treatment Methods and Investigation Processes on the Influencing Parameters on HIPS Polymer *Polymers* **13** 6 p. 901

[10] Hlinka J et al 2020 Wetting properties, recrystallization phenomena and interfacial reactions between laser treated Cu substrate and SAC305 solder *Applied Surface Science* **501**

IOP Conf. Series: Materials Science and Engineering 1246 (2022) 012028 doi:10.1088/1757-899X/1246/1/012028

Surface treatment of polymer matrix nanocomposites for adhesion enhancement by cold plasma

M Berczeli[1*], B Hatoss[1] and E Kókai[1]

[1] Department of Innovative Vehicles and Materials, GAMF Faculty of Mechanical Engineering and Computer Science, John von Neumann University, Kecskemét, HU-6000, Hungary

*berczeli.miklos@gamf.uni-neumann.hu

Abstract. Innovative surface technology processes offer a wide range of surface preparation options, allowing engineers to make major changes to the properties of products. Surface modification can also improve the quality and the strength of the bond created by technologies such as bonding, painting, coating, sealing and other adhesion. By activating molecular groups on the surface, wetting properties can be improved, while altering the microtopography of the surface can create roughness and texture. Surface modification suitable for industrial conditions in cold plasma surface modification. The the wetting boundary angles are measured with ethylene glycol and distilled water, and the results are used to calculate the surface free energy of the modified surface in mN/m according to Fowkes surface free energy theory. The demonstration of adhesion improvement of specimens with good wetting surfaces should be carried out by a series of bonding experiments, showing the changes in strength due to surface treatment. In addition, the technical advantages obtained by modifying the topography can be characterised by examining the separated surfaces.

1. Introduction

Polyamide-based nanocomposites are of great importance in many industries, but they are difficult to attach to other surfaces and their application is limited. Their manufacturing processes are difficult too [1-2]. Although polyamide has polar moieties, further increasing their proportion, increasing the polarity of the surfaces will lead to a shift from weak dispersion interactions to stronger interactions such as H-bridging or even covalent bonding. To this end, there are many surface treatment processes that leave the array properties intact [3]. The most prominent of which is cold plasma surface treatment. The surface preparation gives the opportunity to make hybrid material joints which expand the application of the polyamide based nanocomposites [4-5]. In the literature, we have seen that plasma jet surface treatment has been successful in activating the surface of many polymers, achieving large positive changes in wetting edge angle, but there are few reports in the literature that address the strength enhancement potential of adhesion bonding technology that can be achieved by improving wetting.

Homola et al performed similar measurements on poly(ethylene terephthalate) (PET) and poly(methylene methacrylate) (PMMA). Atmospheric pressure DCSBD-generated plasma was applied to the surface of the plastics and then wetting contact angle was measured. Wetting angle measurements were performed with water, ethylene and diiodomethane, from which surface energy was calculated.

In the case of PMMA, the wetting angle measured with water was reduced from 85° to 44° by a treatment of 3 seconds at a distance of 0.5 mm with a power of 400 W. XPS measurements showed that the oxygen

content at the surface increased, which contributed to the increase in the polar component of the surface free energy [7,8].
For PET, the initial 78.4° edge angle was reduced to 40.1° with a 1 s treatment time at 0.3 mm spacing and 400 W power. By increasing the treatment time, the wetting contact angle decreases to a limit, which starts to increase again after the end of the treatment, but the data obtained show that it does so only to a limit which is smaller than the initial contact angle. The surface free energy changes in accordance with this [8].
Zhao et al treated a mesh of PA-6,6 fibres with plasma to improve the preparation of coatings by increasing the specific surface area. The process was carried out in two steps, during which the specific surface area increased by one and a half times and the roughness by more than 20 times. The plasma treatment was carried out at 100 mTorr for 5 minutes at 200W power and the coating was subsequently prepared from a solution containing ammonia, dopamine hydrochloride and cellulose. The coating was tested with various oils and was able to measure a contact angle of around 155° under water from an initial contact angle of around 130°. FT-IR analysis was performed on samples before and after the two-step process [9]. This shows that the biggest change during the treatment is the change in the peak at 3300 cm-1, which widened and increased in intensity compared to the other peaks during the modification. This was caused by the appearance of the coating due to the high hydroxyl (-OH) group content of the cellulose.
Kuzminova and co-workers treated PA-6,6 with dielectric barrier-induced atmospheric pressure plasma. The wetting contact angle was successfully reduced from the initial 64° to 30° with a 0.5 s treatment, with a minimum of 25°, which was essentially maintained for the first 6 days after treatment [10]. As a result of X-ray photoelectron spectroscopy, it was determined that the longer the treatment time, the higher the proportion of oxygen and nitrogen atoms on the surface, while the lower the proportion of carbon [11]. Objectives of the research is the investigation of surface treatment of polyamide matrix nanocomposites by atmospheric pressure plasma jet method. Improvement and optimization of wetting properties. Testing the effect of treatment by the still drop method. Determination of the variation of surface free energy as a function of treatment parameters. Investigation of bond strength of specimens with best wetting.

2. Materials and methods

2.1. Polyamide-6 based nanocomposites

PA-6 base nanocomposites with matrix of polyamide-6 (Grilon BS: manufacturer: EMS Chemie AG) and nanoalloy: Montmorillonite (MMT); Nanomer I.30P (manufacturer: Nanocor Corporation). Carbon nanotube (CNT): PLASTICYlTMPA1503 (manufacturer: Nanocyl SA). The different mixtures were prepared on an IDMX mixer fed with a Brabender extruder with a fluid temperature ranging from 210-250 °C with a rising temperature profile. The composition of the mixtures we use is shown in Table 1.

Table 1. The composition of the mixtures used in the experiment.

Type of mixture	PA-6 content (%w)	MMT content (%w)	CNT content (%w)
0	100	0	0
A	94,.9	5	0.1
B	94.5	5	0.5
C	94	5	1

The effluent was discharged into a water bath maintained at 30-45 °C, which was drawn off and shredded using a Collin-type grinder. After drying at 75 °C for 4 hours on a Wittmann-Battenfeld injection moulding machine, ISO 3176-compliant test specimens were produced in a two-cavity mould, one of the cavities of which was closed due to quality problems. The injection moulding process was carried out at a temperature profile of 260-285 °C, with a mould temperature of 80°C, an injection speed of 50 mm/s, a torque pressure of 50 bar and a changeover pressure of 450 bar, with a cooling time of 10 seconds. The polyamide test specimens used were subjected to tensile testing at a tensile rate of

1.0 mm/min for modulus and 50 mm/min for tensile strength as in previous research. The results of the measurements on air-conditioned specimens are presented in Table 2 [9].

Table 2. Mechanical properties of the investigated mixtures.

Type of mixture	Tensile strength (MPa)	Elongation (mm)	Young's Modulus (MPa)
0	50.7	162.65	1534
A	51.7	3.19	2709
B	44.0	3.13	2672
C	44.2	2.3	2760

2.2. Cold plasma surface treatment

For the surface treatment, we used a PlasmaTreat OpenAir® atmospheric pressure system, which can treat the surface without introducing a large amount of heat. It can clean and activate the surface at the same time. The plasma was generated by an FG5001 plasma generator with a maximum output power of 1000 W and an output frequency between 19-23 kHz. This was coupled to a RD1004 dynamic plasma head and an HTR12 high voltage transformer. In use, the equipment operated at a frequency of 21 kHz with a current of 17.5 A. The plasma-emitting head was moved by a Yaskawa robotic arm, which allowed the distance to the surface to be maintained accurately along its entire line. During the surface treatment, the plasma parameters were left unchanged, instead the movement of the robotic arm was controlled, allowing the intensity of the treatment to be varied sensitively. The surface treatments were carried out at different speeds of movement and at different distances between the plasma head and the surface. The procedure involved washing the test specimens with alcohol before treatment, after which the surface was treated once with a plasma beam, moving the head parallel to its longitudinal side while it was attached to the underlying metal plate with a magnet to prevent the plasma from moving the specimen. The markings on the metal plate underneath made it possible to keep the specimen in the same position under the plasma each time.

2.3. Surface energy measurement

Immediately after the surface treatment, we measured the wetting angle on the specimens using the stilling-drop method. During the measurement, the test specimen containing 5 μl of liquid is placed on the movable table top under the automatic pipette. The drop must be applied to the test specimen without dropping it, as this will change the wetting angle and falsify the measurement result. This can be avoided by bringing the movable stage closer to the pipette. Once the drop has been placed on the specimen, the surface of the specimen on the movable stage should be moved back into the plane of the camera and the light source, and the photograph can then be taken with the Nikon D5300. Exposure times of 1/100 s, ISO-200 sensitivity, 100 mm focal length and f/10 depth of field were used. As the camera vibrations were not reduced when touching the camera before taking the picture, we controlled the camera from a computer, which also greatly facilitated data storage. We took measurements with five to five drops of both distilled water and 99.9% pure ethylene glycol, which we subsequently evaluated using software, producing a total of ten edge angles for each parameter. From the obtained boundary angles, we calculated the surface free energy using the Fowkes method. From this, the dispersion component of the solid surface is immediately obtained.

2.4. FT-IR investigation

The measurement was performed with a Jasco FT-IR 4600. In order to measure the true condition after surface treatment, we transported the sample from the plasma treatment room to the FT-IR instrument in sealed packaging. The measurement was performed in the wavelength range 4000 - 500 cm^{-1} with 64 sweeps.

2.5. Adhesive shear strength test

The adhesion of the nanocomposites with different filler contents was investigated by bonding using Loctite 9466 adhesive on the stem of the specimen on a 20 x 10 mm^2 surface using a template, using the parameters considered ideal. Since the aim is to characterize the bonded surface, the amount of adhesive was chosen as the smallest possible amount to be handled, which was 0.10 g. Five replicate measurements of each composition were performed. After each sample was glued together, it was placed in a press and left for 72 hours. The bonded joints were tensile tested using an INSTRON 3366 tensile testing machine at a tensile speed of 10 mm/min. The tensile testing machine has manual clamping jaws. To avoid lateral tearing, counter pieces were used during clamping.

3. Results

3.1. Results of wetting tests on untreated polymer

The wetting contact angles of the untreated specimens are given in Table 3, together with the resulting surface free energy values.

Table 3. Wettability results of untreated polymer mixtures.

	Wetting contact angle (°)		Surface energy (mN/m)		
Type of mixture	Distilled water	Ethylene glycol	Polar	Disperse	Total
0	73	43	11	26	37
A	71	47	16	19	35
B	81	62	11	15	26
C	71	44	14	6	21

In evaluating the results of the research, we take the wetting and surface free energy values for the materials shown here as the untreated values, which we use in the subsequent diagrams. From the wettability results measured on the untreated polymer surface, it can be seen that basically an unfavourable, poor wetting condition is observed for the injection moulded product. With such a surface it would not be appropriate to immediately implement an adhesion bonding technology. Looking at the polymer types, the distilled water contact angle of 71-81° and the low surface free energy of 25-40 mN/m are expected to show poor bonding performance and low adhesion energy. Compared to the wetting of the unfilled polyamide 6 based specimen, the composite with 0.5% CNTs exhibits significantly poorer wetting phenomena at higher and lower contents, with no change in the resolution tested.

3.2. Effect of plasma surface treatment on wetting phenomena

As a first step, the aim is to investigate the effect of polyamide wettability as a function of treatment distance and speed on polyamide samples not containing MMT and CNT. For this purpose, a relatively broad but reasonable scale was chosen, i.e. small enough to require relatively few measurements but large enough to find a possible minimum in wettability or maximum in surface energy. On the advice of the plasma system manufacturers, the closest recommended distance was 4 mm as a lower limit. On this basis, treatment distances were chosen to be 4; 10; 15 and 20 mm, and treatment rates 0.5; 1.0; 2.5; 5.0 and 10.0 m/min (Figure 1.).

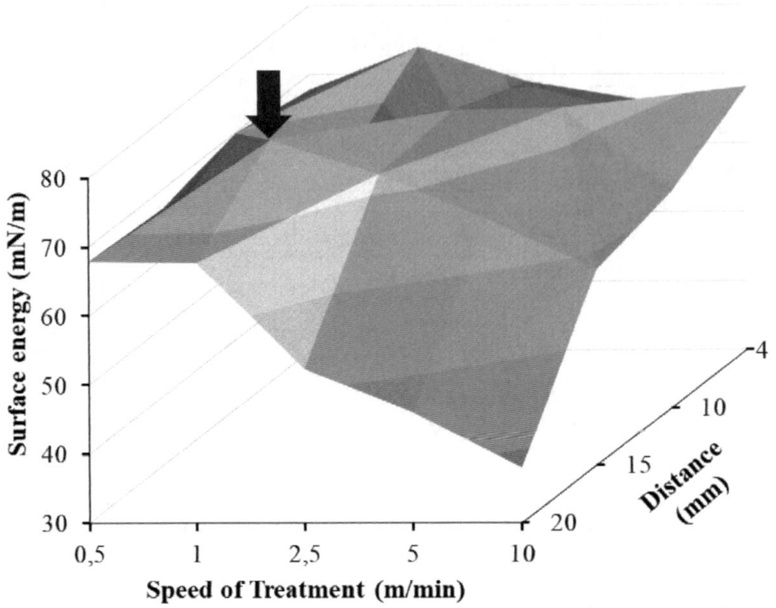

Figure 1. Effect of cold plasma technology parameters to surface energy improvement.

A clear minimum was observed for the sample treated at 1 m/min; 15 mm spacing, for the latter, as the ethylene glycol was completely spread over a wider range, this cannot be determined. The total interfacial energy of the polar component of the surface free energy calculated by the Fowkes method is shown in Figure X. These confirm that for the sample treated at 1 m/min; 15 mm spacing, both the polar component of the surface free energy and the total surface free energy are larger than for any other parameter pair. If the plastic surface is treated for too long or too close, this is detrimental to wetting. In some of these cases (e.g. feed rate of 1.0 m/min for 4 mm), the specimen has visibly warped while the surface has visibly deteriorated. At high feed rates or long treatment distances, the treatment intensity is not such that it shows a significant change, as for example at a treatment distance of 20 mm and a feed rate of 10 m/min.

In order to be able to declare that the polymer was not significantly degraded, we used Fourier transform infrared spectroscopy to check the sample treated at the optimum treatment distance of 15 mm and at a feed rate of 1 m/min, the results of which are shown in Figure 2.

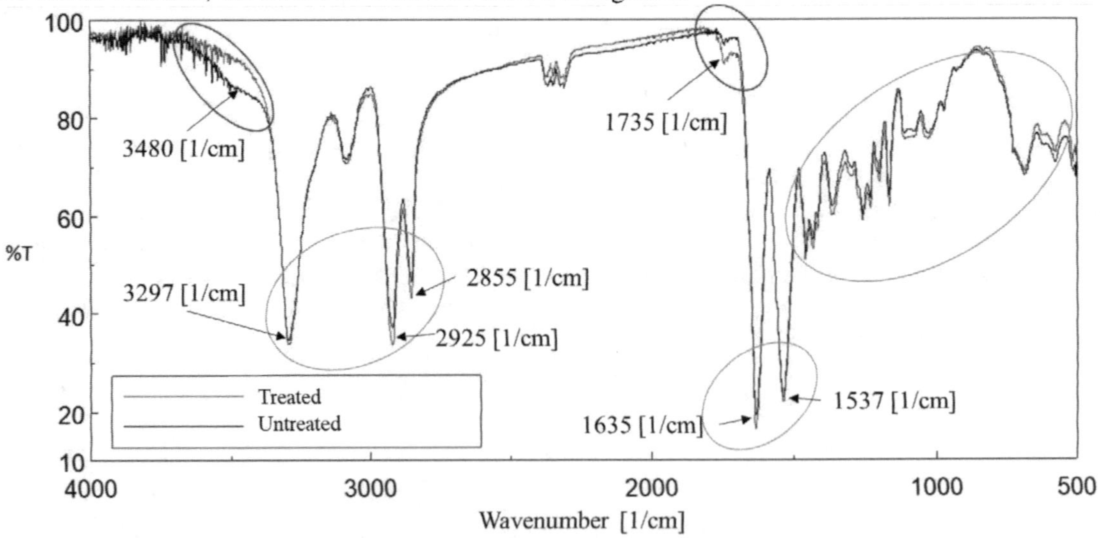

Figure 2. FT-IR analysis of untreated and hydrofoil PA-6 samples.

In the spectra shown, the curves before and after treatment are overlapping, and after 1600 cm^{-1} they are almost perfectly identical, which indicates that no significant degradation has occurred during the surface treatment, with only two minimal differences. At 3500 cm^{-1}, one signal appears to have disappeared as a result of the treatment. This was probably due to evaporation of water bound by the polyamide. At 1737 cm^{-1}, a small shoulder appeared as a result of the treatment, showing the vibration of a carbonyl group that is not bound to acid amide. This can be explained by the fact that the plasma causes a radical reaction to take place and an oxygen from the air is attached to the polyamide carbon chain. Another explanation could be that some of the -CO-NH- bonds formed during polymerisation are broken, but the spectrum does not show any trace of the amine remaining. Thus, we can see from the FT-IR spectra that plasma treatment has succeeded in forming additional polar groups (-CO-) in the carbon chain of the polyamide, which has improved wetting. With the optimum parameters obtained, we then treat the samples with different CNT contents, which are also first wetted, then bonded and torn, taking 5 parallel measurements for each of them. The wetting of the sample treated at 1 m/min; 15 mm spacing was found to be the best, and the FT-IR analysis ensured that if degradation occurred, it would be minimal, so this pair of parameters was chosen to measure wetting before and after treatment on nanocomposites of compositions A, B and C (Figure 3).

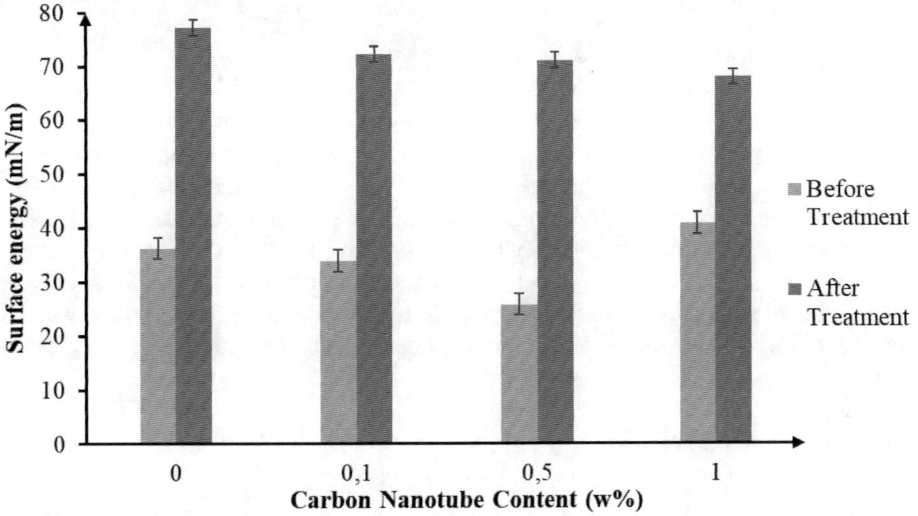

Figure 3. Total surface free energy component before and after treatment as a function of CNT content.

Somewhat surprisingly, at 0.5% CNT content, there appears to be a maximum wetting angle for both distilled water and ethylene glycol. After treatment, the contact angle disappears completely for ethylene glycol, but only the maximum disappears for distilled water. The most likely reason for this is that there was contamination on the surface of the test specimen that could not be washed off with alcohol, but was removed by the plasma surface treatment.

These results follow expectations based on the wetting contact angle, however it is worth pointing out that the total interfacial tension decreases as a function of CNT content. The polar component also decreases, which implies a decrease in the number of polar groups on the surface, which can be explained by the fact that more fibres are deposited on the surface with increasing fibre concentration. In parallel, the disperse component, i.e. the strength of the non-site-specific interactions, increases as more fibres are present that are able to dissipate the site-specific interactions in a larger surface area. Although the fibres are arranged parallel to the flow direction on the surface of the injection moulded specimens, it is easy for a fibre to turn into the block phase in the case of relatively flexible CNTs. In this case, a contact is created between the surface and the array phase, which reduces the energy surplus of the surface relative to the array level, i.e. the surface free energy.

3.3. Investigation of the shear strength of samples surface treated with optimum parameters

The results of tearing samples 0, A, B and C treated and then bonded at 1 m/min; 15 mm spacing and reference samples bonded without treatment are shown in Figure 4. It can be seen that the reference specimens already separated along the bond at relatively low loads.

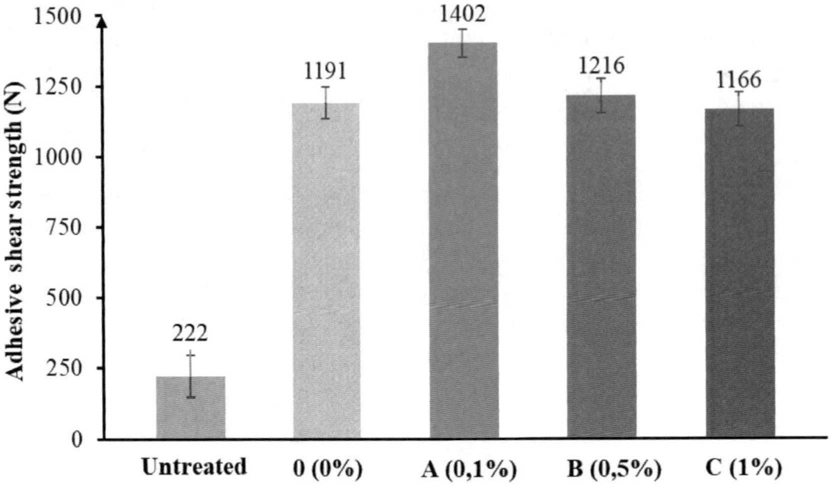

Figure 4. Stress at fracture of nanocomposites of different compositions.

As the treated specimens fractured in their material after surface treatment and the bonded surfaces did not separate, and in addition to the fact that the bonding effort was directed towards the smallest amount that could be treated, the increase in adhesion can be considered clearly successful, as the initial load of 222 N (1.1 MPa) was increased to 1402 N (7 MPa), almost seven times. The specimens fractured as the CNT content increased, so that the 0.5% and 1% CNT nanocomposites, we are not able to actually test the shear strength developed on the bonded surface. In the tensile tests, the mixtures were found to be homogeneous, but increasing the CNT content deteriorated the mechanical properties of the nanocomposite.

3.4. Investigation of the time dependence of samples surface treated with optimal parameters

The total interfacial energy as a function of time for samples treated with the optimum parameters is shown in Figure 5.

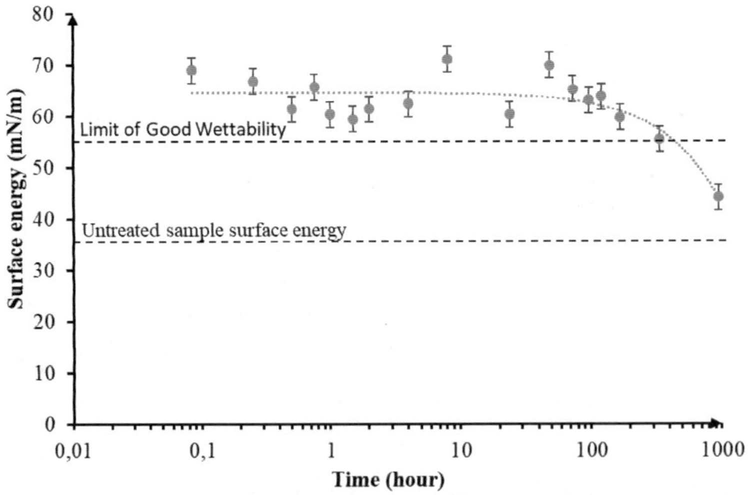

Figure 5. Time dependence of surface free energy.

In parallel, we also treat an independent sample and measure wetting at 5; 15; 30; 45; 60; 90; 120 minutes, 4; 8; 24; 48; 72; 96; 120; 168; 336 and 936 hours after treatment, thus obtaining information on the time dependence of the treatment. As long as the contact angle does not reach 45°, the surface is well wetted. This shows that the surface starts to lose its activation only after about 100 hours, i.e. 4-5 days, but reaches the good wetting limit of 55 mN/m only after about 300 hours.

4. Summary

Results of the research:

- Atmospheric pressure plasma jet surface treatment was successfully used to treat the surface of polyamides and polyamide matrix nanocomposites.
- In the still drop method, the contact angle with water decreased from 73° to 21° and the contact angle with ethylene glycol decreased from 43° to 0°, with a concomitant increase in the total surface free energy from 36.3 mN/m to 77.2 mN/m. To achieve these results, the optimum distance used was 15 mm and the optimum treatment speed was 1 m/min.
- Based on the FT-IR spectra, these improvements were not accompanied by significant degradation and it is likely that the changes seen are the result of surface activation.
- The results show that wetting properties can be improved by treatment with the right parameters and equipment, which is related to adhesion and cohesive bonding techniques.
- Bonding and tearing of all specimens treated with optimal parameters showed an improvement of bond strength of nearly 630% compared to the specimens.

5. References

[1] Kun K and Weltsch Z 2021 Effect of Femtosecond-Laser-Structured Injection Molding Tool on Mechanical Properties of the Manufactured Product *Polymers* **13** 2187

[2] Kun K and Weltsch Z 2021 Research of the Effect of Macrogeometric Structures on the Melt Front Using Simulation *In Advances in Manufacturing Engineering and Materials II: Proceedings of the International Conference on Manufacturing Engineering and Materials* **1** 282

[3] Béres G J, Végvári F and Danyi J 2021 Nagy szilárdságú acéllemezek és hegesztett terítékek alakítási jellegzetességei *ACTA MATERIALIA TRANSYLVANICA* **4** 18

[4] Kovács P, Sztupinszky Zs, Patkás S, Béres G J and Hareancz F 2021 Lítiumion cellák hegesztett kötéseinek fejlesztése *GRADUS* **8** 195

[5] Kovács Zs F and Hareancz F 2021 A kavaró dörzshegesztés alkalmazási lehetőségei alumínium kompozitok esetén *GRADUS* **8** 294

[6] Kelar J, Shekargoftar M, Krumpolec R and Homola T 2018 Activation of polycarbonate (PC) surfaces by atmospheric pressure plasma in ambient air *Polym. Test.* **67** 428

[7] Homola T, Matoušek J, Hergelová B, Kormunda M, Wu L and Černák M 2012 Activation of poly(methyl methacrylate) surfaces by atmospheric pressure plasma *Polym. Degrad. Stab* **97** 886

[8] Homola T, Matoušek J, Hergelová B, Kormunda M, Wu L and Černák M Activation of poly(ethylene terephthalate) surfaces by atmospheric pressure plasma *Polym. Degrad. Stab* **97** 2249

[9] Zhao P, Qin N, Ren C L and Wen J Z 2019 Surface modification of polyamide meshes and nonwoven fabrics by plasma etching and a PDA/cellulose coating for oil/water separation *Appl. Surf. Sci* **481** 883

[10] Kuzminova A, Shelemin A, Kylián O, Choukourov A, Valentová H, Krakovský I, Nedbal J, Slavínská D and Biederman H 2014 Study of the effect of atmospheric pressure air dielectric barrier discharge on nylon 6,6 foils *Polym. Degrad. Stab.* **110** 378

[11] Çoban O, Akman E, Bora M Ö, Genc Oztoprak B and Demir A 2019 Laser surface treatment of CFRP composites for a better adhesive bonding owing to the mechanical interlocking mechanism *Polym. Compos.* **40** 3611

Acknowledgments

This work was supported by the Hungarian Government and co-financed by European Social Fund (GINOP-2.3.4-15-2016-00001).

IOP Conf. Series: Materials Science and Engineering 1246 (2022) 012029 doi:10.1088/1757-899X/1246/1/012029

Investigation of segregation in the runner system during injection moulding

D. R. Fris, F. Szabó

Department of Polymer Engineering, Faculty of Mechanical Engineering, Budapest University of Technology and Economics, H-1111 Budapest, Műegyetem rkp. 3., Hungary

E-mail: szabof@pt.bme.hu

Abstract. The goal of this study is to investigate segregation in the runner system during injection moulding. Segregation is a phenomenon which occurs during the injection moulding of filled or reinforced polymer systems. The effects of segregation can be so severe that the process can lead to waste products. Even though segregation is an important problem, so far it has been little researched. However, it would be very beneficial if segregation during injection moulding could be predicted through simulation software. In order to achieve that, we must first understand what causes segregation. We injection moulded several specimens using polypropylene, a transparent polymer, and used 10.8 V% ceramic beads as filler with different injection moulding parameters. We investigated the effect of melt temperature and injection rate, which were set to 180, 220, 260 °C and 5, 20, 80 cm^3/s. We observed segregation with an optical microscope and used a furnace to burn the PP and measured the weight of the leftover beads in order to determine the bead content in various regions.

1. Introduction

Although plastic products have been manufactured only since the second half of the 20th-century, nowadays there is virtually no industry that is not using polymers. Injection moulding is one of the most widely used and cost-effective polymer processing technologies, with which practically any kind of product can be manufactured. Injection moulding is fast and waste is minimal. It is essential, however, to optimise the parameters of the process.

One of the most sensitive aspects of the quality of injection moulded products is homogeneity. Proper mixing and setting the injection moulding parameters correctly are key to achieving good homogeneity [1, 2]. Inhomogeneity can take the form of colour streaks, which is a relatively common problem during injection moulding. It is caused by the irregular distribution of the colour pigments, the masterbatch.

In order to enhance or alter the properties of polymer products, engineers often use fibre reinforcement or filler materials. The proper distribution of these materials is more important than the distribution of the masterbatch; improper distribution of the reinforcement or filler materials can make the product defective. When reinforcements or fillers are used during injection moulding, segregation must be taken into account. A segregated system or mixture is different from an inhomogeneous system—the latter is a perfectly randomised distribution of two phases, while the former means that the probability of finding a particle in a certain region differs from the overall average of particle content [3, 4]. The purpose of reinforcement in polymers is to strengthen the structure, but due to segregation and the unfavourable orientation of the fibers, considerably weaker regions may form across the weld lines [5-7]. Although segregation is important in designing the product [8], only a few studies

investigate the reason why and how segregation occurs during injection moulding. Since the exact mechanism of segregation is unknown, industrial simulation software cannot calculate it properly.
Our long-term goal is to develop a method that would be capable of simulating segregation by either using a FEM-DEM coupled system or developing a new material model. But first, it is necessary to investigate segregation. In this study, we investigate how different injection moulding parameters affect segregation, and attempt to find what causes segregation. We injection moulded a product with a geometry that resembles the runner system. We suspect, that segregation somehow correlates with the thickness and formulation of the frozen layer during injection moulding. We intend to investigate segregation by comparing simulation results describing the frozen layer to the rate of segregation.

2. Materials, machinery and methods

In order to examine segregation, we injection moulded specimens using a mould with several branches to represent the runner system. The injection moulding machine was an Arburg Allrounder Advance 270S 400-170. The matrix was polypropylene (PP), MOL TIPPLEN H145 F and the filler (10,8 V%/32 m%) was ceramic beads, Saint-Gobain ZirPro Microblast® B120. We used PP because it is a polymer not prone to shear-induced fill imbalances, therefore the lengths of how the polymer melt fills each cavity is more uniform and the measurements can be done more easily. We used ceramic beads, since they are opaque, and can be seen more clearly under a microscope. We investigated the effect of two injection moulding parameters: melt temperature and injection rate. Melt temperature was 180, 220 and 260 °C and injection rates were 5, 20 and 80 cm^3/s. Mould temperature was set to a fixed 50 °C. We selected 5 regions from the eight-branched geometry and examined the surface of the cross-section with a Keyence VHX-5000 optical microscope. Then we burnt off the matrix near the relevant regions with a Denkal 6B furnace to determine filler content in m%. We injection moulded specimens and ran simulations in Autodesk Moldflow with the same injection moulding settings, to examine how the thickness of the frozen layer correlates with segregation.

3. Experiments

We examined the specimens in certain regions with an optical microscope and burnt off the matrix. We compared the results of segregation to the simulation results.

3.1. Simulation results

We ran simulations with the same injection moulding settings with which the specimens were manufactured. The temperature results of the simulations can help predict the thickness of the frozen layer. Table 1 shows the simulation results of the volume ratio of the frozen layer with different injection moulding settings. The left column shows the notation of the specimens. As the table suggests, the higher the melt temperature, the thinner the frozen layer on the surface of the 50 °C mold. The explanation is quite simple, the hotter melt cannot cool down so easily and freeze. The results indicate that with a lower injection speed, the frozen layer will be thicker. This is because the frozen layer has more time to form.

Table 1. The results of the volume ratio of the frozen layer with different injection moulding parameters.

Notation of the specimen	Melt temperature (°C)	Injection speed (cm^3/s)	Volume ratio of the frozen layer (%)
A	180	20	19,13
B*	220	5	37,54
C	220	20	13,57
D	220	80	5,28
E	260	20	10,47

3.2. Optical Microscope

Before the experiment, we decided where we would examine the cross-section in the specimens. We examined the runner in the first branch (figure 1, B), which carries the melt towards the junctions directly from the sprue and in the "1st" and "3rd" junction at both the top (figure 1, F1 and F3) and the bottom (figure 1, A1 and A3), so overall there are 5 regions to check. We cut the specimens at these areas and produced images of the cross-sections using an optical microscope. Figure 1 shows the

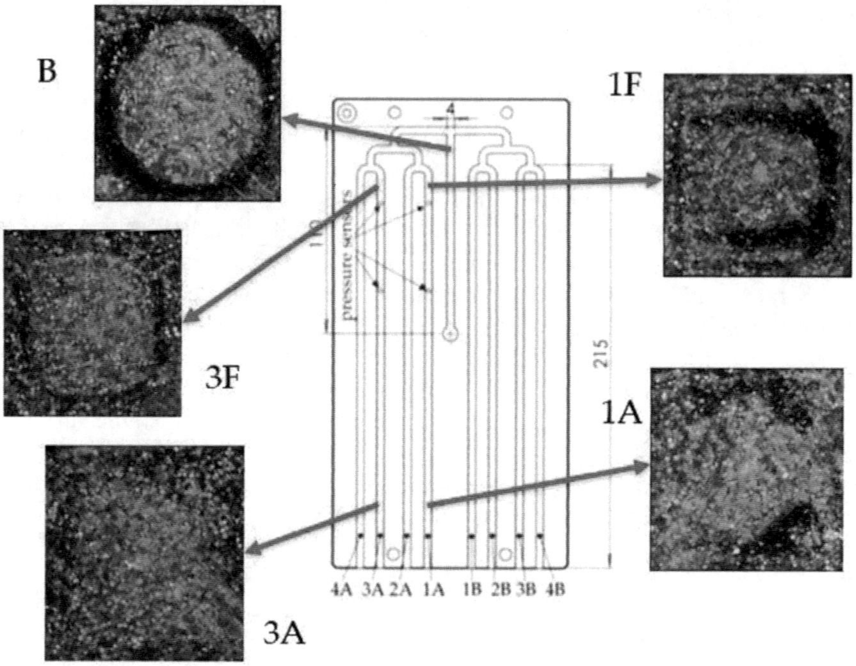

Figure 1. The characteristic microscope image of each cross-section and their notations.

discussed regions and their notations, and a characteristic microscope image of the cross-sections. Figure 2 displays all the microscopic images with different injection moulding settings at the locations we selected. The white areas are the ceramic beads and the dark region is the PP matrix material with only a few beads. Sometimes a dark spot can be observed at the core region of the melt (e. g. 3F/E, figure 2); these are probably caused by air bubbles or vacuum during the injection moulding process. As figure 2 suggests, the 5 different cross-sectional regions each have their own characteristic segregation profile. The way the injection moulding settings affect these profiles can be observed for example in region B (figure 2, column B), where with setting B*, which is the slowest injection rate, the darker area near the surface of the mould is the largest Because of table 1, we know that with these parameters, the frozen layer is the thickest there.

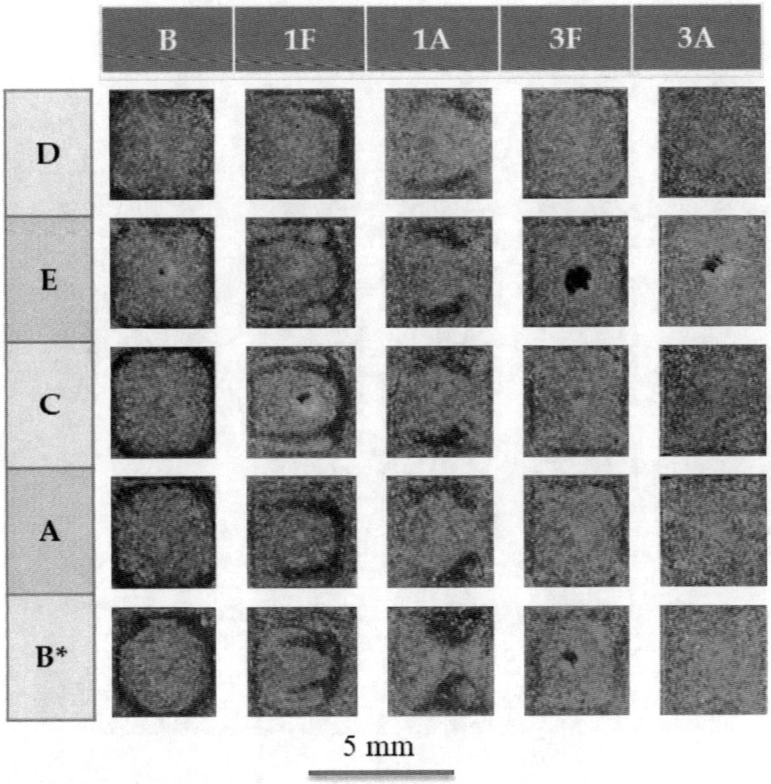

Figure 2. The microscopic images created from the cross-sections of different regions with different injection moulding settings.

3.3. Burnoff

To determine the filler content of the examined regions, we burnt off the matrix in 3 specimens with each set of settings. With the weight of the leftover ceramic beads the filler content can be easily calculated. The nominal bead content of specimens is 32 m%. We introduced a notation system in order to easily plot the results. The first part of the name of the specimen suggests the location of the cross-section (B, 1F, 1A, 3F and 3A, figure 1) and the second part is the injection moulding setting (A, B*, C, D, E, table 1). The burnoff results of cross-section B are shown in figure 3. The red line marks the nominal 32 m% value. Figure 4 shows all the microscopic images of cross-section "B" lined up according to the thickness of the frozen layer. A comparison of figure 3 and 4 indicates that the segregation of the system is only local, does not cause a quantitative change—figure 3 indicates that filler content is roughly the same. We compared the burnoff results (figure 5) in all the cross-sections of the specimens produced with injection moulding setting B*, which causes the thickest, and setting D, which causes the thinnest frozen layer. With both settings, cross-section B is low in ceramic beads, but while setting B* results in high filler content in cross-section 1F, 1FD stays below the nominal value.

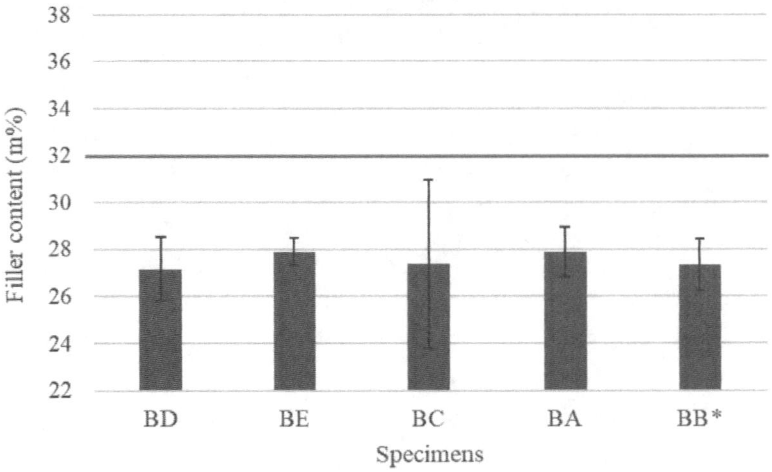

Figure 3. The burnoff results of cross-section B.

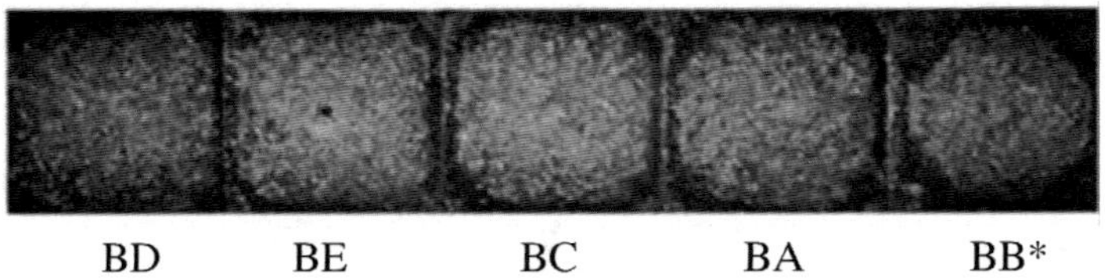

Figure 4. Microscopic images of cross-section B.

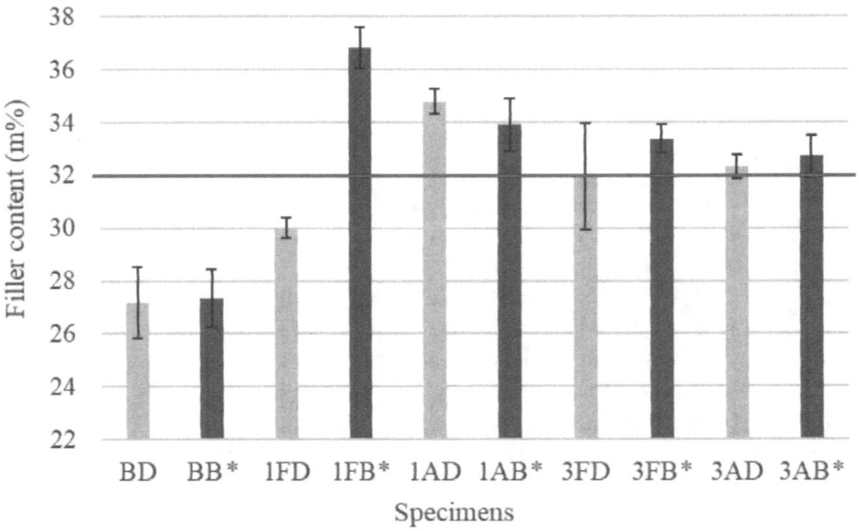

Figure 5. The burnoff results of setting B* and D.

4. Results and discussion

According to the literature, segregation during injection moulding causes an increase in filler content at the end of the flow path and a decrease at the beginning, near the sprue. We assume that this is because the flow is able to tear out beads or fibre, which are not fully embedded in the frozen layer. Hence, a correlation between the frozen layer and the rate of segregation can be assumed as well.

Microscopic images in different cross-sections show different bead distributions. We assume that these segregation profiles are related to the viscosity profile during injection moulding. If we compare the temperature profiles of the simulation results (figure 6) with the microscopic images, a similar distribution can be observed. The temperature profiles are asymmetric due to shear induction. Figure 2 shows that segregation, the distribution of the beads, is also asymmetric. The profile of the simulation results and segregation are similar in the case of 1F. This suggests a similar phenomenon to the one discussed with cross-section B (figure 4), where the frozen layer might be the reason behind the segregation profile. However, based on shear induced fill imbalances, the right side of cross-section 1F is filled with a hotter melt, therefore the frozen layer is theoretically considerably thinner there. The reason why this hot melt region is absent in the simulation results (figure 6) is that even industrial simulation software is not able to correctly calculate the effects of shear induction. The segregation

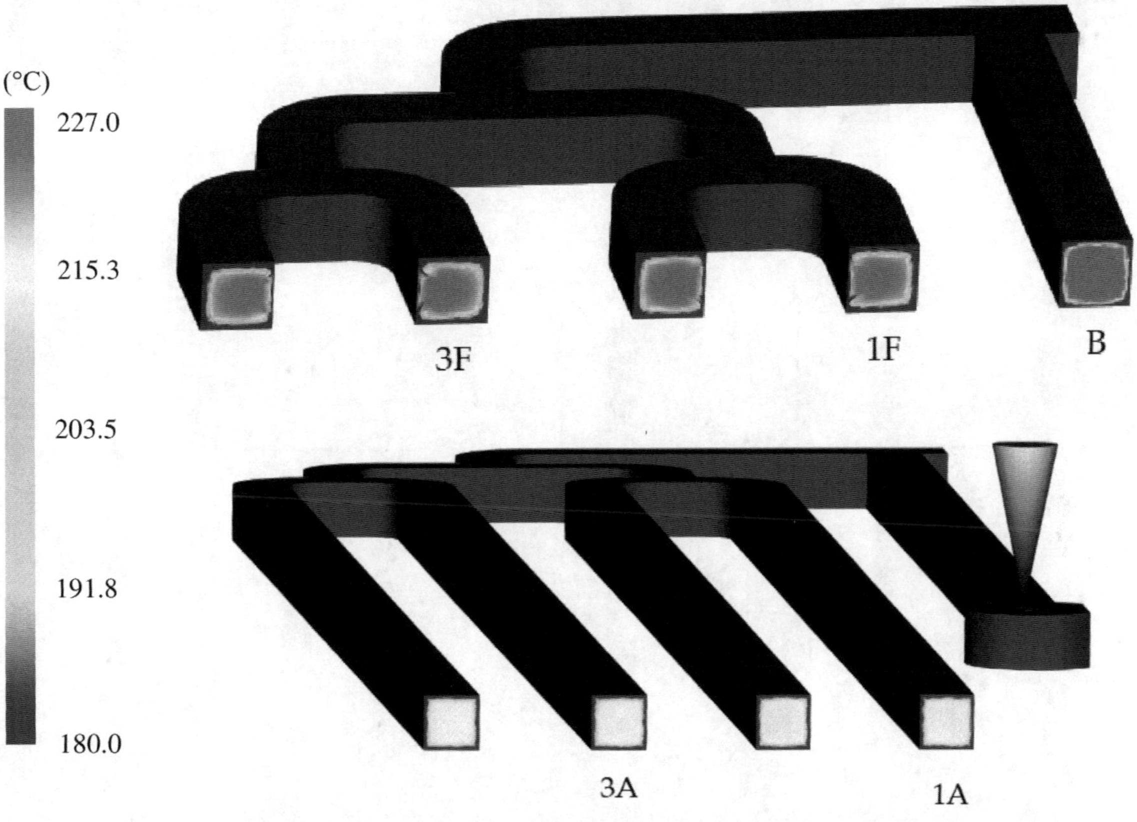

Figure 6. The temperature profiles of the simulation results.

profile of 1F (figure 2) cannot be explained by the frozen layer—further studies are necessary.What we expected from the burnoff results was a tendency in segregation: increased filler content at the end of the flow path (cross-section 1A and 3A, figure 7) and lower filler content at the beginning of the flow path (cross-section B, figure 3). Cross-section 1A behaves according to this theory (figure 7) with every injection moulding setting. Segregation is shown by the amount of beads in a given location. The microscopic results indicate that the distribution of the ceramic beads is more even than in cross-section 1F (figure 2). In the case of cross-section 3A, the increase is less significant or non-existent (figure 7, 3AE). We also examined the filler content in cross-section 1F and 3F (figure 8). 1F has high filler content, except with setting D, which produces the thinnest frozen layer, while with setting B*, the thickest frozen layer shows the greatest increase. This pattern cannot be observed in cross-section 3F. It would seem that Junction 3 is more uniform in bead content than Junction 1. Both the burnoff results (figure 7 and 8) and the microscopic images (figure 2) show more even distribution of the beads. This might also be connected to shear induction, but further studies are necessary for a definitive explanation.

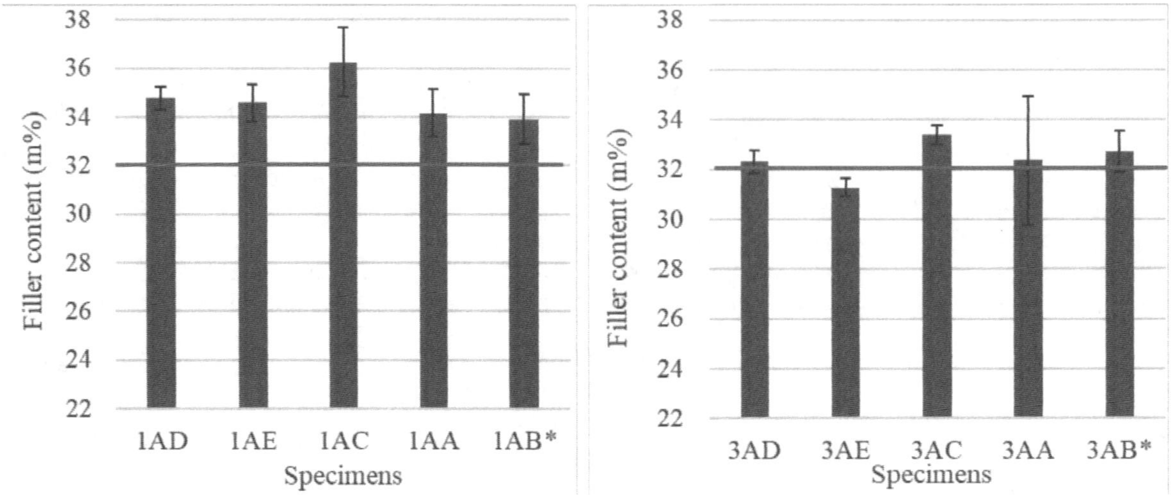

Figure 7. Burnoff results of cross-sections 1A and 3A.

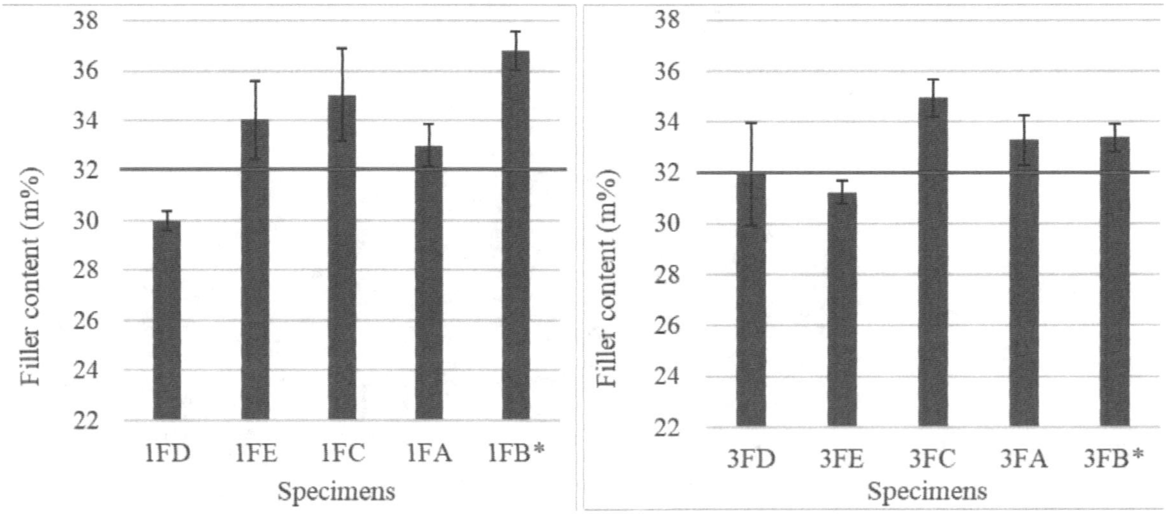

Figure 8. Burnoff results of cross-sections 1F and 3F.

5. Conclusion

Although the segregation of the filler or reinforcement during injection moulding is a considerable problem, it is little researched. Our long-term goal is to develop software that can simulate segregation during injection moulding. To do that, first we have to understand the phenomenon better.

We injection moulded specimens with different injection moulding settings using PP as matrix material and ceramic beads as filler. We chose 5 regions where we cut the specimens and observed the cross-sections with an optical microscope to observe the distribution of the beads. We also burnt off the matrix around the 5 examined cross-sections to determine filler content. We searched for a correlation between the thickness of the frozen layer and segregation. In the case of cross-section B, it seems that the frozen layer does have an effect on segregation, but only the distribution of the beads changed, their quantity did not. Junction 1 shows more segregation than Junction 3; we assume that the reason might be shear induction. For this, further experiments are needed and in the future, it might be expedient to use a polymer as matrix which is prone to extreme shear-induced fill imbalances.

6. References

[1] Zsíros L and Kovács J G 2018 Surface homogeneity of injection molded parts *Periodica Polytechnica Mechanical Engineering* **62** 284-291

[2] Wang M-V, Fatahul A and Vu V-H 2019 The Study of Optimal Molding of a LED Lens with Grey Relational Analysis and Molding Simulation *Periodica Polytechnica Mechanical Engineering* **63** 278-294

[3] Rhodes M 2008 Introduction to Particle Technology New Jersey (John Wiley & Sons, Ltd)

[4] Kovács J G 2011 Shrinkage alteration induced by segregation of glass beads in injection molded PA6 Experimental analysis and modeling *Polymer Engineering & Science* **51** 2517–25

[5] Albrecht K, Baur E, Endres H-J, Gente R, Graupner N, Koch M and Müssig J 2016 Measuring fibre orientation in sisal fibre-reinforced, injection moulded polypropylene – Pros and cons of the experimental methods to validate injection moulding simulation *Composites Part A: Applied Science and Manufacturing* **95** 54–64

[6] Hegler R P and Mennig G 1985 Phase separation effects in processing of glass-bead- and glass-fiber-filled thermoplastics by injection molding *Polymer Engineering and* Science **25** 395–405

[7] Wu H, Zhao G, Wang J, Wang G and Zhang M 2019 A kind of special weld lines with high specific strength and elongation obtained by core-back chemical foam injection molding *Express Polymer Letters* **13** 1041-1056

[8] Jozo M, Cui L and Pukanszky B 2020 Processing induced segregation in PLA/TPS blends: Factors and consequences *Express Polymer Letters* **14** 768-779

Acknowledgments
This research is funded by the (NKFIH) National Research, Development and Innovation Office (FK138501 - The utilization of the rheological and structural characteristics of polymer solutions and melts for their efficient processing).

Experiment of lithium ion accumulator cells welding with diode laser

P Kovács, F Hareancz

John von Neumann university, Kecskemét, Hungary

Abstract. Accumulator cells have an important role in today's industry. It produces a need to connect them, and there are several ways to make the bonds. We examined in this experiment already made resistance projection welded bonds, and we created laser beam welded bonds with various geometries. We tried to find out, which geometry produces the greatest bonds and which parameters will be the best solution.

1. Introduction

Here in our department we examine a lot of technologies, adhesive bonding and various type of welding or brazing technologies. We use surface treatments to improve the strength of bonds, or we remove material from metals or even cut those. [1-2]

We started to examine and improve laser beam welding on lithium ion cells. In this experiment we welded connectors on the already separated end of the lithium ion cells. It was necessary to find the right parameters before we could try to weld on a complete cell.

2. Literature of laser beam welding technology

In order to achieve great welds, we examined a lot of articles, which shows great examples for us. Jingquan Zhang and his team made laser beam welded bonds on copper. They achieved bonds in a wide range of conditions, welding speed from 5 m/min to 15 m/min, 400N pressure and laser power from 3 kW to 6 kW [3]. Eric Punzel and his team investigated different system technologies for continuous laser beam welding [4]. Martin Ruthandi Maina and his team's study two wavelengths of 532 nm and 1064 nm laser welded bonds were examined. They could achieve great welding quality with deep penetration and without porosity [5]. Amirhossein Sadeghian and Naveed Iqbal laser welded dissimilar materials to improve welding technology of batteries. They explored a relationship between process parameters and joint properties [6]. Marc Hummel and his team investigated laser welding of copper welded by 515nm and 1030nm laser beam sources. They selected suitable parameters to weld copper joints, which were independent of the wavelengths of laser beam sources [7].

There is 3 main reason why laser beam welding is great to bond the cell and the connector. It is highly capable of automation, the technology can achieve short cycle times, and the connection makes a stable contact resistance. This is shown in an experiment, that Philipp A. Schmidt and his team made. They achieved great laser welded joint, while using copper as cell connector and they didn't damage lithium ion cells. This solution is great, because when copper is used as connector, electrical resistance can be reduced by 50% [8].

The wavelength of the radiation affects the absorptivity of copper (figure 1.). An Nd:YAG-laser with a wavelength of 1064 nm generates 2-3% absorption in copper, and 98-97% of the energy of the laser is reflected. It looks like higher laser power produces better results while welding, however the absorptivity increases when the material reaches melting temperature. To avoid an unacceptable welding, lower laser power is necessary.

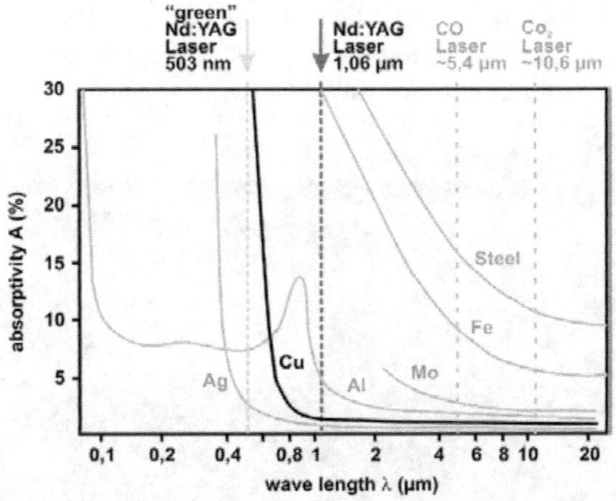

Figure 1.: Absorptivity of different materials [9].

3. Used materials and machines

Used materials were tin coated copper made connectors with a thickness of 0,2 mm. 21700 lithium ion cells were welded with our TruLaser Cell 7020 diode laser. We made a welding device, which holds the connector to the top of the cell (figure 2.), and has a bore in it wherefrom Argon gas flows into the welding area. We prepared it considering the right amount of available distance that the laser beam is in focus where it needs to be. The device is open from the top, therefore the laser beam reaches the welding area.

Figure 2.: Welding device.

We made the welds with a laser welding head attached to our laser device, which was turned 3 degrees, so the reflected radiation does not damage the laser .We cut the top of the cells and joints were cut half then we embedded the joints. After it we examined the welds with a Zeiss Axio Imager. M2m microscope, and we took uptakes from the specimens.

We used fix and variable parameters also for the welding of the lithium ion cells. The Ar gas flow was 9 l/min at the first 4 occasion, after it 6 l/min was used during the welding process. We could change average power, frequency and maximum power of the laser. The welding time was calculated by the machine, from the before mentioned parameters.

4. Laser beam welding experiments

Resistance projection welding provides a great base to compare our laser beam welded joints. figure 3. shows the structure of the welds. It shows that the material of copper and the top of the cell doesn't get mixed. The heating zone causes that, particles in the top of the cell

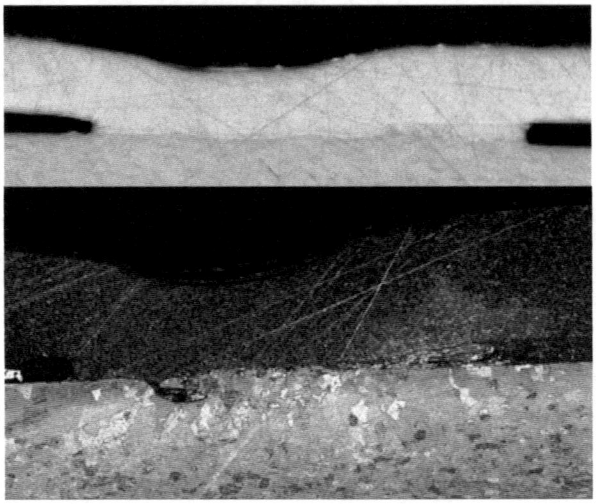

Figure 3.: Resistance projection welded bonds.

In the experiment we changed the maximum power, the average power and the frequency parameters of the laser beam (figure 4). In figures 4 and 5 green colours mean a successful weld, and red colours mean that the laser radiation that the laser radiation reflects or shoots through the materials. In the first experiment we did use 2000 Hz frequency when welding, in the second experiment 2000 Hz and 100 Hz frequency.

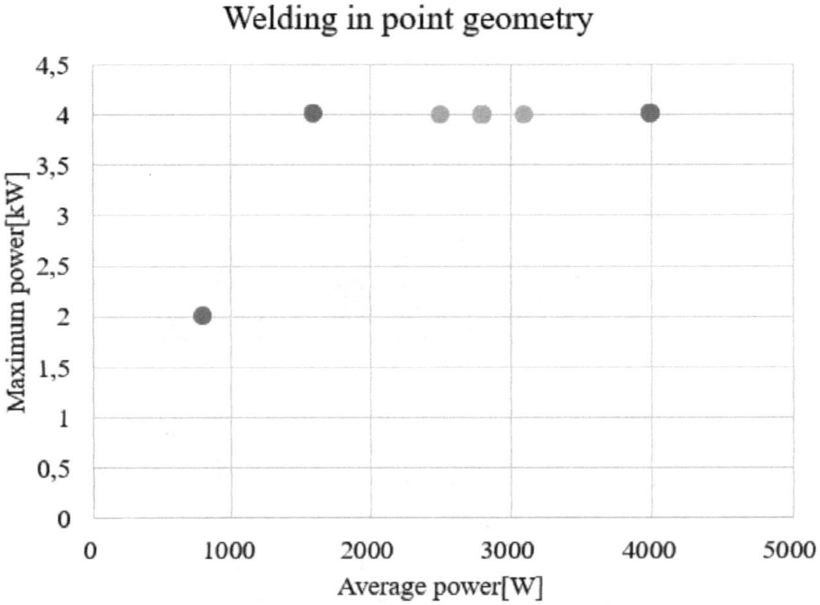

Figure 4.: First experiment, welding in point geometry.

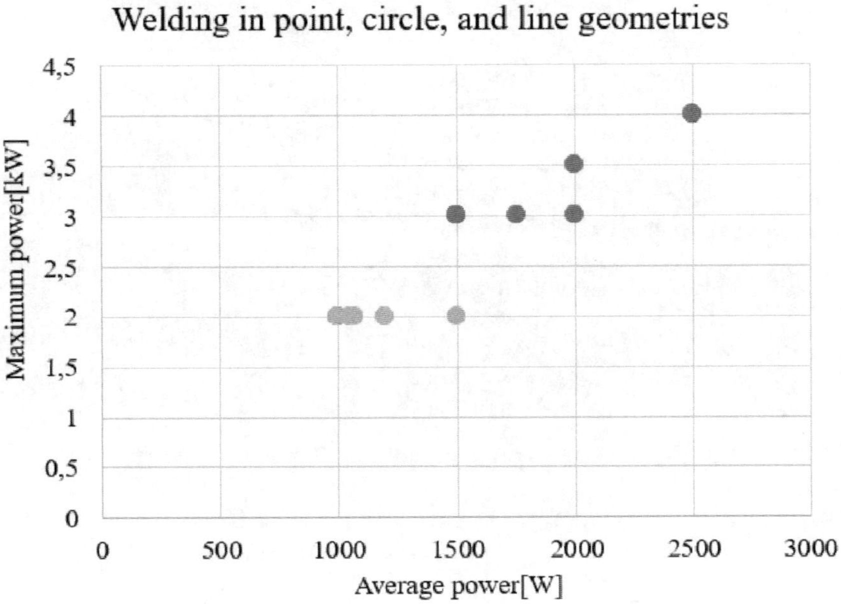

Figure 5.: Second experiment, welding in point, circle and line geometry.

We examined welding joints by a pulling test, if the bonds get unbuttoned than it makes a great bond on top of the cell. In figure 6. welding parameters were 2800 W average power, 2000 Hz frequency, and 4 kW maximum power. It shows a great unbuttoned bonding.

Test 6:

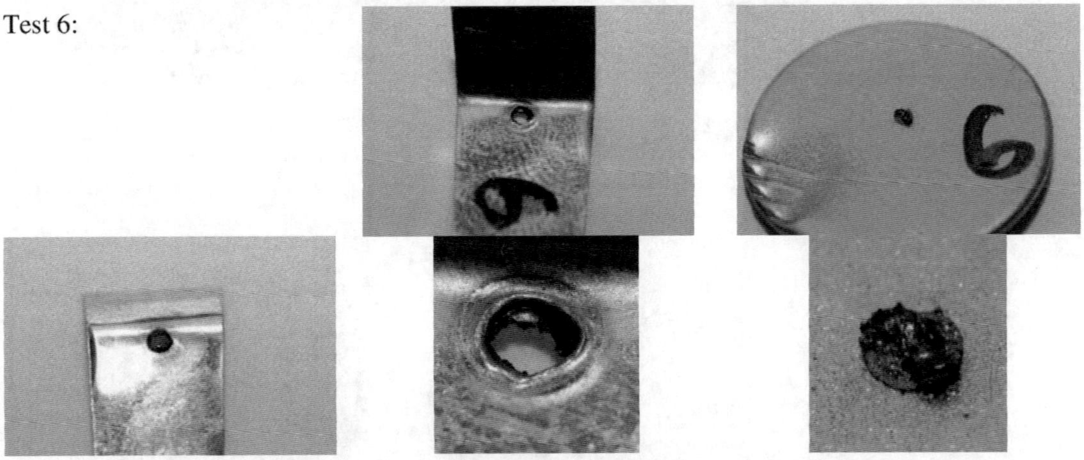

Figure 6.: First experiment, test 6, point geometry.

In figure 7. welding parameters were 2800 W average power, 2000 Hz frequency, and 4 kW maximum power. The laser beam burned through the top of the cell. So this parameter was not acceptable, for the solution. Although the bonds got unbuttoned after the manual pulling test of the connection.

Test 7:

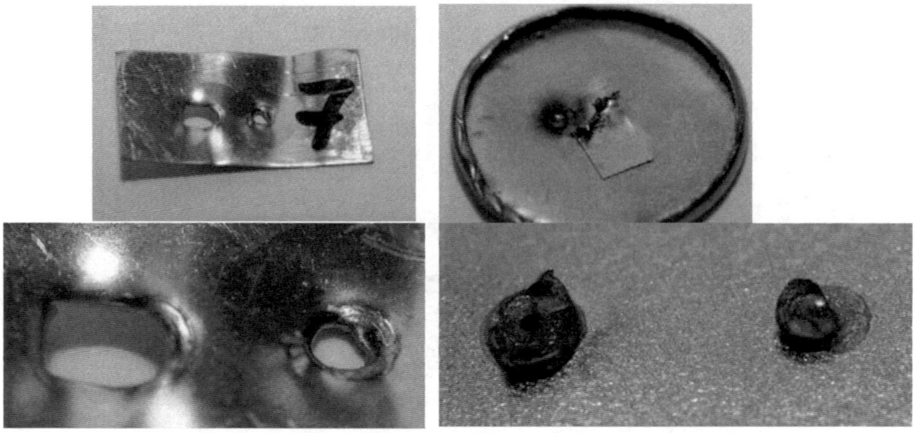

Figure 7.: First experiment, test 7, point geometry.

In figure 8. welding parameters were 2000 W average power, 2000 Hz frequency, and 3,5 kW maximum power. The same problem occurs in this type also, that the laser radiation shoots trough the top of the cell. Makes it also unacceptable.

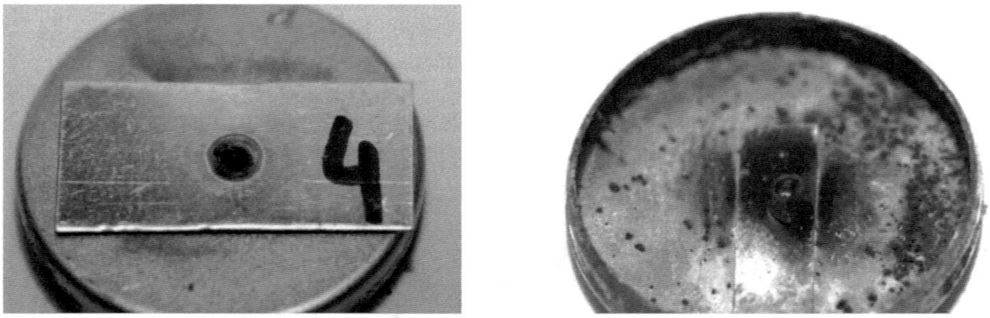

Figure 8.: Second experiment, point geometry, test 4.

In figure 9. welding parameters were 2500 W average power, 2000 Hz frequency, and 4 kW maximum power. In this time, the cell didn't burn through, because the positive side is really different, the material is thicker, where

Test 1:

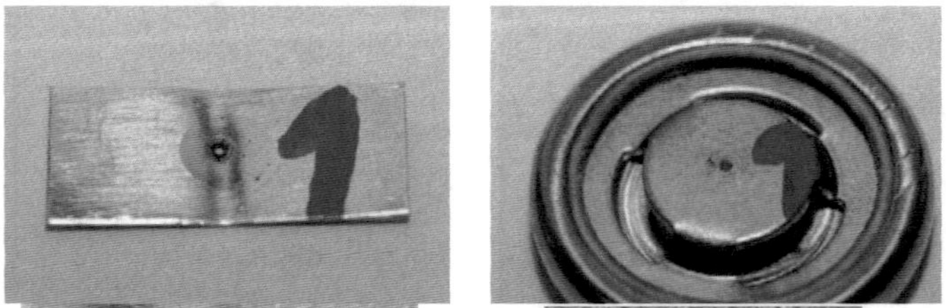

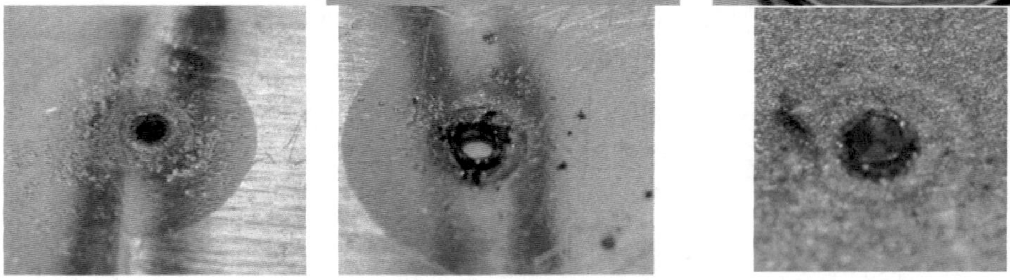

Figure 9.: Second experiment, point geometry, test 1.

In figure 10. and 11. we used 1200 and 1050 average power which caused a totally and a half burn through weld on the connector. After it we tried with lower laser power parameters.

Test 15

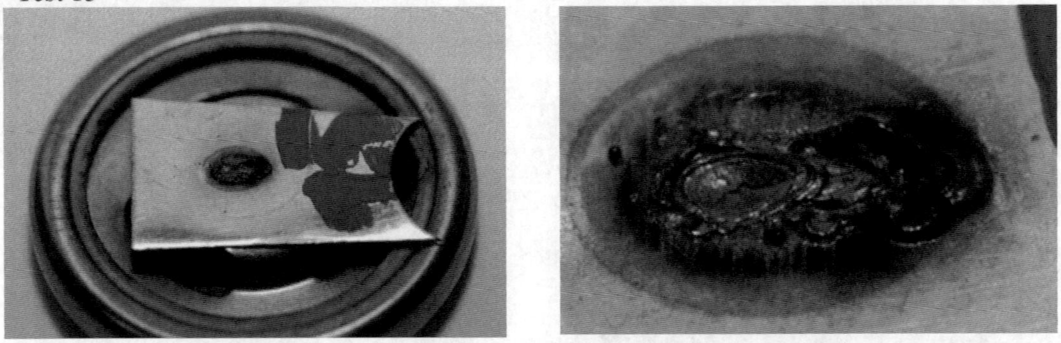

Figure 10.: Second experiment, line geometry, test 15.

Test 16

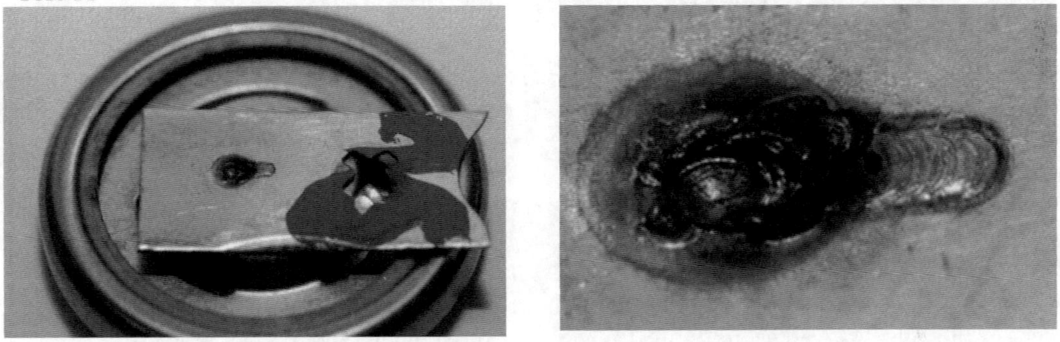

Figure 11.: Second experiment, point geometry, test 16.

We got same results when we welded in a circle geometry, after a little period of time the heat was concentrated in a small area. It caused, that the connector burns through (figure 12.).

Figure 12.: Second experiment, point geometry, test 9.

At this test 1500 W average power, 100 Hz frequency, and 2 kW maximum power were used.

After the test we compared the first experiment, where we achieved a large area with porosity (figure 13.) with the second experiment's bonds. In this time

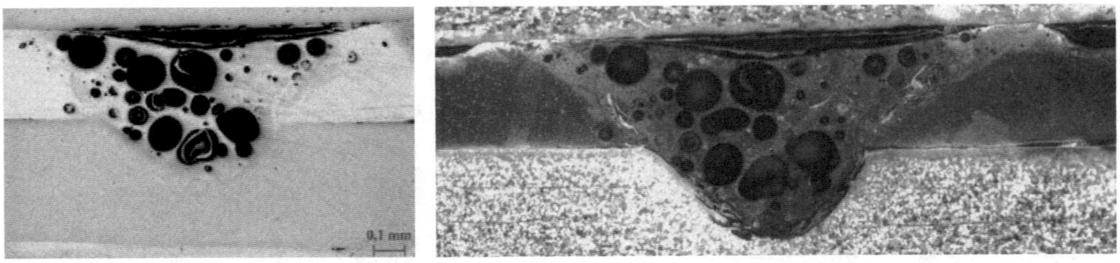

Figure 13.: Early porosity problems.

In figure 14 and 15 the laser beam welded joints had great visual look after the pulling test. This type had greatly unbuttoned bonds

Figure 14.: Welding in line geometry on positive side of the cell.

Figure 15.: Welding in line geometry on negative side of the cell.

Figure 14. shows a little porosity and a crack in the joint, but overall it is an improvement to the structure of the joint. As well as, in figure 15. shows improvement to the earlier welds

5. Conclusions

As result, we achieved visually great welds in the first experiment. After we examined those under the microscope, we found that all of them were porous. Probably the heat intake was so high, that it caused the copper to blow out from the place it was before. Therefore, in the second experiment we used different geometries, because that way heat is absorbed in an increased surface. We examined uptakes from the second experiment, and the geometry or the less laser frequency caused a much better looking welds. There is a little crack in the positive cell side, and no porosity or crack in the negative side.

1. Great welds can be achieved with lower laser power and lower frequency parameters.
2. It should be examined, if changing the laser power during the welding process has a great effect on it.
3. The welds are significantly better if we move the laser beam in a line or circle geometry, but after little time it gets worse. Adjusting laser power might solve this issue.

6. Acknowledgments

The research was supported by 2019-1.1.1-PIACI-KFI-2019-00075. The authors thankfully acknowledge its financial support.

7. References

[1] Berczeli Miklós and Tajti Ferenc: Development of high power femtosecond laser microstructures on automotive stainless steel, IOP CONFERENCE SERIES: MATERIALS SCIENCE AND ENGINEERING 903 p. 012025 (2020) doi:10.1088/1757-899X/903/1/012025

[2] Berczeli Miklós, Weltsch Zoltán: Enhanced Wetting and Adhesive Properties by Atmospheric Pressure Plasma Surface Treatment Methods and Investigation Processes on the Influencing Parameters on HIPS Polymer; POLYMERS 136 p. 901 (2021); https://doi.org/10.3390/polym13060901

[3] Jingquan Zhang, Ting Huang, Sergey Mironov, Dong Wang, Qingwei Zhang, Qiang Wu, Jiejie Xu, Rongshi Xiao: Laser pressure welding of copper; Optics & Laser Technology Volume 134, February 2021, 106645; https://doi.org/10.1016/j.optlastec.2020.106645

[4] Eric Punzel, Florian Hugger, Robert Dörringer, Thorm Lembit Dinkelbach, Andreas Bürger: Comparison of different system technologies for continuous-wave laser beam welding of copper, Procedia CIRP Volume 94, 2020, Pages 587-591; https://doi.org/10.1016/j.procir.2020.09.081

[5] Martin Ruthandi Maina; Yasuhiro Okamoto; Kazuki Hamada; AkiraOkada; Shin-ichi Nakashiba; Norio Nishi: Effects of superposition of 532 nm and 1064 nm wavelengths in copper micro-welding by pulsed Nd:YAG laser, Journal of Materials Processing Technology Volume 299, January 2022, 117388, https://doi.org/10.1016/j.jmatprotec.2021.117388

[6] Amirhossein Sadeghian; Naveed Iqbal: A review on dissimilar laser welding of steel-copper, steel-aluminum, aluminum-copper, and steel-nickel for electric vehicle battery manufacturing, Optics & Laser Technology Volume 146, February 2022, 107595 https://doi.org/10.1016/j.optlastec.2021.107595

[7] Marc Hummel, Mike Külkens, Christoph Schöler, Wolfgang Schulz, Arnold Gillner: In situ X-ray tomography investigations on laser welding of copper with 515 and 1030 nm laser beam sources, Journal of Manufacturing Processes Volume 67, July 2021, Pages 170-176, https://doi.org/10.1016/j.jmapro.2021.04.063

[8] Philipp A. Schmidt, Patrick Schmitz, and Michael F. Zaeh: Laser beam welding of electrical contacts for the application in stationary energy storage devices, Journal of Laser Applications 28, 022423 (2016); doi: 10.2351/1.4943908

[9] E. Beyer and K. Wissenbach, Oberflächenbehandlung mit Laserstrahlung (Springer-Verlag Berlin, 1998)

IOP Conf. Series: Materials Science and Engineering 1246 (2022) 012031 doi:10.1088/1757-899X/1246/1/012031

Analysis of the strength difference effect on tailor welded blanks' springback

G J Béres[1] and **M Tisza**[†2]

[1]Department of Innovative Vehicles and Materials, GAMF Faculty of Engineering and Computer Science, John von Neumann University, Kecskemét 6000, Hungary
[2]Institute of Materials Science and Technology, Faculty of Mechanical Engineering and IT, University of Miskolc, Miskolc-Egyetemváros 3515, Hungary

Abstract. Tailor welded blanks (TWBs) are becoming widespread in the automotive industry due to their advantages, such as the possibility of weight reduction or the material usage optimising, resulting from the application. However, their main disadvantage is that the construction design and the simulation of the mechanical processes (like weld line movement, sheet springback, etc) are more complex. Predicting the mechanical response can only be done by accurately estimating the weld properties, which greatly increases the material testing requirements. Moreover, it often does not coincide with the matter of the field of material forming. In this study we focus on the simplified description of the springback of TWBs, to reveal which parameters have the primary influence on this process. Three kinds of DP steels with different strengths were welded to a mild steel grade one by one. Using these pairs we have varied both the strength difference between each sides and the average strength of the blanks, too. Uniaxial tensile tests were performed for the mechanical characterisation of each material, as well as right-angled bends perpendicular to the weld line were carried out to monitor the springback tendency of the samples. Our results show that a transition zone develops near to the weld line, independently from the strength difference between the TWB's components. This zone is well-defined by linear functions, in which, the parameters are consistent with the physical content of the springback phenomenon.

1. Introduction

Tailor Welded Banks (TWBs) were developed for the most ideal use of materials in the engineering structures, and their application is still spreading today. The TWBs are primary manufactured by laser beam welding, specifically for subsequent forming operations. Namely, the welding is not the final operation in the assembly sequence plan, contrary to the ordinary sense of this technology [1].
The most popular variants of TWBs are made by sheets of different thickness and/or strength to meet the lightweight requirements and retain the structural rigidity at the same time. Besides, different coatings can also be observed in certain cases. In order to distinguish the pre-welded blanks for forming purpose from the traditional sheets, the EN 10359:2015 standard contains certain specifications for the manufacturing conditions.

Figure 1 shows the ideal material distribution of a complex car body panel, both in the sense of the material grade and the thickness. Due to this geometrical and structural complexity, the manufacturing of this part from a simple blank is neither economically nor technically reasonable. It can be seen that if the complete part would be manufactured by the thickest or the highest strength material, the structure will be oversized and the lightweight pursuits are no longer met. On the other hand, if the complete part would be manufactured by the thinnest or the lowest strength material, the structure will be undersized and additional elements should be used for strengthening the critical points. In addition, the geometry of the initial blank can be simplified and accordingly the material usage can be improved if more, simpler banks are included. Another advantage of TWBs is that the welding happens on in-plane sheets, which is easier than for spatial cases [2,3].

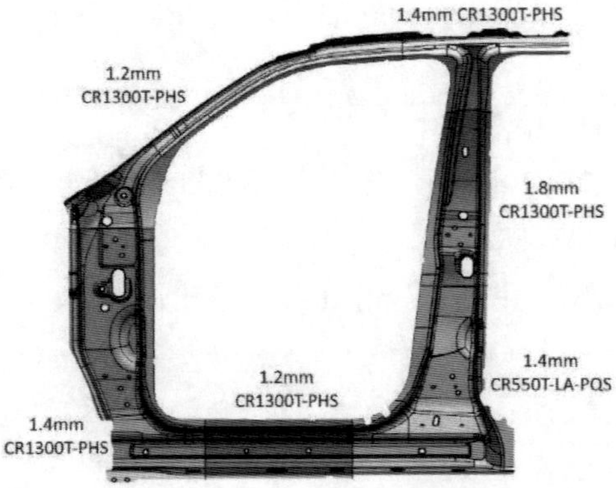

Figure 1. An example for a complex car body panel, made by different grades of materials and thicknesses [2].

Nevertheless, thanks to the interaction of the different elements as well as the weld line, the forming operation is more complicated than for simple blanks. Both the difference of the applied materials' properties (strength, formability, anisotropy, thickness, etc.) and the loading circumstances (the weld line position in relation to the loading or the straining direction) influence the behaviour of the workpiece during forming. In this sense, Korouyeh et al. [4] introduced the limit thickness ratio to describe the formability of TWBs with different thicknesses. The formability ratio was established by Bandyopadhyay et al. [5] for direct characterization of the drawability and the stretchability of TWBs. The deep-drawability was also intensively studied by Padmanabhan et al. in [6,7].

Fundamental springback approximation, which calculates the angle deviation after tools' release (θ^*) is generally defined in the form of equation (1). Marciniak et al. [8] derived that formulation from the moment equilibrium. Lange et al. [9] adopted the fundamentals of mechanic to obtain the resisting moment of the sheet against bending. Pearce [10] and the ASM Handbook of Metal Forming [11] simply communicate the formula, in which expression $\bar{\sigma}$ indicates the flow stress, E refers to the Young modulus and ρ and t_0 are the bending radius and the nominal (initial) thickness of the sheet. The bending angle is $\theta°$.

$$\theta^* = \frac{3\bar{\sigma}}{E} \cdot \frac{\rho}{t_0} \cdot \theta. \tag{1}$$

Due to the complexity of the TWBs, authors certainly do not know any simplified formula for analytical evaluation of the TWBs' springback. Although, it is worth noting that the present results only applies to this case of bending, and work on how this can be generalised is still ongoing.

This study presents the experimental results and the theoretical analysis on the springback behaviour of TWBs, considering the interaction of each sides made by different strength steels. The results show that the interaction can be described by a transition zone, which strength difference dependent mathematical description meets the physical content of the springback phenomenon.

2. Applied materials

Four grades of commercial, automotive, thin sheets in 1 mm nominal thickness (t_0) were used in this research work. The basic mechanical properties of the applied dual phase (DP) steels and the DC04 mild steel were determined by tensile tests, according to the ISO 6892-1:2019 standard's prescription. Five parallel experiment were done on each material, with 20 mm/min constant crosshead speed. The measurement of the r-values were carried out by a non-contact Instron AVE video-extensometer at Ag-1 (%) engineering strain level. The mean values in each direction are listed in table 1.

The applied three types of DP steels, DP600, DP800 and DP1000 have contained varying amounts of the ferrite-martensite phases. Basically, the higher the martensite ratio, the higher the tensile strength, which is indicated in the materials' number. The hard martensite and the soft ferrite together provide the relatively high strength and good formability of these materials at the same time [3]. Such as, DP steels are widely used in the automotive industry for car body panels, thanks to the abovementioned benefits as well as their favourable weldability and solderability [12,13].

The DC04 is also a widespread material for both the automotive and the kitchen utensils industry, since it has excellent formability and high anisotropy coefficient (i.e. good resistance to thinning). Its microstructure is ferritic for the most part, and thus the deformation is relatively homogenous on the microscopic scale [14].

Table 1. Basic, average material properties and the r-values of the applied materials.

	YS (MPa)	UTS (MPa)	Ag (%)	r_0	r_{45}	r_{90}
DC04	238	336	20.4	1.79	1.33	2.39
DP600	444	656	12.8	0.80	0.91	1.12
DP800	570	879	10.2	0.64	0.79	0.77
DP1000	758	1099	6.7	0.74	0.71	0.79

In the case of the TWBs, the DC04 mild steel was constantly coupled to each grade of DP steels by laser beam welding. Thereby, three variable strength ratios were available. Next to it, the average strength of the welded blanks was also varied with the usage of the different pairings.

The TWBs were manufactured by laser beam welding using a Rofin DY 027 Nd:YAG machine. The shielding gas was Ar 99.996 on the crown side of the weld. The welding has happened without additives. There was no particular requirement for the weld line, except that it should not open up under the bending stress.

3. Bending experiments

Simple, rectangular, free-bending in V-die was performed on square shaped TWBs, the illustration of which can be seen in figure 2 (right). The illustration contains the relation of the weld line to the rolling- and the bending directions, too. The final angles after the tools' release were measured by a workshop angle meter on the edges of the blanks. With increasing the length of the initial blanks (100…600 mm), the springback (θ*) could be determined in the function of the distance from the weld line. All the bending experiments were repeated three times.

30° bending tools (figure 2 left and mid) made by the Eurostamp GmbH were installed on an AMADA HFE 50-20 CNC controlled bending machine. The final angle (90°) was adjusted by the punch movement, which had a constant 20 mm/min speed. The sheets were not lubricated.

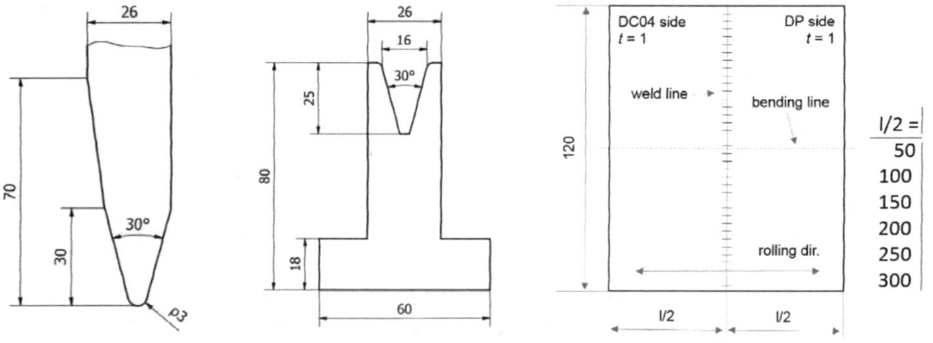

Figure 2. Schematic view of the applied bending tools and the initial blank, indicated the rolling-, the bending-, and the welding directions.

4. Results and discussion

4.1. Experimental springback results

The average difference between the measured and the nominal bending angle (90°) has given the value of free springback and is depicted in figure 3 for the three cases of TWBs. The angle deviation was measured in every 50 mm from the weld line. Therefore, six measurement points depending on the distance from the weld line were determined on each sides.

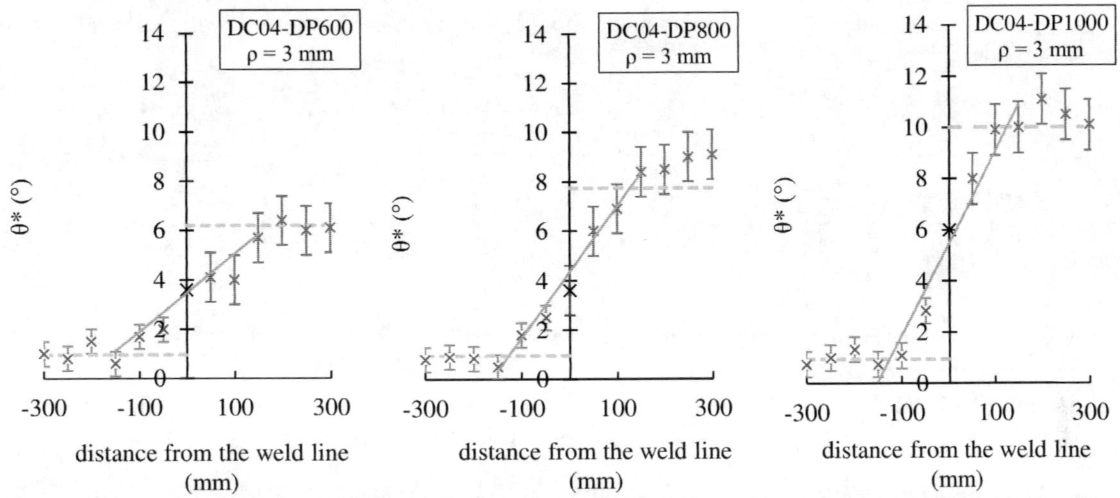

Figure 3. Experimental springback angles of the DC04-DP600, DC04-DP800 and DC04-DP1000 pairs.

Seeing the components' interaction in the figure above, it can be observed that a transition zone is developed near to the weld line, independently from the strength difference of the creating elements. Before and after the transition zone, the springback fits the components' separately characteristic values. The character of the transition zones, which can be properly described by linear functions is varying with the change of the strength differences ($\Delta\sigma$). Increasing the strength difference enhances both the steepness (m) and the point of intersection with the vertical axis (b) of the linear functions. The mentioned two parameters are indicated in table 2.

Table 2. Characteristic values (steepness and the point of intersection with the vertical axis) of the transition zones.

	m (-)	b (-)
DC04-DP600	0.0157	3.67
DC04-DP800	0.0267	4.25
DC04-DP1000	0.0361	5.59

4.2. Theoretical analysis

According to equation (1), the expected springback values only depend on the strength difference between theoretical *A* and *B* blanks in a steel-steel TWB, since the other influencing parameters are the same on both sides. Our assumption was that two theoretical cases should be existed in this manner:

- case one (i): the components have symmetrical influence on the springback, i.e. the welded blank behaves in the same way in both sides;

IOP Conf. Series: Materials Science and Engineering 1246 (2022) 012031 doi:10.1088/1757-899X/1246/1/012031

- case two (ii): the components have asymmetrical influence on the springback, i.e. the effect of one side is more pronounced.

Figure 4-6 show the appearances of both cases theoretically, if the strength difference is infinitely large, or infinitely small, or if it is between them. (The strength difference is indicated by $\Delta\sigma_{ps}$ in the following figures, where *ps* abbreviation refers to the plane strain state.) Provided that the strength difference is infinitely large (figure 4), the linear function should lie into the vertical axis in (i), since the *m* and *b* parameters should be infinitely large, too. In (ii), when the effect of the sides is inhomogeneous, the higher strength component should distort the TWB, since it validates its effect on the softer side.

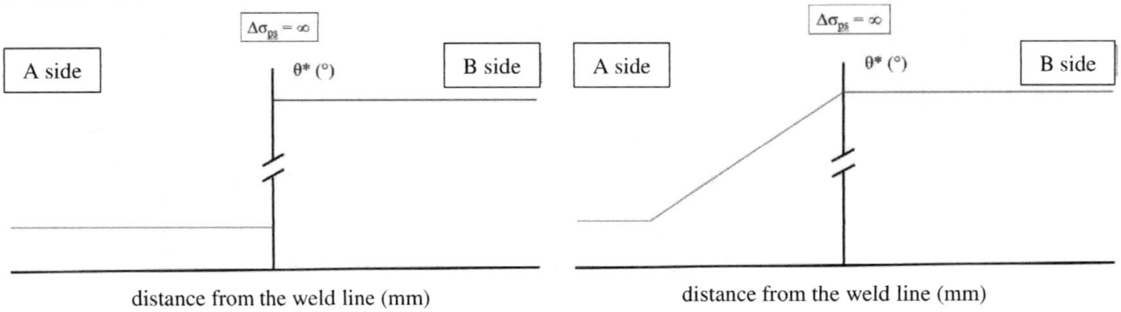

Figure 4. Theoretical interaction between the TWB's sides with infinitely large strength-difference: the effect of each components is symmetric (left) or asymmetric (right).

As long as the strength difference is infinitely small (figure 5), there is no discrepancy between case (i) and (ii), and the blank behaves like a non-welded sheet.

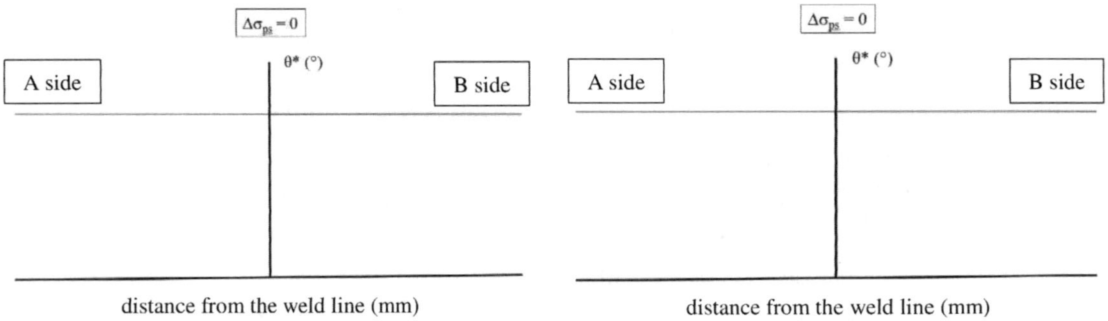

Figure 5. Theoretical interaction between the TWB's sides with infinitely small strength-difference: the effect of each components is symmetric (left) or asymmetric (right).

In the cases we investigated practically, when the strength difference is neither infinitely large nor infinitely small (figure 6), the transition zone should develop symmetrically (i), or asymmetrically (ii). We suppose that this kind of strength difference could have the most common occurrence in the practice. Our results (figure 3) show that the symmetrical behaviour (i.e. case (i)) is an acceptable approximation in the tested range, since there is no observable shift of the linear functions in figure 3.

IOP Conf. Series: Materials Science and Engineering 1246 (2022) 012031 doi:10.1088/1757-899X/1246/1/012031

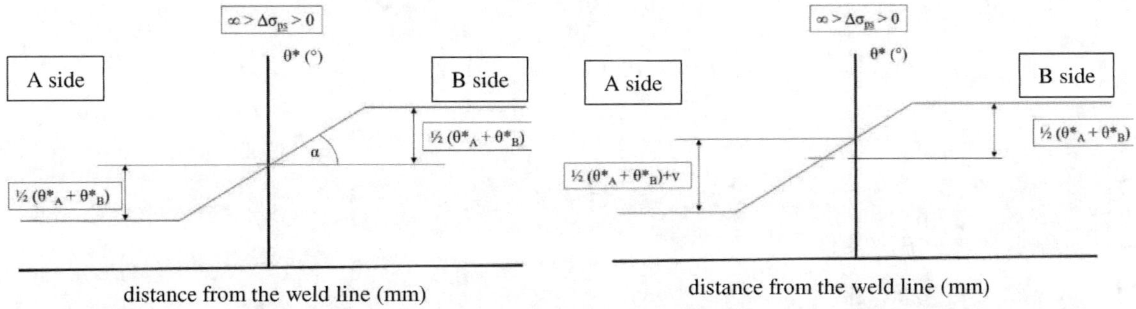

Figure 6. Theoretical interaction between the TWB's sides with a given (practical) strength-difference: the effect of each components is symmetric (left) or asymmetric (right).

If we accept the assumption that the TWBs behave as shown in figure 6 left side, then we can consider influencing parameters, which affect the steepness and the intersection point of the linear functions. The fact that both the steepness and position of the intersection point have grown in figure 3 is explained schematically by figure 7. It becomes visible from this figure that the steepness only depends on the strength difference itself, however, the intersection point is influenced by both the strength difference and the average strength. In other words, if $\Delta\sigma_{ps}$ is increasing, both the *m* and *b* parameters increase, too (figure 7 left). On the other hand, the average strength only has an effect on the intersection point, as it can be seen in figure 7 right, thus the average strength does not change the steepness.

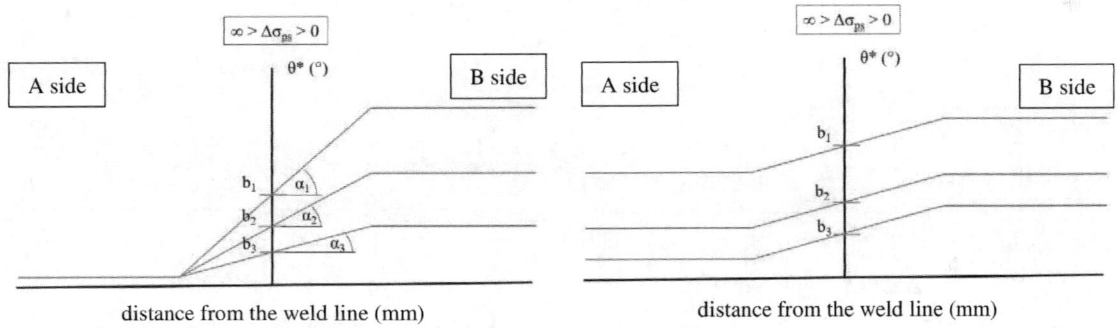

Figure 7. Theoretical interaction of given strength-difference sides, considering the effect of the strength difference (left) and the average strength (right).

The above mentioned facts can be mathematically described in equations (2) and (3)

$$m = f(\Delta\sigma_{ps}) \tag{2}$$

$$b = f\left[\left(\frac{\sigma_A+\sigma_B}{2}\right);\ (\Delta\sigma_{ps})\right] \tag{3}$$

as well as depicted in figure 8 and 9.

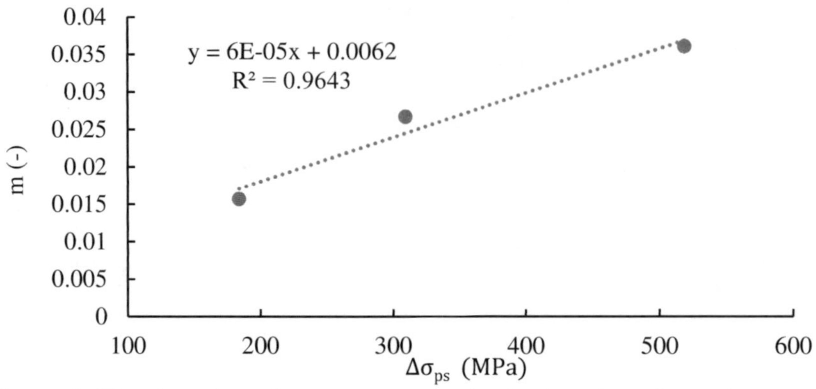

Figure 8. The changing of the m parameter in the function of the plane strain yield strength difference.

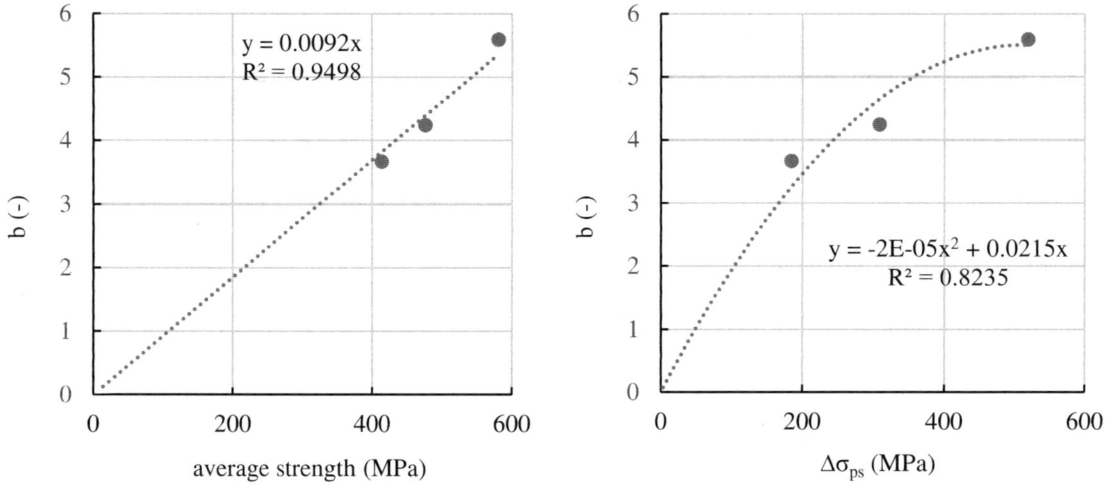

Figure 9. The changing of the b parameter in both the function of the yield strength difference in plane strain state and the average yield strength.

5. Summary

In this study we investigated the springback tendency of tailor welded blanks. Different pairs were manufactured by using constantly the same DC04 mild steel jointed to different kinds of DP steels. This have given three different types of blanks regarding the strength difference between the components and the average strength. In this way, the effect of the constituent materials' strength on the springback was even better revealed, since the elastic modulus, the sheet thickness and the bending geometry were the same in both sides of the blanks.

From the results it can be observed that a transition zone develops near to the weld line, in which the springback can be defined by linear functions. Theoretical analysis showed that parameters of the obtained functions have physical meaning in line with the phenomenon of springback, and is clearly aligned with the strength difference as well as the average strength of the components.

6. References

[1] Merklein M, Johannes M, Lechner M and. Kuppert A 2014 A review on tailored blanks - Production, application and evaluation *Journal of Materials Processing Technology* **214** 151 164

[2] https://ahssinsights.org/forming/tailored-products/tailored-products/

[3] Keeler S, Kimchi M 2017 *Advanced High-Strength Steels Application Guidelines Version 6.0.* 3S-Superior Stamping Solutions, LLC., WorldAutoSteel 1-17

[4] Korouyeh R S, Naeini H M, Torkamany M J and Liaghat G 2013 Experimental and theoretical investigation of thickness ratio effect on the formability of tailor welded blank *Optics and Laser Technology* 24-3

[5] Bandyopadhyay K, Panda S K, Saha P and Padmanabham G 2015 Limiting drawing ratio and deep drawing behavior of dual phase steel tailor welded blanks: FE simulation and experimental validation *Journal of Materials Processing Technology* 48-64

[6] Padmanabhan R, Oliveira M C and Menezes L F 2008 Deep drawing of aluminium-steel tailor-welded blanks *Materials and Design* 154-160

[7] R. Padmanabhan R, Baptista A J, Oliveira M C and Menezes L F 2007 Effect of anisotropy on the deep-drawing of mild steel and dual-phase steel tailor-welded blanks *Journal of Materials Processing Technology* 288-293

[8] Marciniak Z, Duncan J L and Hu S J 2002 *Mechanics of Sheet Metal Forming* (Butterworth Heinemann, Oxford) pp 82-107

[9] Lange K 1985 *Handbook of Metal Forming* (SME Dearborn:Michigan) pp 19.1-19.23.

[10] Pearce R 1991 *Sheet metal forming* (Adam Hilger Series on new manufacturing processes and materials IOP Publising Ltd.) pp 79-88.

[11] Metals Handbook 1969 *Vol. 4. Forming* (8th ed. American Society for Metals, Ohio: Metals Park)

[12] Hareancz F and Kiss N 2018 Szakaszos energiabevitel hatása nagyszilárdságú autóipari vékonylemezek ellenállás-ponthegesztésekor *29. Nemzetközi Hegesztési Konferencia* 213 p ISBN: 9789633581605

[13] Berczeli M and Weltsch Z 2018 Experimental studies of different strength steels MIG brazed joints *Periodica Polytechnica Transportation Engineering* **46** 63-68

[14] Thyssenkrupp: Steel Product Information, deep-drawing steels DD, DC and DX, Issue: July 2018

IOP Conf. Series: Materials Science and Engineering 1246 (2022) 012032 doi:10.1088/1757-899X/1246/1/012032

The plastic strain and the rolling direction dependence of the traditional and the biaxial r-values

Á Zs Hoffmann[1] and **G J Béres**[1]

[1]Department of Innovative Vehicles and Materials, GAMF Faculty of Engineering and Computer Science, John von Neumann University, Kecskemét 6000, Hungary

Abstract. The Lankford number (i.e. r-value) has long been used as a key parameter in the science of sheet metal forming, for the mathematical description of plasticity. The traditional way of thinking is that this value is a constant in a given direction, and the definition is recommended at Ag-1 % plastic strain. In this study we focused on the evolution tendency of r-value in the function of the plastic strain, in different loading directions. Two automotive thin sheets, DC04 and DP1000 were examined by uniaxial tensile tests and compression tests to investigate both the traditional and the biaxial r-values. The axial and the transverse deformations during the tensile tests were applied under continuous process monitoring with a touchless video extensometer. The compression tests were carried out intermittently, i.e. the tests were stopped after given forces, and the thickness as well as the diameters were measured parallel and perpendicular to the rolling direction. Our results show that the evolution of the traditional r-value is clearly shown for both materials, and its value is given here by linear and polynomial functions. Besides, the changing characteristic of the r-value as a function of deformation does not show a trend with respect to the rolling direction. The biaxial r-value can be well-defined by compression tests despite the small thickness, and the results are less dependent on friction.

1. Introduction

The theoretical and practical definition of the yield surface for isotropic materials is one of the state-of-the-art research topic in the sheet metal industry. In the cold forming processes, it is of paramount importance to understand and describe the behaviour of the sheet to be formed as accurately as possible. A major advancement was the description of anisotropy of the cold rolled sheets by Lankford [1], in the last century. Hill was another pioneer on this subject, who made the first mathematical formulation of anisotropic yield surface, the Hill'48 theory [2]. Since then, several improvements have been made on the description of plasticity functions, nevertheless the r-value is further used in its original form. However, by observing the r-value more closely than is usual in the practice (e.g. defining it as a variable), it is hopefully possible to predict the deformation (and other resulting forming characteristics) in a more realistic way during the calculations.

Recently, a number of theories have been proposed to determine anisotropic behaviour of sheet metals using the r-value, such as Hill'90 [3], Hosford [4], Barlat and Lian [5], without claiming completeness. What these theories have in common is that they all treat the r-value as a constant, and thus lead to moderately complicated calculations. Lian et al. [6] have already used the r-value as a variable and defined by a linear function of the plastic deformation ($\bar{\varepsilon}$) to make calculations of forming limits, including the Hill'48 formula. Hou et al. [7] proposed the description of the evolution of the r-value similarly, but with polynomial function in the following form. They have used the calculated r-values related to the yield loci determination using associated and non-associated flow rules.

$$r_\alpha = A_\alpha \bar{\varepsilon}^2 + B_\alpha \bar{\varepsilon} + C_\alpha \quad (1)$$

Here A_α, B_α and C_α mean the parameters of the polynomic function mentioned above, for the specimens machined at an α angle to the rolling direction.

In this paper we report the investigation of the r-value for two different sheet materials, DC04 and DP1000. These were subjected to uniaxial tensile tests supported by non-contact video extensometer for strain measuring and intermittent compression tests. The r-vaule was measured as a function of the equivalent plastic strain and the rolling direction. From the obtained results it can be seen that the change in traditional r-value as a function of deformation is clear, and does not show a trend with respect to the rolling direction. The biaxial r-value obtained from the compression tests is well-defined despite the small thickness. Moreover, the lubrication mode did not affect the results significantly.

2. Applied materials

DC steels are Al-killed, non-alloyed steels with favourable deep-drawing properties. They are cold-rolled materials intended for further cold forming, and as such, have their r-value above unity. DC04 is primarily used by the automotive industry for exterior body panels due to its excellent formability, i.e. its ability to form complex shapes [8]. It has the advantage of good weldability, too [9]. At the same time, it is not used as skeletal elements as its strength is not sufficient, due to its purely ferritic microstructure.

DP (Dual Phase) steels are so-called second generation, advanced, high strength steels, which are widely used in the automotive industry. They are named after their microstructure: dispersed martensite islands in ferrite matrix, exploiting the advantageous properties of both phases. The soft ferrite provides improved formability, while the hard martensite is responsible for superior strength properties. Therefore, this kind of steels can be relatively well-formable and can also be high strength by adjusting the ferrite-martensite ratio: their tensile strength is typically between 500-1200 MPa, with a corresponding 5-35% strain without fracture. For DP steels no yield point in the strict sense is present, due to the free movement of dislocations in the ferrite and the dislocation mechanisms at the ferrite-martensite grain boundaries [10,11,12].

In general, DP steels have been developed in response to the need for weight reduction in the automotive industry. They have certain formability, and can also be joined by bonding, welding and soldering [13,14]. The average mechanical properties, such as ultimate tensile strength (UTS), yield strength (YS), uniform (A_g) and total (A_{total}) elongations for both materials are summarize in table 1.

Table 1. Basic, average material properties of the applied materials.

	YS (MPa)	UTS (MPa)	Ag (%)	A_{total} (%)
DC04	238	336	20.4	37.9
DP1000	758	1099	6.7	10.6

3. Measurement methods

3.1. Tensile tests

The experimental method for determining the traditional r-value, is the tensile test. These kind of tests were carried out on standard specimens (ISO 6892-1:2019) machined in several different orientation as shown in figure 1. Although, specimens were manufactured in every 15°, due to the limitations of the paper, only the results in 0°, 45°and 90° to the rolling direction are reported here.

Contrasting, regular, circular marks were applied on the specimens for monitoring the deformation. Using a non-contact video extensometer, both the marks in the axial and the transverse direction were tracked continuously during the investigations. The non-contact measuring equipment has been calibrated according to ISO 9513:2013, with +/- 2.5 μm measurement accuracy in both directions. The operating principle is based on the reflection: a high-intensity LED array illuminates the workpiece, on which the contrast marks are reflected differently from the surrounding background.

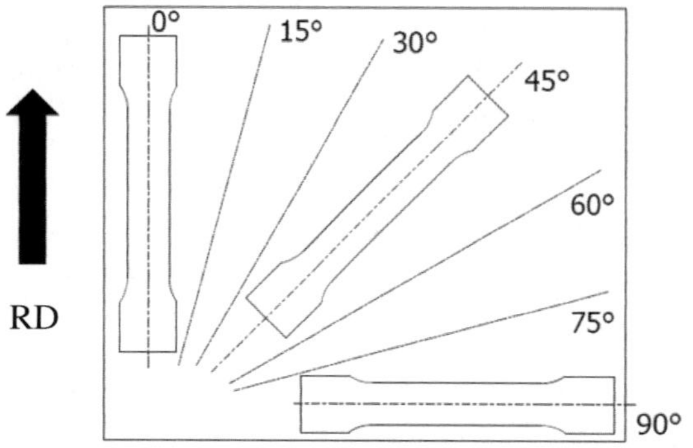

Figure 1. Machining orientations of the specimens to the rolling direction (RD).

The Instron AVE 2663-821 video extensometer we use has a Basler A622f monochrome camera, which has a maximal resolution of 1.3 Megapixels (1280x1024). It is coupled with a frame rate of 25 fps, which is equivalent with a movie film rate. The corresponding crosshead speed was 4 mm/min. The schematic view of the set-up of the measurement method is depicted in figure 2.

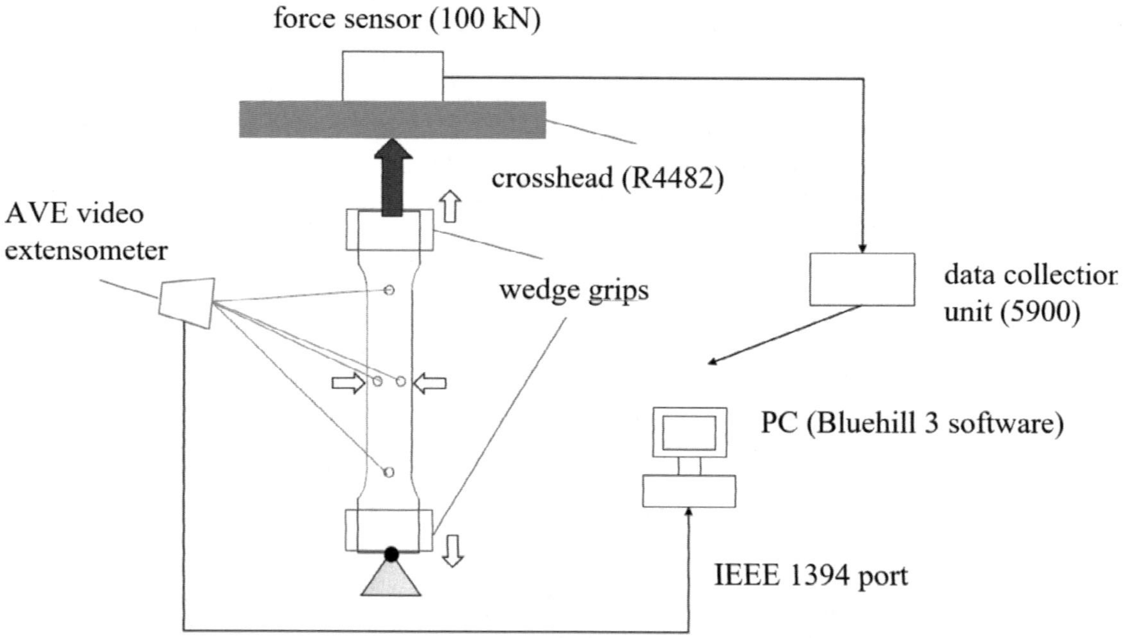

Figure 2. Schematic view of the tensile tests, showing the equipment and the additional units.

3.2. Compression tests

The compression tests were carried out on 10 mm diameter discs, with axial symmetry, between two, flat, tool surfaces. The blanks were separated from the tools with different lubrications: oil as well as nylon and PTFE foils. The foils had 0.05 nominal thickness.

The disks were tested individually by stopping the test at a certain force and then measuring the diameters parallel and perpendicular to the RD and the thicknesses. The tests were repeated three times for each force and each lubricant. The biaxial anisotropy (r_b) is obtained by dividing the plastic strain rate perpendicular to the RD ($d\varepsilon_{90}$) by the plastic strain ratio parallel to the RD ($d\varepsilon_0$):

$$r_b = \frac{d\varepsilon_{90}}{d\varepsilon_0}. \tag{2}$$

In this context, the strain ratios are the results of changes of the diameters during forming [15]:

$$d\varepsilon_0 = ln\frac{D_0}{D_0^0}; \quad d\varepsilon_{90} = ln\frac{D_{90}}{D_{90}^0}. \tag{3}$$

4. Results and discussion

4.1. Tensile tests' results

Figure 3 shows the r-values of the DC04 steels in three different rolling directions as a function of the equivalent plastic strain. The results show that the r-value for this material varies significantly with the progression of the deformation, and the specimens machined in different orientations have different characteristics. In each case, the equation of the approximate polynomial and linear trend line is plotted on the figures. From these, it is visible that the linear approximation is considered to be adequate for specimens machined in the 0° and 45° orientations, but this is no longer the case for the 90° orientation. Based on the results, it is therefore recommended that the variation of the r-value should also be monitored during the characterisation of this kind of material.

The results of the DP1000 material are shown in figure 4, in which the linear approximation is only acceptable for specimens machined in 45° to the rolling direction. In the 0° and 90° directions, the evolution of the r-value tends to an upper limit, which always reaches its maximum value at approx. 0.05 plastic strain. This approaches the value of the limit of the uniform elongation (A_g). In any case, the change of the r-value with the progress of the plastic deformation is also striking for this material, and the orientation dependence is clearly evident.

4.2. Compression tests' results

The results of the compression tests are depicted in figure 5. The diagrams contain the data of DC04 steel only. The data points represent the average of the results obtained with different lubricants and compression forces. Each measurement was stopped at certain force values and then the sheet thickness and the diameters were measured. From the diameters, deformations parallel and perpendicular to the rolling direction were calculated using equation (3). The resulting strains are on the horizontal and vertical axes, and the biaxial anisotropy represents the relationship between these two factors.

The diagrams show that the biaxial anisotropy parameter can be expressed by linear functions, i.e. the steepness refers to the r_b-value. The different lubricants do not have a significant effect on the steepness. Using lube oil and PTFE foil have led to almost the same results (0.92), even though the deviation (and thus the uncertainty) is higher when using oil. The biaxial r-value is slightly higher when using nylon foil, approx. 0.96. It means that the DC04 material behaves similarly in the 0° and the 90° orientations during biaxial deformation. These results are in line with the tensile tests' results, too, if one sees the characteristic of the traditional r-values in these directions (figure 3). Although, the stress state is different, similarities can be observed with the uniaxial tensile tests.

IOP Conf. Series: Materials Science and Engineering 1246 (2022) 012032 doi:10.1088/1757-899X/1246/1/012032

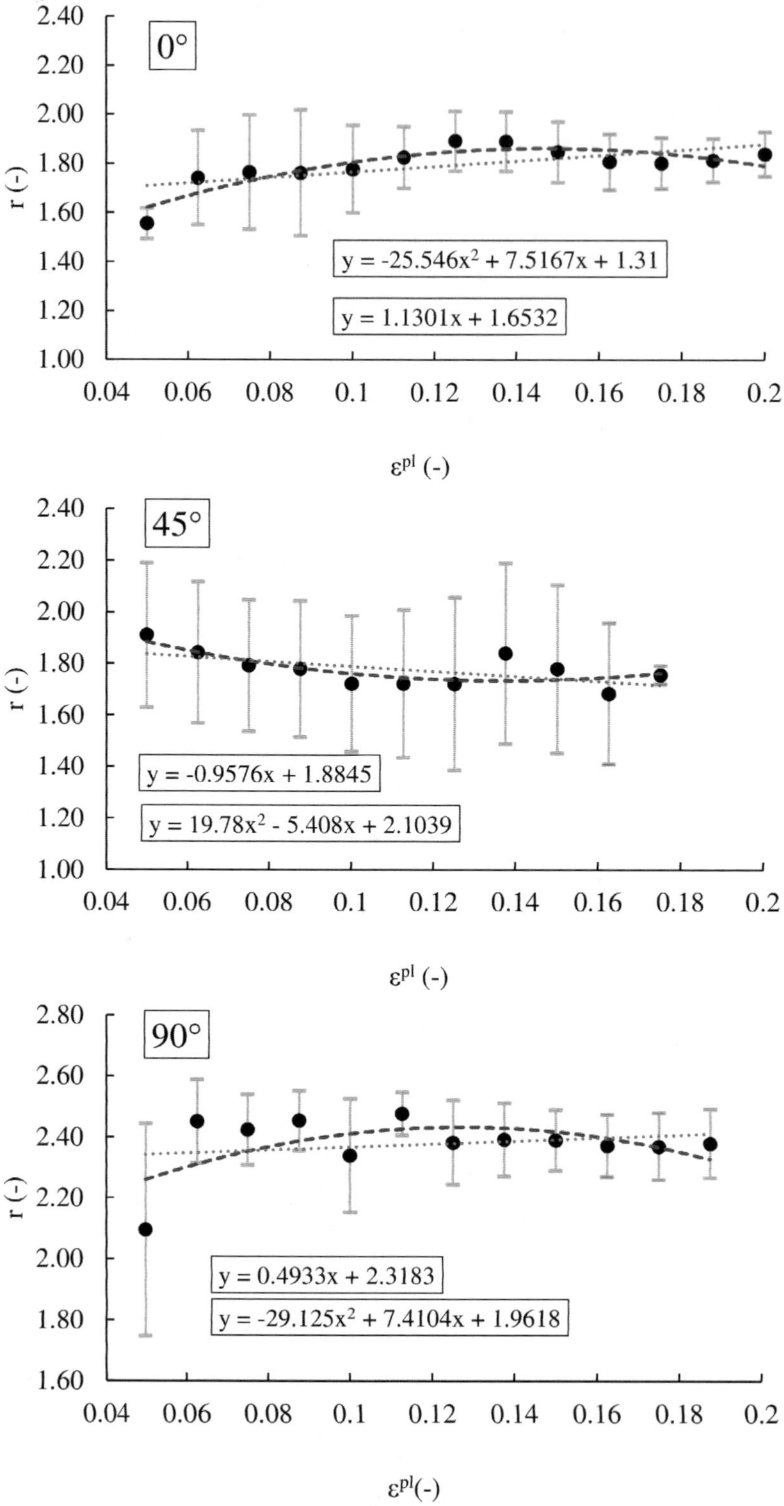

Figure 3. Variation of the r-value as a function of plastic strain in three different directions for DC04 steel.

IOP Conf. Series: Materials Science and Engineering 1246 (2022) 012032 doi:10.1088/1757-899X/1246/1/012032

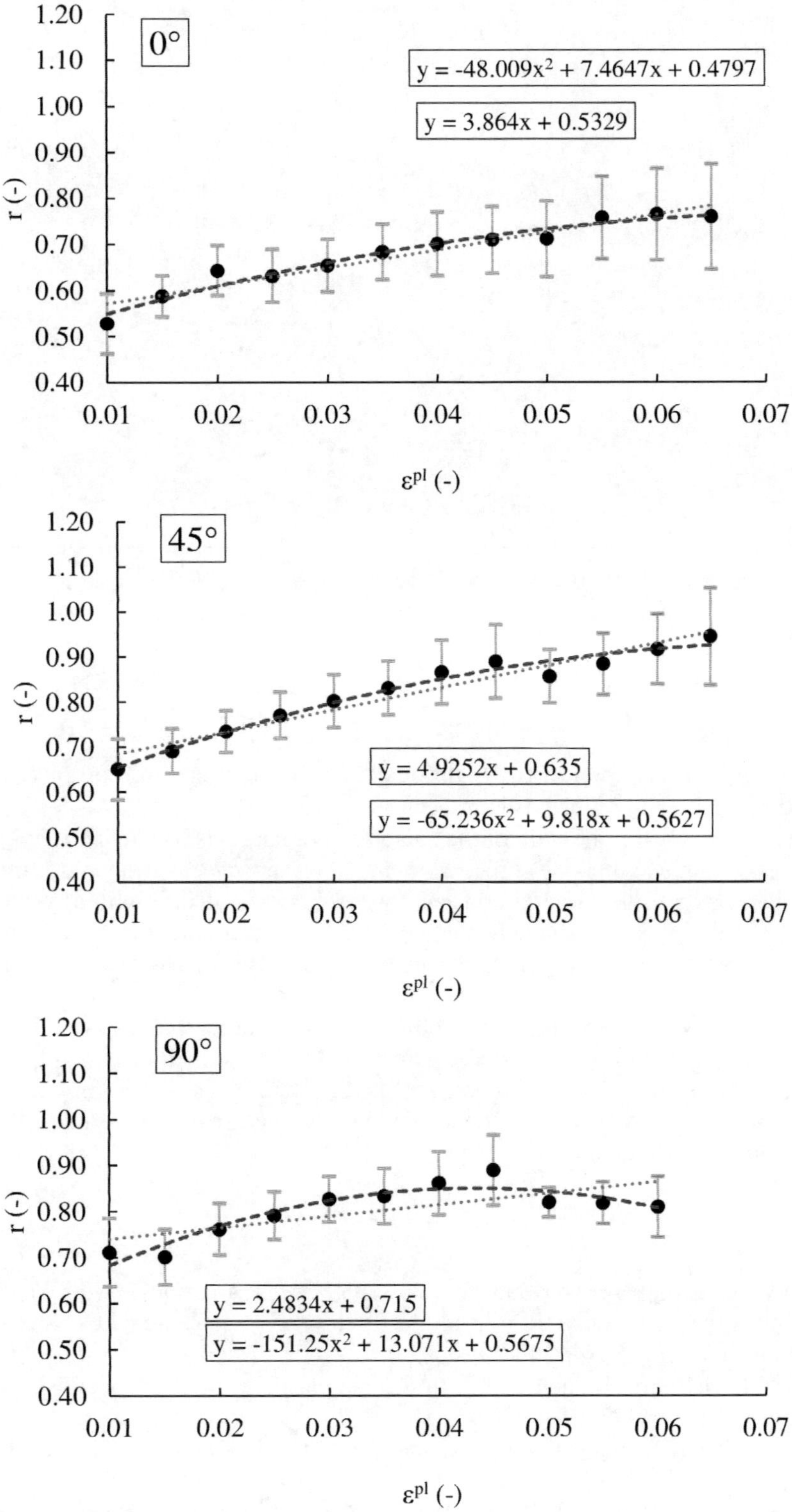

Figure 4. Variation of the r-value as a function of plastic strain in three different directions for DP1000 steel.

IOP Conf. Series: Materials Science and Engineering 1246 (2022) 012032 doi:10.1088/1757-899X/1246/1/012032

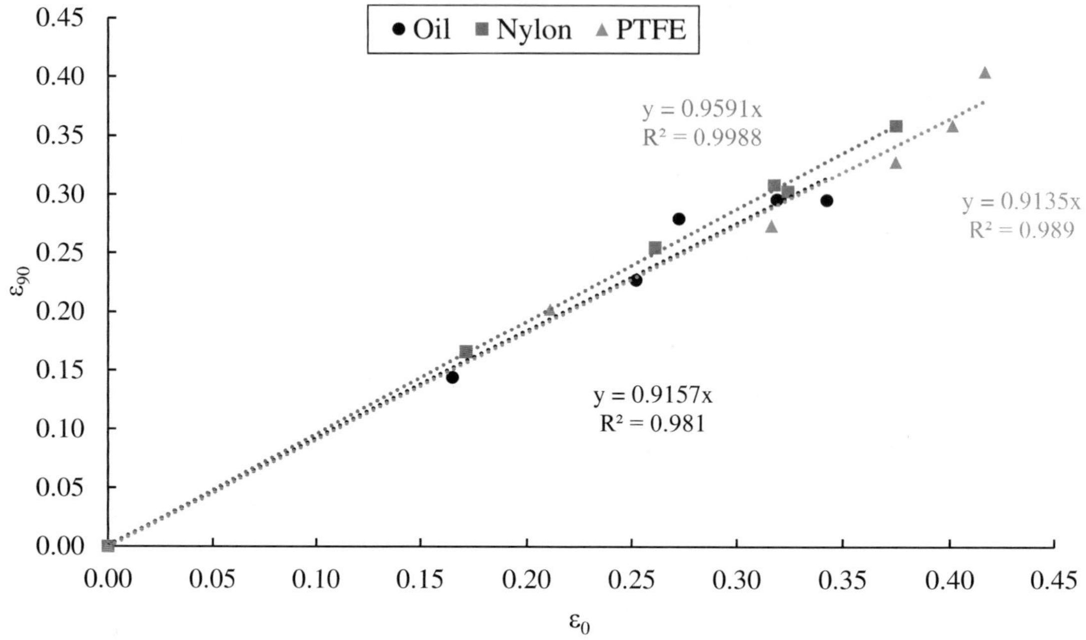

Figure 5. Biaxial r-value with three different lubricants, for DC04 steel.

5. Summary

This study deals with the variation of the traditional, and the determination of the biaxial r-values for DC04 and DP1000 steels. This value has long been a key parameter in the sheet metal forming industry, and will remain so until a better measure of the anisotropy of sheets is found. The applied tests included uniaxial tensile tests coupled with an in-process, non-contact strain measuring system and axisymmetric compression experiments. The compression tests were carried out intermittently, i.e. the tests were stopped at certain loads, where the diameters and the thickness were measured. During the tensile tests, the r-value's evolution in the function of the plastic deformation as well as the specimen orientation was monitored.

The results depicted above show that the evolution of the r-value is clearly visible at both materials, and there is no tendency between the evolution trend and the specimen orientation. In the case of the compression tests, the biaxial r-value was well-defined despite of the small thickness, in addition, the parameter has no significant dependency on the applied lubricant. In this stress state, the DC04 material showed nearly isotropic behaviour.

6. References

[1] Lankford W T, Snyder S C and Bauscher J A 1950 *Trans. Am. Soc. Metals* **42** 1197

[2] Hill R 1948 A theory of the yielding and plastic flow of anisotropic metals *The hydrodynamics of non-Newtonian fluids. I* 281-297.

[3] Hill R 1990 Constitutive modelling of orthotropic plasticity in sheet metals *J. Mech. Phys. Solids Vol. 38* 405-417

[4] Hosford W F 1985 Comments on anisotropic yield criteria *Int. Journ. Mechanical Sciences, Vol. 27* 423-427.

[5] Barlat F and Lian J 1989 Plastic behaviour and stretchability of sheet metals: A yield function for orthotropic sheet under plane stress conditions *Int. Journ. Plasticity, Vol. 5* 51-56.

[6] Lian J, Shen F, Jia X, Ahn D-C, Chae D-C, Münstermann S and Bleck W 2017 An evolving non-associated Hill48 plasticity model accounting for anisotropic hardening and r-value evolution and its application to forming limit prediction *International Journal of Solids and*

Structures doi: 10.1016/j.ijsolstr.2017.04.007
[7] Yong H et al. 2019 Investigation of evolving yield surfaces of dual-phase steels *Journal of Materials Processing Tech.* doi.org/10.1016/j.jmatprotec.2019.116314
[8] Thyssenkrupp: Steel Product Information, deep-drawing steels DD, DC and DX, Issue: July 2018
[9] Prém L and Balogh A 2013 Hagyományos és korszerű autóipari acéllemezek ponthegeszthetőségének vizsgálata, *Hegesztés-Technika XXIV* **4** 41-49.
[10] Xiong Z P, Kostrythev A G, Stanford N E and Pereloma E V 2015 Microstructures and mechanical properties of dual phase steel produced by laboratory simulated strip casting *Materilas and Design Vol 88* **2** 537-549
[11] Surajit K P 2013 Real microstructure based micromechanical model to simulate microstructural level deformation behavior and failure initiation in DP 590 steel *Materials and Design* **44** 397-406
[12] Sodjit S and Uthaisangsuk V 2012 Microstructure based prediction of strain hardening behaviour of dual phase steels *Materials and Design* **41** 370–379
[13] Hareancz F and Kiss N 2018 Szakaszos energiabevitel hatása nagyszilárdságú autóipari vékonylemezek ellenállás-ponthegesztésekor *29. Nemzetközi Hegesztési Konferencia* 213 p ISBN: 9789633581605
[14] Tajti F, Berczeli M and Weltsch Z 2019 Improvement of high strength automotive steels wettability properties using CO2 laser surface treatment *Int. J. of Eng. and Man. Sci.* **4** 422-427
[15] Barlat F et al. 2003 Plane stress yield function for aluminium alloy sheets – part 1: theory *Int. J. of Plasticity* **19** 1297-1319

Investigation of nickel coated steel tapes welded joints

F Hareancz and T Sinka

John von Neumann University, Kecskemét, Bács-Kiskun, Hungary

Abstract. Nickel-plated steel tapes are primarily developed for battery production. The aim of these developments was to provide all battery cell cases with good contact resistance and high corrosion resistance, thus increasing their durability. When individual battery cells are packaged, the electrical connection is also mostly made with nickel-plated plates. Welding and soldering are the bonding technologies used to make the connections between the plates, but the advantages mentioned above may vary when making these connections. To our knowledge, lasers with diode beam sources are the most efficient and are becoming more and more suitable for a wide range of applications as the technology develops. The aim of our investigations is to investigate the quality of joints made with low power diode lasers with different parameters from a strength and metallographic point of view.

1. Introduction

Foils and thin sheets are found in many areas of industry. Nowadays, the increasing penetration and development of electric cars is affecting and influencing many industrial areas. The impact of these trends and developments can also be felt in Hungary, where battery assembly plants are being set up in many parts of the country to meet the growing demand for batteries. The batteries, which can be used in electric cars, come in 3 different designs: cylindrical, flat and brick. However, a battery pack will have the right parameters for use in electric cars if several smaller battery cells are connected together. There are several solutions for interconnecting these cells in industry. The conductive materials used between each cell can also be made of different alloys. One of these materials used as conductive plates is nickel-plated steel plates. Our investigations explore the weldability of this material using a low power 300 W diode laser beam source [1-7].

2. State of the Art

Welding of thin films is currently one of the most actively researched areas due to its automotive relevance. Different materials have different physical and chemical properties. In order to weld copper and aluminium guide rails well, for example, radiation sources have had to evolve. It is well known that laser welding of copper was difficult due to the high reflection of the laser beam. However, by choosing the right wavelength, the absorption on the copper substrate can be increased. Laser beam source manufacturers have adapted their developments to this and are now producing increasingly powerful blue and green laser beam sources. Green laser beam sources are particularly advantageous for welding copper plates. Optimisation and finding the right parameters are still the basis of a considerable amount of research [8-17].

3. Objectives

The objective of the present study is to determine the parameters for welding these 8x100 mm nickel coated steel plates with a thickness of 0.2 mm, which are also used as guide bridges, with a suitable quality. The weld formed is a C weld with a 3 mm diameter not fully closed circular track. There is a distance of 1 mm between the start and the end of the circular path. Joints made with different performances and movement speeds are examined by visual inspection, tensile testing, metallography and hardness measurement, looking for the parameter window that the weld geometry is also the most suitable and a cohesive connection is established between the two layers in the desired cross section.

4. Experimental setup and method

For the welded joints, we used a Compact-Evolution 980 diode beam source with a maximum power of 300W and a laser beam quality of 22 mm*mrad, which is fed through a 200 μm diameter light cable into a Coherent PHFQ-C0 fixed optic at a focal length of 94 mm to obtain a focal spot diameter of 538 μm. The optics are protected by a cross yet nozzle. The smoke that may be generated during welding is kept as far away from the optics as possible. The optics are moved by a Yaskawa robot in figure 1. The shielding gas used for welding was 4.6 argon, with a flow rate of 6-7 l/min.

Figure 1. Yaskawa robot equipped with optics.

Once the robot and its environment were built and programmed, we had to carry out preliminary experiments to determine the parameters that would allow us to successfully create a bond. First, for each of the three seams, we recorded a scale to move within the range we had defined. For power, we applied a 5% variation, which translates into a difference of 15 W. For the welding speed of the welds, a step of 5 mm/s was taken for speed. The resultant sets of these experiments are shown in figure 2.

IOP Conf. Series: Materials Science and Engineering 1246 (2022) 012033 doi:10.1088/1757-899X/1246/1/012033

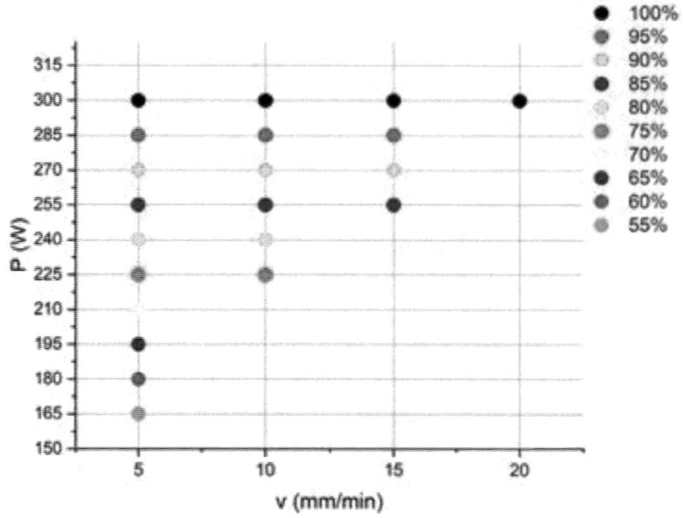

Figure 2. Welding parameter combinations.

5. Results and discussion

5.1. Visual inspection
The completed joints were first checked by visual inspection. The defects looked for were absorption cavities, incomplete fusion, surface cracks and porosity. Joints that were found to have a defect were not examined further. The most common defect is incomplete cross-sectional melt-through.

5.2. Tensile test
The uniaxial tensile tests were performed on an Instron type universal material testing machine. The results with different welding parameters (laser beam power and welding speed) are plotted on a common graph shown in figure 3.

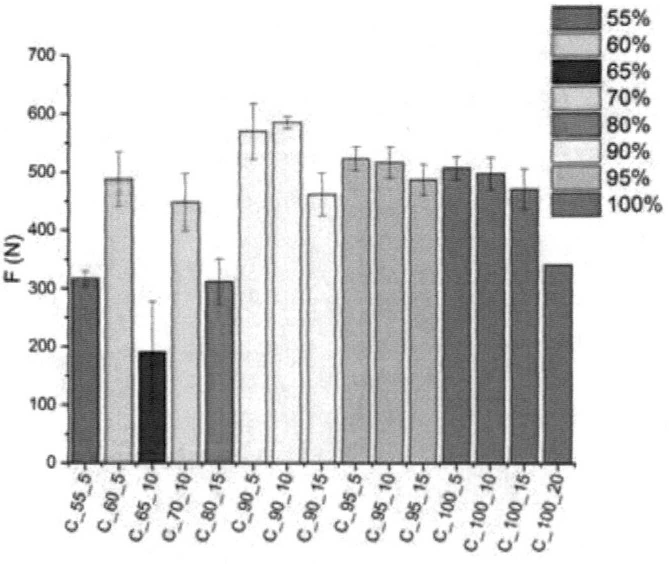

Figure 3. Tensile test results.

5.3. Metallography

The picture was taken halfway around the arc so that I could see both parts of it. The resulting micrograph is shown in figure 4.

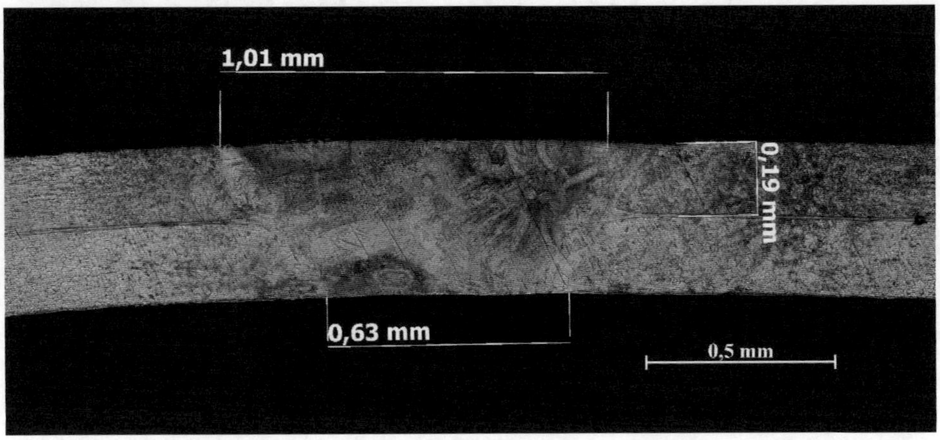

Figure 4. Welding parameter combinations.

The picture clearly shows the bond between the two plates. The laser beam, guided along a small arc, heated the plates by thermal conduction, and the fusion depth increased from thermal expansion and the same energy coupling. The resulting micrograph is shown in figure 5.

Figure 5. Heat-affected zone.

5.4. Hardness test

Following metallographic tests, hardness was also measured on the embedded pieces.

I used a Wilson Wolpert MVD type machine to carry out the test. The test procedure was Vickers hardness measurement with low loading force, justified by the fact that the pieces to be measured were 0.2 mm thick and that the small width of the joints formed would otherwise have made the impression too large in table 1. During the measurement, two parameters can be set on the machine, namely load force and load time. Here, the 0.1 kp and 12 seconds considered optimal through preliminary experiments were used.

Table 1. Hardness test results.

Base material	*BM HAZ*	Fusion zone	*BM HAZ*	Base material
123 HV	158 HV	310 HV	186 HV	140 HV

6. Summary and outlook

In summary, it is possible to weld nickel strip with a low power radiation source with certain parameters. Based on the conclusions drawn from the results of our experiments, the welding windows set up by the preliminary experiments have to be narrowed down since for some parameters no weld is formed, only surface fusion or the quality of the joint is not satisfactory.

The resulting diagrams give a good approximation of how the technology can be used. However, further modifications and tests are needed to fully calculate the appropriate parameters.
As regards the design of the clamping, a special clamping device would be appropriate in order to prevent possible imperfect fit (air gap). Another structural element that could be improved is the gas injection system. The use of a possible root gas in the case of multi-side argon injection would probably reduce the oxidation rate.
Further parameters to be investigated include the gas outflow rate and the deviation from the focal point, as these are typical of the technology and have a significant influence on the formation of the bond.

The technology is viable and can be applied on an industrial scale, as the system we have created can be further developed to create an automated production line capable of producing high volumes and high quality joints, making it a viable alternative for any company involved in this field.

Acknowledgements
We are grateful for the support of this research, which was carried out within the framework of the grant 2019-1.1.1-PIACI-KFI-2019-00075 "Development of efficient machining technologies for structural materials with increased strength at Ferropatent Zrt." The project is supported by the National Research, Development and Innovation Fund.

References

[1] Berczeli M, Buza G. The effect of laser beam welding parameters onto the evolving joints geometry. Mater Sci For 2017;885:178–83. https://doi.org/10.4028/www.scientific.net/msf.885.178.
[2] Sadeghian A, Iqbal N. A review on dissimilar laser welding of steel-copper, steel-aluminium, aluminium-copper, and steel-nickel for electric vehicle battery manufacturing. Opt Laser Technol 2022;146:107595. https://doi.org/10.1016/j.optlastec.2021.107595.
[3] Grabmann S, Tomcic L, Zaeh MF. Laser beam welding of copper foil stacks using a green high power disk laser. Procedia CIRP 2020;94:582–6. https://doi.org/10.1016/j.procir.2020.09.080.
[4] Eskander SMSU, Fankhauser S. Reduction in greenhouse gas emissions from national climate legislation. Nat Clim Chang 2020;10:750–6. https://doi.org/10.1038/s41558-020-0831-z.
[5] Shaw-Stewart J, Alvarez-Reguera A, Greszta A, Marco J, Masood M, Sommerville R, et al. Aqueous solution discharge of cylindrical lithium-ion cells. Sustain Mater Technol 2019;22:e00110. https://doi.org/10.1016/j.susmat.2019.e00110.
[6] Lee SS, Kim TH, Hu SJ, Cai WW, Abell JA. Joining technologies for automotive lithium-ion battery manufacturing: A review. ASME 2010 International Manufacturing Science and Engineering Conference, Volume 1, ASMEDC; 2010.
[7] Saariluoma H, Piiroinen A, Unt A, Hakanen J, Rautava T, Salminen A. Overview of optical digital

measuring challenges and technologies in laser welded components in EV battery module design and manufacturing. Batteries 2020;6:47. https://doi.org/10.3390/batteries6030047.
[8] Brand MJ, Schmidt PA, Zaeh MF, Jossen A. Welding techniques for battery cells and resulting electrical contact resistances. J Energy Storage 2015;1:7–14. https://doi.org/10.1016/j.est.2015.04.001.
[9] Li J, Zhou S, Han Y. Advances in battery manufacturing, service, and management systems. Standards Information Network; 2016.
[10] Yang J, Chen J, Zhao W, Zhang P, Yu Z, Li Y, et al. Diode laser welding/brazing of aluminium alloy to steel using a nickel coating. Appl Sci (Basel) 2018;8:922. https://doi.org/10.3390/app8060922.
[11] Das A, Fritz R, Finuf M, Masters I. Blue laser welding of multi-layered AISI 316L stainless steel micro-foils. Opt Laser Technol 2020;132:106498. https://doi.org/10.1016/j.optlastec.2020.106498.
[12] Hu Y, He X, Yu G, Ge Z, Zheng C, Ning W. Heat and mass transfer in laser dissimilar welding of stainless steel and nickel. Appl Surf Sci 2012;258:5914–22. https://doi.org/10.1016/j.apsusc.2012.02.143.
[13] Das A, Li D, Williams D, Greenwood D. Joining technologies for automotive battery systems manufacturing. World Electric Veh J 2018;9:22. https://doi.org/10.3390/wevj9020022.
[14] LeBozec N, LeGac A, Thierry D. Corrosion performance and mechanical properties of joined automotive materials: Corrosion performance and mechanical properties of joined automotive materials. Mater Corros 2012;63:408–15. https://doi.org/10.1002/maco.201005888.
[15] Shui L, Chen F, Garg A, Peng X, Bao N, Zhang J. Design optimization of battery pack enclosure for electric vehicle. Struct Multidiscipl Optim 2018;58:331–47. https://doi.org/10.1007/s00158-018-1901-y.
[16] Weil M, Ziemann S, Peters J. The issue of metal resources in Li-ion batteries for electric vehicles. Behaviour of Lithium-Ion Batteries in Electric Vehicles, Cham: Springer International Publishing; 2018, p. 59–74.
[17] Pereirinha PG, González M, Carrilero I, Anseán D, Alonso J, Viera JC. Main trends and challenges in road transportation electrification. Transp Res Procedia 2018;33:235–42. https://doi.org/10.1016/j.trpro.2018.10.096.

IOP Conf. Series: Materials Science and Engineering 1246 (2022) 012034 doi:10.1088/1757-899X/1246/1/012034

Surface roughness prediction applying artificial neural network at micro machining

U L Adizue[1], B Z Balazs[1] and M Takacs[1]

[1]Budapest University of Technology and Economics, Faculty of Mechanical Engineering, Department of Manufacturing Science and Engineering, H-1116 Budapest, Műegyetem rkp. 3, Hungary

E-mail: takacs.marton@gpk.bme.hu

Abstract. Micro machining forms an integral part of modern manufacturing technology. In order to satisfy the most important demands, like cost reduction, quality improvement, and production efficiency, this process has to be implemented into the framework of Industry 4.0, too. Defect reduction of products can be realized by beneficial trends of Industry 4.0, such as application of sensors driven AI based predictive models. Surface roughness is one of the main indicators, which can be used for evaluation of the machined surface. The aim of this research is to develop a predictive surface roughness model in the case of micro milling of hard materials. Most important scientific knowledge related to micro milling, and application possibilities of AI based predictive models in the case of chip removal processes were summarised. Micro milling experiments on hot work tool steel (AISI H13) were carried out in order to determine the effect of different cutting parameters on surface roughness. The effects of the cutting parameters were investigated by the analysis of variance (ANOVA). Two-layer feedforward artificial neural network (ANN) based model was trained, validated and tested by means of experimental results. It was found that the trained model has an average MSE, RMSE and R values of 0.00063850, 0.019116835 and 0.9948, while the additional validation test provided 0.00038148, 0.019531513 and 0.8545, respectively, which are statistically significant. The developed model can effectively predict the surface roughness under the investigated circumstances.

1. Introduction

Two of the main trends and demands in current manufacturing technology are digitalisation and miniaturisation. Micro milling offers a very promising cutting process for flexible machining small parts of devices of any materials with high material removal rate. The kinematic of micro milling is actually the same as that of conventional sized one, but the process characteristics cannot be adapted directly into the small size ranges because of the size effect. The main important phenomena, which have to be considered in the case of micro milling, are as follows: importance of minimum chip thickness, relative small tool stiffness, relative large tool run-out, and burr formation [1]. In the case of micro machining tool wear plays an essential role in the geometric accuracy and surface quality of the product. Micro machining is referred to as setting uncut chip thickness in the range of 0.1-200 μm [2]. Micro machining can also be considered as mechanical cutting of features by geometrically defined tool, where the tool engagement is less than 1 mm, and tool applied is geometrically defined[3]. J. Chae used the term micro machining for fabrication of miniature devices and parts with features ranging from tens of micrometres to a few millimetres [4]. Micro machining offers an efficient way to produce small and complex components with tight tolerance in both plastics and metals. The process is also ideal for creating prototypes, and reducing lead time [5]. The special characteristics of micro milling make the assurance of uniform product quality challenging. Application of different artificial intelligence based models can improve the possible prediction on tool life and surface quality, which can be resulted in reduced defect manufacturing.

Machine learning (ML) is an extension of computer science that can be defined as exploiting algorithms and data in a way that allows a machine to learn similar to human learning. Different algorithms are trained to develope predicting models using statistical approaches, revealing crucial insights in data mining initiatives [6]. Voluminous data can be efficiently analized by machine learning algorithms in order to identify complex patterns, and extract conclusions [7]. The traditional machine learning models such as Bayesian networks, linear regression, logistic regression, support vector machine (SVM), radial basic function, and single-layer artificial neural networks cannot scale effectively, when the amount of training data is excessive. However, Multi-layer neural networks and deep learning approaches perform better for learning and prediction as the amount of training data grows. Therefore, it's critical to consider the amount of data utilized to train a machine learning model [8].

ML can be efficiently used for the prediction of process parameters, quality features, and remaining useful tool life in the case of micro machining. Tool life prediction is a prerequisite for application of a decision support system (DSS) to determine the optimal time for tool change. DSS can be part of tool condition monitoring (TCM). The objective of TCM is to use sensor signal processing techniques to monitor the cutter status to prevent losses caused by tool wear or damage. TCM technique estimates tool condition and tool health in real-time using prior feature-based knowledge or some empirical relationship. A set of recorded data is specifically utilized to predict the state of the situation in real-time [9]. Artificial neural network (ANN) is one of the most sophisticated ML algorithms, which is used in form of supervised learning to identify complex relationships between inputs and outputs [10]. They are mathematical models inspired by the behaviour of biological neurons [11]. The ANN and deep belief networks were employed to predict the wear state along with the remaining useful life of the tools [12].

The contemporary advancement in ML and its pertinency in tool condition monitoring approach has created awareness amongst researchers and the manufacturing industry. A robust TCM system could boost production while also ensuring product quality, which has a significant impact on machining efficiency. Therefore, TCM plays a significant role in the industrial business [13]. There are various machine learning approaches available for developing predictive models regarding micro machining processes. S S Patil et al. summarized the different deep learning algorithms, which can be applied for tool condition monitoring at milling, such as artificial neural network, convolutional neural network (CNN), and recurrent neural network (RNN). Depending on the characteristics of the measured signals the appropriate technique for tool condition monitoring can be selected [14].

Aims of this research work is the development of an artificial neural network supported predictive model related to the surface roughness of a micro milled workpiece of high hardness. Systematic experiments were carried out on AISI H 13 workpiece material. The effect of feed per tooth (f_z) and depth of cut (a_p) on average surface roughness (R_a) are investigated in detail. ANOVA test was conducted to explore the effects of cutting parameters on surface roughness. An artificial neural network based model was developed to ensure a better prediction of the surface roughness, which can be applied for sensor based process optimization in the future.

2. Experimental design

The micro milling experiments were performed on a 5-axis micro machining centre of the type of VHTC 130 with a main spindle of a maximum rotational speed of 60 000 (figure 1). Two flutes, AlTiN coated end milling cutter of the diameter of 500 μm was applied. The end mill is prepared with a corner radius of 50 μm to ensure larger edge stability. High concentricity is provided by a 6 mm precision shank with h5 tolerance. Tool was clamped by means of a high precision collet directly in the main spindle. Tool run-out was checked by a dial test indicator on the shank, and a digital microscope (500x magnification) at the tip while it was rotated manually. No measurable run-out of the tool was found. Grooves of the length of 5 mm with full diameter were micro milled during the experiments. Contact surface roughness tester of type of Mitutoyo SJ-400 was applied to measure the average surface roughness (R_a) values. Three measurements were carried out in the case of each parameter combination. The workpiece after

Figure 1. 5-axis micro milling machine with micro end mill and workpiece.

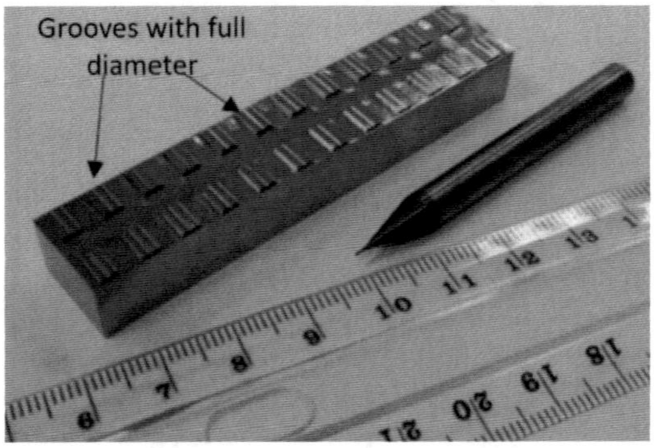

Figure 2. Workpiece and end mill of micro milling experiment [15].

the cutting experiments together with the micro end mill is shown in figure 2. The grooves were milled with different cutting parameters in the middle of the islands of the workpiece.

Hot work tool steel of type AISI H13 (table 1) was applied as experimental material, because it is commonly used for conventional sized dies of injection moulding [16]. One of the main application fields of micro milling is the production of injection moulds for miniature sized plastic pieces.

Table 1. Chemical composition of the experimental material [16].

	C (%)	Si (%)	Mn (%)	Cr (%)	Mo (%)	Ni (%)	V (%)
AISI H13	0.39	1.10	0.40	5.20	1.40	-	0.95

Altogether 48 micro milling experiments based on the full factorial design of experiments were carried out according to the parameters given in table 2.

Table 2. Design of experiments.

Factors	v_c (m/min)	f_z (μm)	a_p (μm)	Milling strategy	Material
Levels	1	8	6	1	1
Values	90	1, 2, 4, 6, 8, 10, 12, 14	25, 50, 75, 100, 125, 150	Milling with full diameter	AISI H13

3. Results and discussion

3.1. Surface roughness

Table 3. Experimental values with measured surface roughnesses.

Nr. of experiments	v_c (m/min)	f_z (μm)	a_p (μm)	$R_{a\text{-measured Nr 1}}$ (μm)	$R_{a\text{-measured Nr 2}}$ (μm)	$R_{a\text{-measured Nr 3}}$ (μm)	$R_{a\text{-average}}$ (μm)
1	90	1	25	0.046	0.055	0.057	0.053
2	90	2	25	0.052	0.049	0.050	0.050
3	90	4	25	0.062	0.063	0.064	0.063
4	90	6	25	0.076	0.065	0.067	0.069
5	90	8	25	0.146	0.126	0.106	0.126
6	90	10	25	0.126	0.124	0.123	0.124
7	90	12	25	0.130	0.151	0.194	0.158
8	90	14	25	0.149	0.154	0.153	0.152
9	90	1	50	0.059	0.083	0.061	0.068
10	90	2	50	0.080	0.079	0.091	0.083
11	90	4	50	0.082	0.085	0.083	0.083
12	90	6	50	0.080	0.089	0.081	0.083
13	90	8	50	0.117	0.115	0.118	0.117
14	90	10	50	0.145	0.144	0.140	0.143
15	90	12	50	0.124	0.123	0.124	0.124
16	90	14	50	0.123	0.123	0.123	0.123
17	90	1	75	0.072	0.065	0.085	0.074
18	90	2	75	0.104	0.091	0.095	0.097
19	90	4	75	0.110	0.109	0.117	0.112
20	90	6	75	0.092	0.089	0.089	0.090
21	90	8	75	0.106	0.089	0.106	0.100
22	90	10	75	0.118	0.099	0.102	0.106
23	90	12	75	0.128	0.115	0.121	0.121
24	90	14	75	0.102	0.105	0.101	0.103
25	90	1	100	0.057	0.057	0.050	0.055
26	90	2	100	0.077	0.076	0.077	0.077
27	90	4	100	0.130	0.136	0.128	0.131
28	90	6	100	0.084	0.096	0.087	0.089
29	90	8	100	0.091	0.089	0.079	0.086
30	90	10	100	0.114	0.149	0.096	0.120
31	90	12	100	0.114	0.120	0.119	0.118
32	90	14	100	0.129	0.125	0.123	0.126
33	90	1	125	0.048	0.083	0.048	0.060
34	90	2	125	0.096	0.093	0.085	0.091
35	90	4	125	0.072	0.079	0.080	0.077
36	90	6	125	0.122	0.114	0.103	0.113
37	90	8	125	0.084	0.101	0.082	0.089
38	90	10	125	0.087	0.085	0.087	0.086
39	90	12	125	0.113	0.106	0.116	0.112
40	90	14	125	0.143	0.135	0.138	0.139
41	90	1	150	0.038	0.037	0.048	0.041
42	90	2	150	0.093	0.092	0.091	0.092
43	90	4	150	0.089	0.089	0.094	0.091
44	90	6	150	0.082	0.087	0.082	0.084
45	90	8	150	0.073	0.085	0.081	0.080
46	90	10	150	0.085	0.084	0.087	0.085
47	90	12	150	0.115	0.110	0.117	0.114
48	90	14	150	0.152	0.139	0.128	0.140

Results of surface roughness measurement can be found in table 3. In order to get more detailed information related to the effects of the investigated cutting parameters, an ANOVA test was applied. The main effects plot for surface roughness is shown in figure 3. Feed per tooth (f_z) seems to affect the surface roughness in a significant way, the larger the feed the worse the surface quality. There is a remarkable region of the main effect plot of f_z between 4 and 6 μm, where the strong linear relationship seems to be disturbed. This can be considered as a local minimum of the surface roughness. One possible reason for this is the fact that the value of the minimum chip thickness under certain conditions falls within this range, which is equal to the size of the cutting edge radius of the micro end mill. In that case the optimal ratio of ploughing effect and shear process at material separation is realized. The depth of cut (a_p) has a limited effect on the surface roughness. Within the investigated range, there is a slight improvement in the surface quality at depth of cut larger than 100 microns.

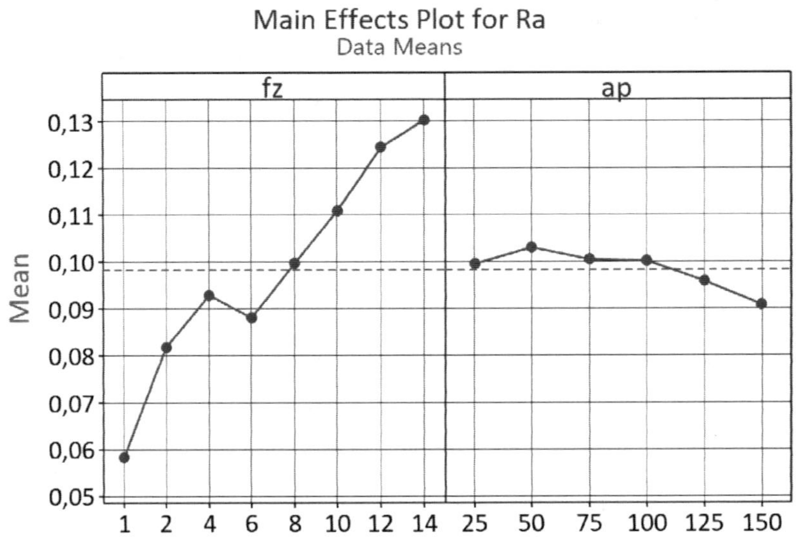

Figure 3. Main effects plot for surface roughness.

3.2. Artificial neural network

In the frame of the analysis, an ANN model was developed with a two-layer feedforward backpropagation neural network by means of MATLAB. Levenberg-Marquardt algorithm (LMA) was applied for the training procedure. The data set was divided randomly in the ratio of training set (90%), validation set (5%), and testing set (5%). The training was carried out by the "trainlm" function in the software with 25 hidden layers. Cutting speed (v_c), depth of cut (a_p), and feed per tooth (f_z) were used as input parameters to predict 1 output, which was the average surface roughness (R_a).

Results of model development are summarized in table 4. It can be seen that the values of RMSE (Root Mean Square Error) for the training, validation and test sets are low enough, and statistically significant with their average value of 0.019. This signifies that the root mean square error between the measured values and predicted values are minimal. Also, the average value is minimal and very close to the RMSE-value of the additional test data (0.0139) for verifying the developed ANN model. The MSE (Mean Square Error) of the training, validation and test sets are closed to each other and tend to zero. The MSE for the best validation performance is 0.0018338 at epoch 3.

Table 4. ANN model results summary.

	Observations	MSE	RMSE	R
Training	33	0.00002492	0.004991593	0.9845
Validation	2	0.00180000	0.042426407	1
Test	2	0.00009059	0.009517825	1
Additional Test	11	0.00038148	0.019531513	0.8545

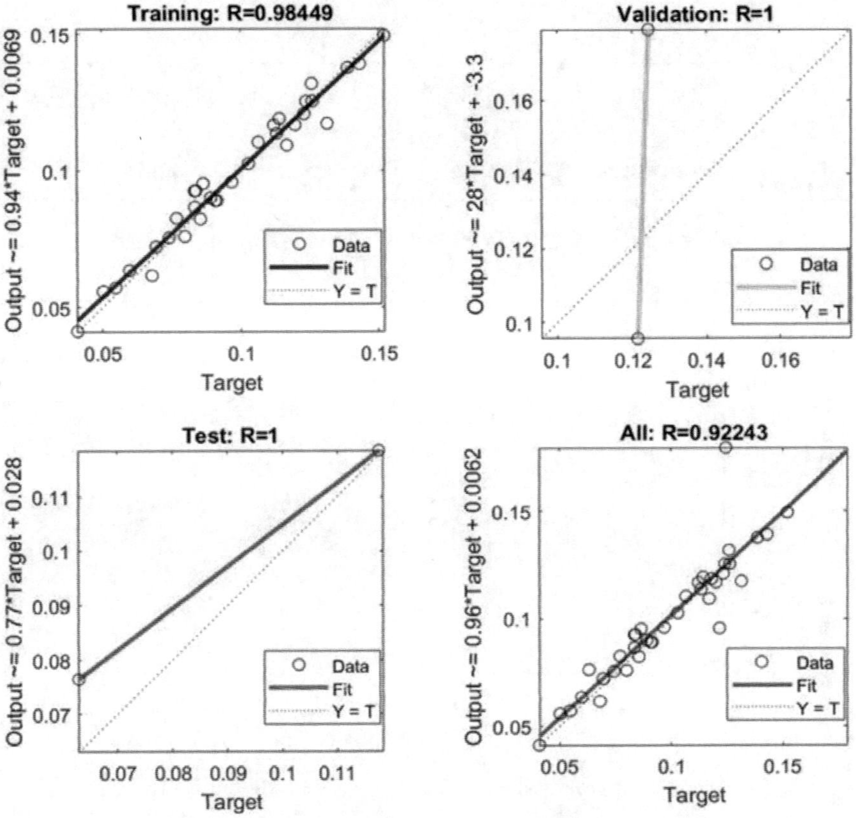

Figure 4. Regression plots of the trained model.

More so, the R-values of trained data are near perfect (0.9845), and that in the case of validation data set and test data set equals to 1 (figure 4). The additional test data give a confirmation of the developed LMA ANN model, in that a high R-value (0.8545) signifies a strong correlation between the measured and predicted R_a-values.

The comparative plots of the surface roughness based on both training data (figure 5) and additional test data (figure 6) show that the error of measured and predicted values was minimal and correlated. The numbers on the horizontal axis of the diagram in figure 6 shows the experiments, which were selected to carry out the additional test by the previously trained model. It can be stated that the developed ANN based predictive model can be efficiently used for estimation of surface roughness (R_a) at micro machining of AISI H13 using two flutes micro end mill. The responses of the model can be applied for process monitoring, and realizing cutting parameter optimization, too.

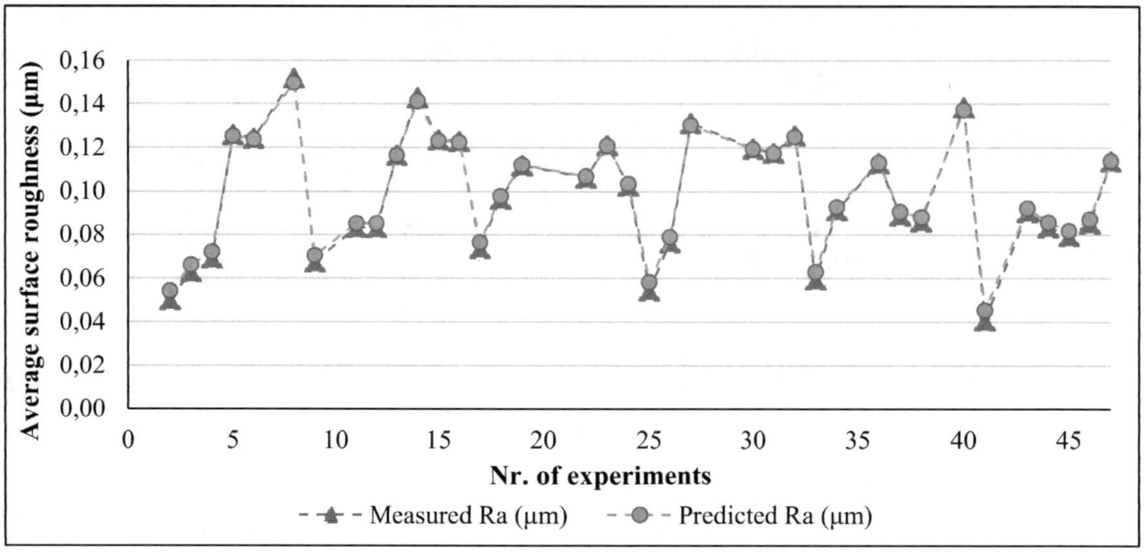

Figure 5. Comparison of measured and predicted R_a-values.

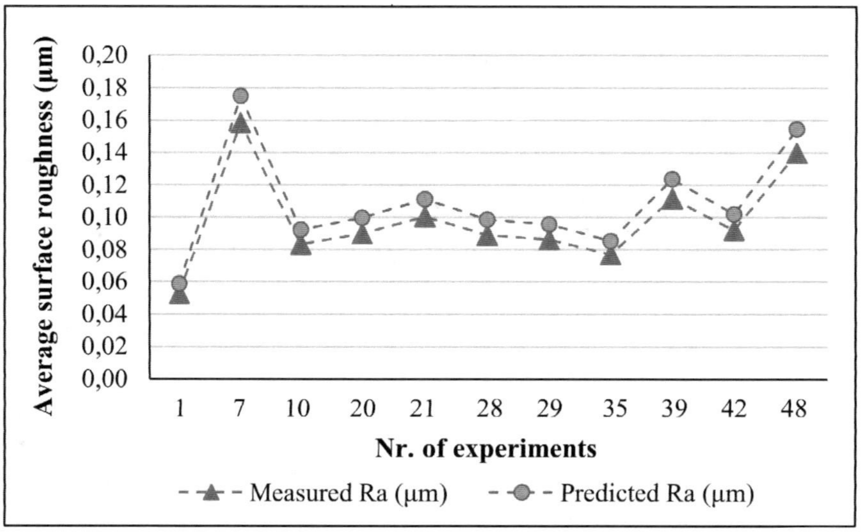

Figure 6. Comparison of measured and predicted R_a-values in the frame of additional validation test.

4. Conclusion

Main aim of this research work was to prepare predictive model applying artificial neural network for determining surface roughness at micro machining. Micro milling experiments on hot work tool steel (AISI H13) were carried out to determine the effect of different cutting parameters on surface roughness, and to provide data for training, validation and testing of predictive model.

The developed LMA ANN model for prediction of the surface roughness at machining AISI H13 is effective considering the results from the investigated data, where MSE and RMSE values were very low signifying minimal error between the measured and predicted surface roughness. Similarly, a high R-value (0.8545) in the case of checking the trained model with additional experimental data implies a strong correlation between the measured and predicted values.

The developed predictive surface roughness model at micro milling can serve as the basis for a decision-making system in order to ensure better product quality.

5. References

[1] Balazs BZ, Geier N, Takacs M, Davim JP. A review on micro-milling: recent advances and future trends. International Journal of Advanced Manufacturing Technology. 2021; 112 (3-4): 655-684

[2] Masuzawa T, Tonshoff H. Three-dimensional micromachining by machine tools. CIRP Annals – Manufacturing Technology. 1997; 46: 621–628

[3] Dornfeld D, Min S, Takeuchi Y. Recent advances in mechanical micromachining. CIRP Annals – Manufacturing Technology. 2006; 55: 745–768

[4] Chae J, Park SS, Freiheit T. Investigation of micro-cutting operations. International Journal of Machine Tools and Manufacture. 2006; 46: 313-332

[5] Inverse Solutions, Inc. The Evolution of Micromachining and Its Advantages. Available from: https://inversesolutionsinc.com/evolution-micromachining-advantages/ [Accessed 7th December 2021]

[6] IBM Cloud Education. Machine Learning. Available from: https://www.ibm.com/cloud/learn/ machine-learning. [Accessed 15th November 2021]

[7] Taoufik N, Boumya W, Achakb M, Chennouk H, Dewil R, Barka N. The state of art on the prediction of efficiency and modelling of process pollutants removal based on machine learning. Science of The Total Environment. 2022; 807: p. 16

[8] Serin G, Sener B, Ozbayoglu AM, Unver HO. Review of tool condition monitoring in machining and opportunities for deep learning. The International Journal of Advanced Manufacturing Technology. 2021; 109: 953–974

[9] Mandal S. Applicability of Tool Condition Monitoring Methods used for Conventional Milling in Micromilling. Journal of Industrial Engineering. 2014; 2014: p. 8

[10] Le vT, Dragoi EN, Almomani F, Vasseghian Y. Artificial Neural Networks for Predicting Hydrogen Production in Catalytic Dry Reforming: A Systematic Review. Energies. 2021; 14(10), 2894: p. 11

[11] Nicola GD, Coccia G, Pierantozzi M, Tomassetti S, Grifoni RC. Artificial neural network for the second virial coefficient of organic and inorganic compounds: An ANN for B of organic and inorganic compounds. Chemical Engineering Communications. 2018; 8: 1077-1095

[12] Knittel D, Makich H, Nouari M. Milling diagnosis using artificial intelligence approaches. Mechanics & Industry, 2019; 20, 809: p. 9

[13] ZhangC, Yao X, Zhang J, Jin H. Tool Condition Monitoring and Remaining Useful Life Prognostic Based on a Wireless Sensor in Dry Milling Operations. Sensors. 2016; 16(6):795: p. 20

[14] Patil SS, Pardeshi SS, Patange AD, Jegadeeshwaran R. Deep Learning Algorithms for Tool Condition Monitoring in Milling: A Review. Journal of Physics Conference Series. 2016; 1969 012039: p. 14

[15] Balázs BZ. Edzett acélok mikromarás útján történő megmunkálásának elméleti és kísérleti vizsgálata, PhD dissertation. Budapest: Budapesti Műszaki és Gazdaságtudományi Egyetem, Pattantyús-Ábrahám Géza Gépészeti Tudományok Doktori Iskola; 2021.

[16] voestalpine BÖHLER Edelstahl GmbH & Co KG. Product description, property features and technical specifications of AISI H13. Available from: https://www.bohler-edelstahl.com/en/products/w302isobloc/ [Accessed 31th January 2022]

Acknowledgements

This research was partly supported by the National Research, Development and Innovation Office (NRDIO) through the projects No. OTKA-K-132430 (Transient deformation, thermal and tribological processes at fine machining of metal surfaces of high hardness), and No. ED_18-2018-0006 (Research on prime exploitation of the potential provided by the industrial digitalisation). The research reported in this paper and carried out at BME has also been supported by the National Laboratory of Artificial Intelligence funded by the NRDIO under the auspices of the Ministry for Innovation and Technology.

IOP Conf. Series: Materials Science and Engineering 1246 (2022) 012035 doi:10.1088/1757-899X/1246/1/012035

The effect of Argon Shield Gas and Cold Air for Friction Stir Welding of Ceramic-reinforced Aluminium Matrix Composite and Aluminium Alloy

Zs. F. Kovács [1]

[1] John von Neumann University, GAMF Faculty of Engineering and Computer Science, Dep. of Innovative Vehicles and Materials, Manufacturing Technology Research Group, Kecskemét H-6000, Hungary

kovacs.zsolt@gamf.uni-neumann.hu

Abstract. A dissimilar joint between friction stir welded ceramic-reinforced C35 aluminium matrix composite and AA7075-T6 aluminium alloy has some difficulties. For fusion welding of aluminium, a shielding gas always has to be used to achieve the required welding joint quality, but in the case of Friction Stir Welding (FSW) no shielding gas or filler material is required for welding aluminium alloys. To achieve a good welding joint between super strength aluminium alloys and ceramic-reinforced aluminium matrix composites some sort of support is needed. In this study, the effect of blowing argon gas and cold air (0 °C) was investigated with regards to the quality of the welding. For the experiment a cylindrical pin tool, a different feed and a constant rotation speed were utilized along with a marble table to avoid harmful heat conduction. For evaluations, hardness and tensile tests examinations were carried out to classify the mechanical properties of the joint. Experimental results showed that the blowing of gas and cold air having a positive effect on the welded joint quality and also for the weld face. Additionally, the optimal welding traveling speed and the temperature of the air blown was determined.

1. Introduction

In 1991, a new friction welding process, called "Friction Stir Welding (FSW)", was developed at The Welding Institute (TWI) in England. The process got a European Patent (EP 0 615 480 B1) in 1992. It was adopted by the industry shortly after its appearance because compared to other methods it allows the welding of difficult-to-weld or non-weldable materials. Today FSW is used widely, especially in space technology, aerospace and automotive industries where the weight of the components is crucial, so high-strength aluminium alloy structural elements are preferred, and welding has become much more reliable by the advent of this new process [1] [2] [3] [4].

During the FSW process, the frictional heat is generated by a special geometry tool which is harder than the welded materials. This tool enters the workpieces to be welded during rotation, its shoulders colliding on their surfaces, and when the materials have reached the required temperature and have softened near the associated surfaces, they begin to move along the material boundaries in a linear path, maintaining rotational movement (figure 1.). The softened material flows behind the tool pin due to the combined action of the two movements, and it forms a bond between the workpieces [1] [2].

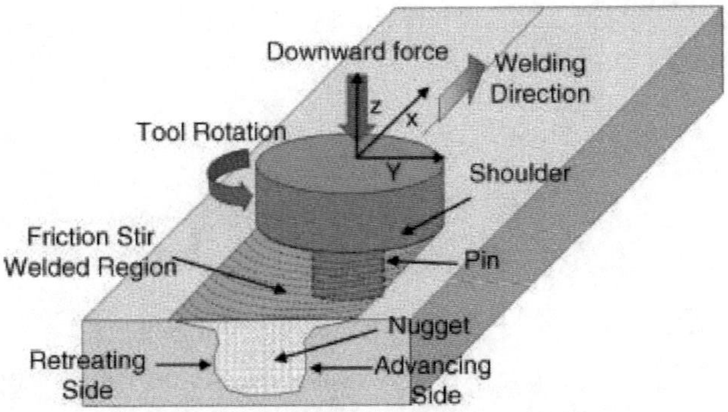

Figure. 1. Principle of FSW process [5]

The welded joint can be divided into four microstructural zones. Along the toolpath, the material recrystallizes after the plastic deformation, resulting in a fine-grained fabric structure; this is the weld nugget (WN). This part of the bond is also called the dynamic recrystallization zone – its circular geometry and mechanical properties depend on the direction of rotation and shape of the tool. Above it is the intensive Flow Zone (FZ). There are two sides, next to them the thermo-mechanically affected zone (TMAZ); in this part the two materials do not mix with each other, but because this zone is subjected to strong impacts, the particles are elongated due to the significant thermal effects and compressive force, thus further plastic deformation takes place, but recrystallization does not occur. Then follows the heat affected zone (HAZ); there is no more pressing force here so plastic deformation does not occur. These zones are surrounded by the raw material. The locations of these zones are illustrated in figure 2. [5] [6] [7].

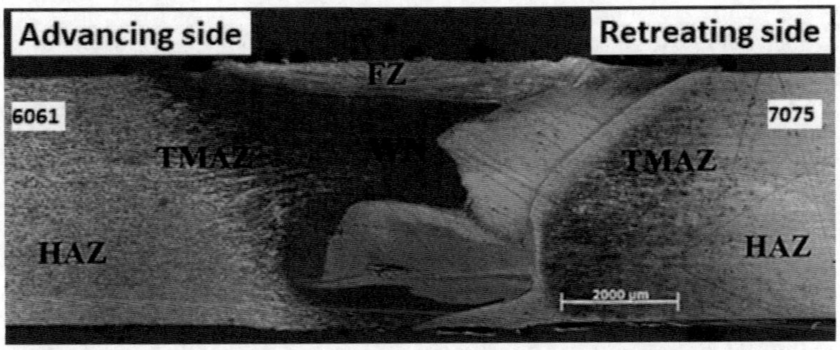

Figure. 2. Microstructural zones of friction stir welded nugget zone [8]

During the welding of mixed joints, the formation of intermetallic compounds is a common phenomenon, which impairs the properties of the joint. These compounds are usually brittle and increase the risk of fracture. They are also present during friction stir welding, so it is generally believed that their formation cannot be avoided, but their extent can be reduced by applying the appropriate welding parameters. For some materials, there are also compounds that improve the properties of the welded bond [9].

To create a proper bond, it is necessary to know the factors influencing welding and their significance. Based on literature, these factors can be divided into three groups. The first one specifies the equipment [10] [11] [12]:

- clamping force
- speed (rpm)
- feed (mm/min)
- angle of inclination from the welding direction (°)
- type of shielding gas or cooling

Second, it is necessary to know the properties of the tool, which would have a fundamental effect on the quality of the joint, so it is advisable to design a tool tailored to the specific task in each situation. Essential parameters of the tool [11] [13] [14]:

- tool material
- shoulder design
- shoulder diameter
- needle design
- needle diameter
- needle length

Other factors to consider when welding [11] [13] [14]:

- distance between workpieces
- wall thickness of the workpieces
- clamping of workpieces
- material characteristics of the workpieces
- mechanical properties of workpieces
- rigidity of the equipment

In general, speed and feed have the greatest effect on the quality of the joint. After all, a significant part of the other factors was constant in most cases, such as the wall thickness of the material, or can be changed in a very small range, such as the angle of the tool [11] [13] [15].

2. Experimental procedure

2.1. Material of the specimens

For the experiments, C355 and AA7075-T6 aluminiums were used. The C355 is a so-called Hybrid Metal Matrix Composite (HMMC), where the aluminium is the matrix that is reinforced with ceramic particles, in our case SiC particles. This aluminium composite is well-known as it is the raw material of aluminium foams, in which SiC particles are responsible for foaming. Its composition contains 91.7-94.1% Al, 4.5-5.5% Si, 0.4-0.6% Mg and 1.0-1.5% Cu which significantly increases the strength, toughness and hardness, and in turn, reduces corrosion resistance and impairs weldability. Furthermore, the C355 aluminium composite may also contain Fe, Ti, Mg, and Zn, but these are present in a maximum of 0.1-0.2%. [16].

2.2. Welding machine and applied instruments

Friction stir welding was performed on a modified MFP-320 vertical milling machine. Welding requires a machine that has a machine feed and an adjustable angle of the vertical head, which requirements this machine fully satisfies. The used tool for welding was made of HSS end mill. The dimensions of the tool are shown in below table 1.

Table. 1. Dimension of FSW tool

	Parameter
Shoulder diameter	12 mm
Pin diameter	4 mm
Pin length	3.8 mm

During welding, the workpiece was cooled with cold air by an EXAIR 3825 vortex tube (figure 3.). The device is able to cool the air flowing continuously from the compressed air network of the building to a constant temperature between -34 °C and +21 °C.

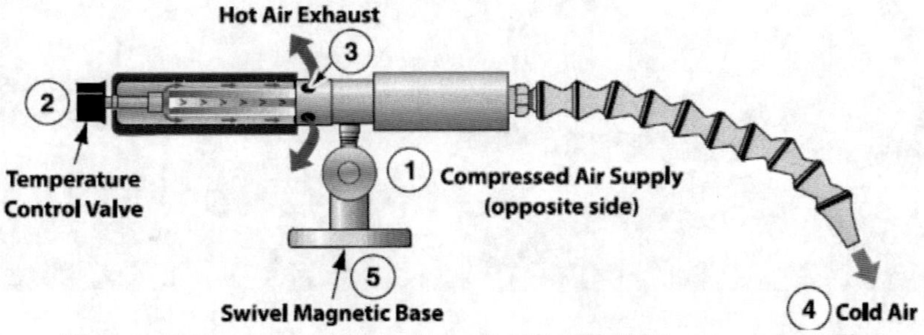

Figure. 3. Conceptual illustration of a EXAIR 3825 vortex tube [17]

The temperature of the cold air flowing out of the vortex tube was set with an SPM LEONOVA Infinity instrument. Also, the tool temperature was determined by a Flir T600 thermal camera.

3. Experimental setup

The aim of the research is to determine the effect of cooling with argon shielding gas or blown cold air on the quality of friction welds. To investigate this, a two-factor and four-level experimental design were made. The factors and levels are shown in table 2., which contains two experiments' setups. In the first one, the cooling was carried out by argon gas and in the second one by cold air. For argon cooling, the flow of gas through the nozzle was constant at 30 l/min.

Table 2. Experimental setups for the first and second experiments

No. of exp.		Feed (mm/min)	Cooling by	
argon	*cold air*		**argon gas (Y/N)** *-first experiment-*	**cold air temperature (°C)** *-second experiment-*
1	1	20	Y	-1
	2	20		3
2	3	20	N	7
	4	20		21
3	5	25	Y	-1
	6	25		3
4	7	25	N	7
	8	25		21
5	9	30	Y	-1
	10	30		3
6	11	30	N	7
	12	30		21
7	13	35	Y	-1
	14	35		3
8	15	35	N	7
	16	35		21

Being aware of the retreating and the advancing side is very important for material placement. Although there is a difference of more than 100 °C (477 °C to 582 °C) between the solidity (the temperature below which only pure solids are present in the material) of the two materials, it is evident that in the cases of this study, the temperature will not reach 477 °C which was confirmed by the thermal camera images taken during welding (figure 4.). With this in mind, the harder material (AA7075-T6) should be placed on the advancing side. The tilt angle of the tool has been set at a constant value of 3°.

IOP Conf. Series: Materials Science and Engineering 1246 (2022) 012035 doi:10.1088/1757-899X/1246/1/012035

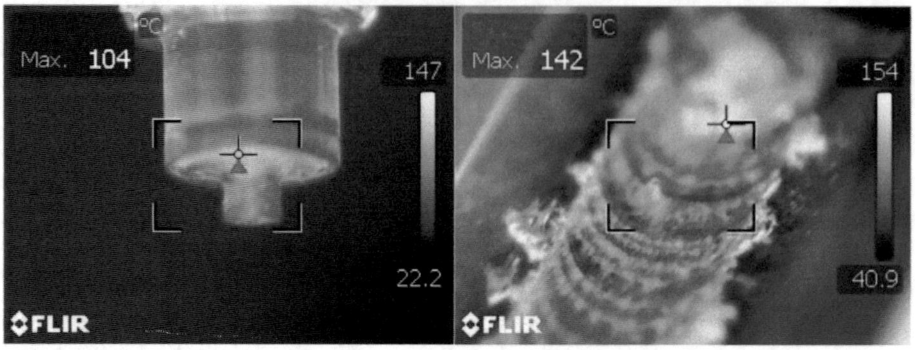

Figure. 4. Thermal camera images of the tool and workpiece immediately after welding

4. Results

4.1. Evaluation of tensile tests

The tensile tests were performed according to ISO 6892. After the tensile tests, the tensile strength of the specimens was determined based on the elongation and force data, the results are shown in table 3.

Table. 3. Results of tensile tests

No. of exp.		Cooled by argon gas *-first experiment-*		Cooled by cold air *-second experiment-*	
argon	*cold air*	**Rm (MPa)**	**Place of rupture**	**Rm (MPa)**	**Place of rupture**
1	1	91	TMAZ	66	WN
	2			88	WN
2	3	111	Base material	121	Base material
	4			110	TMAZ
3	5	128	Base material	110	TMAZ
	6			100	Base material
4	7	124	Base material	114	Base metal
	8			123	TMAZ
5	9	119	Base material	115	Base material
	10			106	TMAZ
6	11	120	TMAZ	112	TMAZ
	12			121	Base material
7	13	114	Base material	122	Base material
	14			109	Base material
8	15	126	Base material	115	TMAZ
	16			108	Base material

Since most specimens were torn in the base material or in the heat-affected zone, they can be considered compliant, while those in the welded nugget are not compliant. In the case of cold air cooling, the specimens No. 1 and 2 were torn in the nugget, so they were not properly rated. Furthermore, it is well observed that in their case, the specimens cracked along the border of NZ and TMAZ. Fragments of these two samples show SiC inclusions and gas pores (figure 5.) that probably caused rupture in the nugget.

Figure. 5. Fracture of sample 1 and 2 (cooled by cold air)

For all other samples (cooled by argon and cold air), the tear occurred outside the nugget on the C355 side. In all cases, greater or lesser SiC inclusions and gas porosity can be observed in the fractures without exception. The presence of inclusions results from the inhomogeneous distribution and inadequate dissolution of the reinforcing material in the matrix. Because the C355 aluminium is used to make aluminium foam, the presence of gas pores is not a problem in the base material, however, in welding cases it weakens the nugget.

4.2. Evaluation of hardness

The $HV_{0.1}$ microhardness measurements were performed on the samples with a Wilson Wolpert 401MVD Vickers hardness tester according to ISO 9015-2.

In the HAZ of the AA7075-T6 plates, a significant decrease can be observed compared to the hardness of the base material 175 $HV_{0.1}$, because the values here fluctuate between 90 and 105 $HV_{0.1}$. Moving towards the centerline, the value of TMAZ fluctuates in the same range, in most cases higher than that measured in the HAZ. The maximum hardness was measured in several cases on the NZ advancing side or in the middle of the nugget, with a value of around 110...115 $HV_{0.1}$. In the middle of the nugget, the hardness values of the crown side and the root side are separated in most cases, because where the two materials are not mixed, there is C355 on the crown side, which has a hardness of around 45...110 $HV_{0.1}$, while the hardness on the root side is around 100...110 $HV_{0.1}$. On this side, the values of TMAZ and HAZ are also in the range of about 40...50 $HV_{0.1}$, in some cases the hardness of HAZ is higher than the values measured in TMAZ and the base material.

Based on the hardness measurement, except for the samples welded at a feed of 30 mm/min, the properties on the AA7075-T6 side and the NZ are more favourable with cooling at 7 °C air. However, from the point of view of the TMAZ and HAZ on the side of the C355, colder air (-1 °C and 3 °C) is more favourable at feeds of 20 mm/min and 30 mm/min.

Comparing the location of the tears with the value of the tensile strengths, there is a significant, visible difference in tensile strength only in the cases where the pieces were torn in the nugget, while in the case of pieces torn in the heat zone tensile strength is higher than in the raw material. Based on this, it is clear that the presence of SiC particles causes the defect in the HAZ and is not associated with an increase in hardness.

5. Summary

Most of the tensile specimens had broken in the base material and the heat affected zone so it can be concluded that these two materials are easily weldable with FSW. It is noticeable that there are voids and porosity in each of them. The metallurgical examinations were looking for common welding defects such as cracks, voids, kissing-bonds and tunnels (figure 6.).

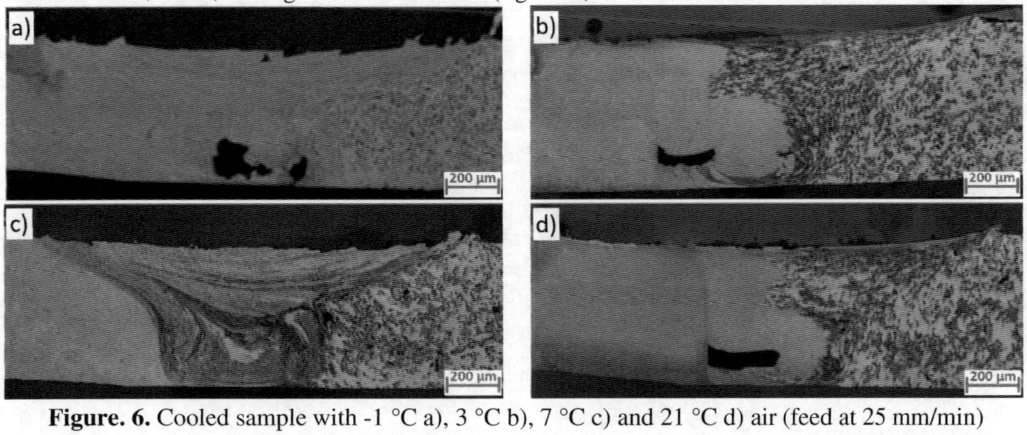

Figure. 6. Cooled sample with -1 °C a), 3 °C b), 7 °C c) and 21 °C d) air (feed at 25 mm/min)

It can be concluded that each of the samples contained tunnel defects, however other regular errors were not as common as that. During the examination, it can be noticed that every second sample is very similar with each other but not with the next one or the one before. The explanation of this phenomenon is that the temperature at the start of the sheets could not reach the equilibrium state yet.

Overall, the effect of cooling on the seam is negligible, but the surface (crown side) of the seam is much smaller in the case of a cooled tool. It should also be mentioned that the effect of argon shielding gas was negligible during the experiment compared to cold air.

6. References

[1] P. B. R. Singh, *A Hand Book on Friction Stir Welding*, First edition. UK: LAP Lambert Academic Publishing (2012)
[2] W. M. Thomas, E. D. Nicholas, J. C. Needham, M. G. Murch, P. Templesmith, és C. J. Dawes, „Improvements relating to friction welding", EP0615480B1 (1995)
[3] E. T. Akinlabi és R. M. Mahamood, *Solid-State Welding: Friction and Friction Stir Welding Processes*, First. Switzerland: Springer International Publishing (2020)
[4] Z. Weltsch, *Comparative study of the joining technologies of vehicle bodywork sheets*, IOP Conference Series: Materials science and engineering Vol. 448 (2018)
[5] J. Hlinka, Z. Fogarassy, Á. Cziráki, Z. Weltsch, *Wetting Properties, Recrystallization Phenomena and Interfacial Reactions between Laser Treated Cu Substrate and SAC305 Solder*, Applied Surface Science Vol. 501 (2020)
[6] Y. Adonyi, J. Rutledge, P. Wang, *Tool and preheat coil offset in dissimilar friction stir welding*, presented at FABTECH, Atlanta, GA (2014)
[7] K. Oyedemi, Y. Adonyi, P. Wang, *Effect of Intermetallics on Aluminum-to-Steel Friction Stir Weld Properties*, STP: Selected Technical Papers, Conference: ASTM E04At: Atlanta, GA (2017)
[8] Zs. Kovács, F. Hareancz, *Joining of non-weldable AA7075 and weldable AA6082 aluminium alloy sheets by Friction Stir Welding,* IOP Conference Series: Materials science and engineering Vol. 448 (2018) Paper: 012001
[9] R. Mishra, K. Nilesh, W. Yuan, *Friction Stir Welding of Dissimilar Alloys and Materials - 1st Edition*. USA: Elsevier (2015) ISBN: 978-0-12-802418-8
[10] A. K. Hussain, S. A. P. Quadri, *Evaluation of parameters of friction stir welding for aluminium AA6351 alloy*, International Journal of Engineering Science and Technology, Vol. 2 (2010) pp. 5977-5984
[11] F. A. Mir, N. Z. Khan, A. N. Siddiquee, S. Parvez, *Joining of aluminium matrix composites using friction stir welding: A review,* Proc IMechE Part L: J Materials: Design and Applications (2022) pp. 1–16
[12] M. W. Safeen, P. R. Spena, *Main Issues in Quality of Friction Stir Welding Joints of Aluminum Alloy and Steel Sheets*, Metals, Vol. 9 (2019)
[13] Y. N. Zhang, X. Cao, S. Larose and P. Wanjara, *Review of tools for friction stir welding and processing*, Canadian Metallurgical Quarterly, Vol. 51 (2011) pp. 250-261
[14] O. P. Abolusoro, E. T. Akinlabi, S. V. Kailas, *Impact of tool profile on mechanical behavior and material flow in friction stir welding of dissimilar aluminum alloys,* Materialwiss. Werkstofftech. Vol. 51 (2020) pp. 725–731
[15] M. D. Sameer, A. K. Birru, *Mechanical and metallurgical properties of friction stir welded dissimilar joints of AZ91 magnesium alloy and AA 6082-T6 aluminium alloy*, Journal of Magnesium and Alloys, Vol. 7 (2019) pp. 264-271
[16] J. Líska, K. Kun, Zs. Kovács, *Attendants at MMC composite milling*, Advances in Manufacturing Engineering and Materials II, Springer International Publishing (2021) pp. 332-342
[17] EXAIR Blog: https://blog.exair.com/2018/11/28/the-adjustable-spot-cooler-cold-air-to-30f-34c-from-your-compressed-air-supply/

AUTHOR INDEX

Institute of Physics
Dirac House, Temple Back
Bristol BS1 6BE UK

ISSN: 1757-8981
ISBN 978-1-7138-6013-6

CURRAN ASSOCIATES INC.
proceedings.com